Huygens: The Man behind the Principle

Huygens: The Man behind the Principle is the story of the great seventeenth-century Dutch mathematician and physicist, Christiaan Huygens (1629–1695). As his first complete biography ever written in English, this book describes in detail how Huygens arrived at discoveries and inventions that are often wrongly ascribed to his younger contemporary Newton. At the same time it paints a vivid picture of Huygens' youth and adulthood, and the many fruits of his science.

Huygens played a key role in the 'scientific revolution' and the 'Huygens Principle' on the wave theory of light helped establish his reputation. The discovery of Saturn's rings and the invention of the pendulum clock made him so famous that he was invited to be the first director of the French Academy of Science, but his busy life as director teetered on the edge of powerlessness.

Despite Huygens' many achievements no complete biography has previously been published in English, a consequence of his Dutch origins and the fact that many important aspects of his life were only ever documented in Dutch. This book gives scientists and historians the opportunity to learn more about all aspects of Huygens' life and work while bringing his story to a wider audience.

C. D. ANDRIESSE studied physical engineering at the Technical University in Delft and gained his Ph.D. in 1969. His research career moved first into astrophysics at the University of Groningen and then materials research, with a prime interest in nuclear safety and electrical network stability. Since 1989 he has been a professor at the University of Utrecht, researching energy conversions in fuel cells and wind turbines. In addition to his research articles, Professor Andriesse's non-research publications include a handbook on problems with nuclear energy, for the International Atomic Energy Agency, essays on the cultural value of technology, diverse historical work and an intellectual autobiography.

Huygens
The Man behind the Principle

C. D. ANDRIESSE

Translated by
SALLY MIEDEMA

CAMBRIDGE
UNIVERSITY PRESS

CAMBRIDGE UNIVERSITY PRESS
Cambridge, New York, Melbourne, Madrid, Cape Town, Singapore,
São Paulo, Delhi, Dubai, Tokyo, Mexico City

Cambridge University Press
The Edinburgh Building, Cambridge CB2 8RU, UK

Published in the United States of America by Cambridge University Press, New York

www.cambridge.org
Information on this title: www.cambridge.org/9780521181433

First published 2005
First paperback edition 2010

A catalogue record for this publication is available from the British Library

Library of Congress Cataloguing in Publication data
Andriesse, Cornelis Dirk.
[Titan kan niet slapen. English]
Huygens : the man behind the principle / C.D. Andriesse ; translated by
Sally Miedema.
    p.   cm.
Includes bibliographical references and index.
ISBN 0-521-85090-8 (hardback)
1. Huygens, Christiaan, 1629–1695.   2. Scientists – Netherlands – Biography.
3. Science – History – 17th century.   1. Title.
Q143.H96A6413   2005
500.2′092 – dc22   2005006464

ISBN 978-0-521-85090-2 Hardback
ISBN 978-0-521-18143-3 Paperback

Contents

The plates are situated between pages 230 and 231.

Plates

Preface

This book describes the life of a great seventeenth-century Dutch mathematician and physicist, in a similar class to Newton. Although his reputation is well established and the Huygens Principle has been a source of enlightenment up until today, the man behind the light has remained in the dark. His Dutch origins, and the difficulties of the Dutch language, in which important facts of his life are documented, may be the reason why no complete biography has yet been published.

Strange as it may seem, this book is the first endeavour to bring to light all aspects of Huygens' life while he was developing his talents. Johan Vollgraff may have had it in mind when he listed the thousand-and-one facts on Huygens' life, which he came across as editor of the *Oeuvres*. He added them to the last volume of the *Oeuvres*, now some fifty years ago, but never wrote that first biography.

Arthur Bell made another attempt, as did Usher Frankfourt together with Aleksandr Frenk, and Alfonsina D'Elia. However, all four were confined to the translations and selections of the *Oeuvres*, since they neither had knowledge of the Dutch language nor access to family documents. For this reason, the books they published in 1947, 1976 and 1985, respectively, cannot be seen as complete, or even reliable biographies. Bell regarded his own attempt as only a beginning, and long cherished the hope to write a large work on this 'great subject'.

Here then is my own endeavour. It is not a large work, and it is written for a wide public, with whom I wished to share my fascination for Huygens, but without encumbering this same public with detailed sources. *Titan kan niet slapen* – the original title of the book – was well received, despite its meagre documentation, and a French translation was quickly published. Yes, Huygens did spend some time in Paris!

The present edition, now in the *lingua franca* that is read all over the world, gives me the opportunity to add the references to my sources. To complete the book still further, I have extended the bibliography with the literature on Huygens that I did not, or was not able to consult initially. Although there is little that is able to throw new light on the man himself, two recent studies prompted me to add further information on Huygens' dioptrics and on his relations with Spinoza. The most substantial addition, however, is the overview with which it begins, the text that I read in Leiden to commemorate Huygens on the 300th anniversary of his death.

My endeavour to write this biography has taken me far from home. As a physicist I had been accustomed to writing papers with a modest number of pages containing clear and logical arguments. But for this book I had to become an historian, and to ponder over the significance of a multitude of lengthy documents that were often obscure and difficult to understand. I would not have got far without the help of scholars.

Therefore I would like to thank Sible de Blaauw, Hendrik Bos, Floris Cohen, Casper Hakfoort and Joella Yoder for their suggestions and comments on an early draft. When the book had appeared Hendrik Casimir, Jan Deiman, Cees Grimbergen, Elisabeth Keesing, Jan Nienhuys and Tim Trachet kindly pointed out some mistakes that I was able to correct in a second Dutch edition. Ad Leerintveld guided me through the world of Huygens' father's poetry, and Jan Nauta through the world of psychoanalysis. I also thank Robert van Gent and Rienk Vermij, who gave me permission to use their extensive Huygens bibliography in the present edition.

Finally I am most grateful to Hans van Himbergen, the dean who requested this English translation for the English-speaking guests of our Faculty, and who provided the means to realise it.

A commemoration[1]

Christiaan Huygens died 300 years ago. Announcing his death, Gottfried Leibniz referred to his teacher and friend as *the Incomparable Huygens*.[2] We would like to use the same phrase today, but if Huygens is to be assured of an important place in history, we feel it is our duty to compare his achievements with those of other scientists. A contemporary colleague, grieving the loss of a close friend, might see that friend's achievements in a more favourable light than would posterity.

It is not difficult to compare him with Galileo Galilei. Huygens was strongly influenced by Galileo and took over many of his ideas. In his youth he had assimilated the ideas of Galileo's *Discorsi* and later in his life he developed many of Galileo's traits and characteristics. In fact, the resemblances were so numerous that when Huygens described Galileo he was almost painting a self-portrait. Shortly before he died Huygens wrote an essay about Cartesianism, in which he distanced himself from that philosophy. In this essay we read the following:[3]

> Galileo had the acuteness of mind and knew all the mathematics necessary to proceed in science. One has to admit that he made beautiful discoveries about the nature of motion, although he left many aspects untouched. He was not so reckless and arrogant as to explain all natural causes, nor was he so vain that he wished to be the leader of a sect. He was modest and truthful. Yet he thought he had acquired eternal fame with his discoveries.

---

[1] Andriesse 3–13   [2] OC **10**, 721   [3] OC **10**, 398 (404)

It was as if Huygens were looking at himself in one of his own carefully polished and silver-plated mirrors.

This comparison with Galileo, although appropriate, is insufficient as an evaluation of Huygens' work and character. The scientific revolution of the seventeenth century involved many scientists, and therefore posterity is able to compare Huygens with many other actors. How, for instance, does he compare with the man who received practically all the honours? The answer to this tricky question is to be found in the background story of Huygens' greatest work.

The most important step on the way to this work, *Horologium oscillatorium*, was the proof that the weight of an isochronous pendulum follows a cycloid and that this path is the evolute of another cycloid. Above the proof Huygens wrote: '*Magna nec ingenijs investigata priorum*' ('This is something great that has not been investigated by geniuses of the past').[4] In other words, he knew he was a genius – there is no doubt about that. He completed the proof on 15 December 1659, and took its motto from Ovid's *Metamorphoses*, which he had read when he was only twelve. Now, having reached the age of thirty, he felt he had made his greatest discovery. Only one week earlier he had told Frans van Schooten in a letter: this is 'certainly the finest thing I have ever come across'.[5]

He had indeed made a great discovery. He had found that the period of a pendulum would be independent of the swing, if the pendulum were deflected by platelets shaped like cycloids. But more important still than the actual finding was the way in which he had arrived at his conclusion. In fact, he had used the method of infinitesimal analysis. Later on, however, he was reluctant to accept formal calculus, for although he perfectly understood its roots, he found the rules obscure. Huygens was a master at summing indefinitely small line elements and an expert at using infinitesimal triangles.

[4] OC **16**, 406   [5] OC **2**, 502

It was quite some time before Huygens' greatest work, *Horologium oscillatorium*, was published. This book about the pendulum clock was his tribute to the Academy of Science in Paris, or rather to the French king. By that time, 1673, Huygens had become the recognised leader of European science. Isaac Newton, who received a copy of the book via the secretary of the Royal Society in London, responded immediately. In a letter to the secretary, Newton commented that the book was very worthy of its author but that he had an easier proof of the isochronous property of the cycloid.[6] 'If he (Huygens) please, I will send it him.'

In this simpler proof Newton used the calculus that he had invented eight years before, but had kept secret. Now he was willing to share his secret with the author of *Horologium oscillatorium*! What greater honour could Huygens receive? But Huygens declined the offer. He did not ask for the proof, probably because he was still offended. Only three months earlier Newton had wrecked their correspondence about light and colour by addressing Huygens like a delinquent schoolboy.[7] Thereupon Huygens had put an end to the exchange of letters with the following polite but icy words:[8] 'In view of the fact that he (Newton) upholds his doctrine with some fervour, I am not interested in continuing this dispute.'

This is drama. Collaboration between these men might have produced great things. Although they clearly did not like each other, each recognised the other's qualities. For instance, in Newton's letter about *Horologium oscillatorium* that ended with theses on centrifugal force, he urged Huygens to publish more about this force, since it 'may prove of good use in naturall Philosophy & Astronomy as well as mechanicks'.[9]

Since Newton himself had discovered the properties of centrifugal force in 1665 (five years after Huygens), he knew what he was talking about. *Horologium oscillatorium* made public what

[6] OC 7, 325   [7] OC 7, 265   [8] OC 7, 302   [9] OC 7, 325

Newton had thought was known only to himself. From that time onwards, therefore, any scientist could infer (and Newton had inferred it in the mean time) that the force that kept planets in orbit round the Sun must become weaker with the square of the distance to the Sun. In addition, any scientist could now generalise the proportionality between force and acceleration (the essence of Huygens' theses) to what is known today as Newton's second law. Huygens had, indeed, hit upon something that could be put to good use in both astronomy and mechanics.

Huygens was urged by Newton to publish more but chose not to do so, even though he had, in fact, written a complete treatise on centrifugal force. The treatise, *De vi centrifuga*, is dated 15 November 1659. This was precisely one month before he completed his proof of the isochrony of the cycloid. He was reluctant to publish his treatise on centrifugal force because he was not satisfied with it. However, after his death Burchard de Volder and Bernard Fullenius decided to publish the treatise, because they believed they were acting in accordance with Huygens' last will and testament. By that time, however, Newton's *Principia mathematica* had made its impact. We can say, therefore, that Huygens' treatise was published twenty years too late. In a sense publication also came 200 years too early. By then, Albert Einstein, not without acknowledging his debt to Huygens, was deducing the ultimate consequences of the relativity of motion.

Why did Huygens not wish to publish *De vi centrifuga*? The most likely reason is that it did not clarify what was relative in circular motion. This was a weak spot in the otherwise brilliant treatment of moving frames of reference, even when accelerated. When the time came for *Horologium oscillatorium* to be printed, he was confused about the relativity of circular motion and thought that it might in fact be absolute. Later on, three or four years before his death, he returned to the problem. Now he rejected the notion that circular motion could be absolute, thereby also rejecting

Newton's idea of absolute space, and attempted to solve the problem in words:[10]

> Rotation is a relative motion of parts driven in different directions
> but kept together by a string or connection. But can one say that
> two bodies move relative to one another if their separation
> remains the same? This is perfectly possible, provided an increase
> in the separation is prevented. In fact, on the circumference (of a
> wheel) there is opposite relative motion.

Huygens believed in the complete relativity of motion, as firmly as he believed in the law of inertia, but he did not fully understand that these two concepts were in conflict with each other. 'Their inconsistency,' wrote Einstein,[11] 'was illuminated very clearly by Mach, but it had already been recognized with less clarity by Huygens and Leibniz.' This praise by Einstein may be excessive, but there is no doubt that Huygens was the first person to take relativity seriously. He wanted to study all its consequences, even if this meant withholding his treatise on centrifugal force, which could have been his greatest contribution to science. He chose a very fitting motto for *De vi centrifuga*. He took it from Horace's letters, which he had also read when he was only twelve years old: '*Libera per vacuum posui vestigia princeps*' ('I was the first to take free steps through emptiness').[12]

By lingering so long at the summit of Huygens' achievements, we are inclined to forget about the rest of the mountain beneath. There is much to be said about his other work, too. One thing we must certainly do is to dispel the myth that the remainder of his work is a loose pile of stones, a collection of casual findings. Another thing we must do is to discard the notion that it must be a monolith, representing one grand idea. Is it not time that we stopped regarding history as a dialectic of grand ideas? For grand ideas are always poorly

[10] OC 16, 233   [11] French 267–268   [12] OC 16, 302

defined, loose constructs that, on close examination, burst like soap bubbles. Huygens, for instance, began by accepting the breathtaking conjectures of René Descartes, but later in life he produced subtle arguments to shatter these conjectures. When Huygens worked with the ideas of others, he elaborated and renewed these ideas, and added important elements. To him, ideas were flexible tools with which he tried to get a grip on the world. His findings were far from casual.

Let us take a look at his early work on collisions that is contained in the unpublished treatise *De motu corporum ex percussione*, dated 1656. In this work he used not only Galileo's idea of relativity, but also an idea of Evangelista Torricelli, namely that the centre of gravity of many connected bodies lies as low as is physically possible. '*Nisi principium ponatur nihil demonstrari potest*' concluded Huygens ('Nothing can be proved unless this principle is laid down').[13]

Recognising the significance of this principle, he rephrased it several times. He first used the principle in 1646, when he tried to prove that the catenary is not a parabola.[14] He rephrased it in 1650 for his extensive study of floating bodies[15] and arrived at the brilliant generalisation of 1652, when he wrote his equations for the conservation of kinetic energy, the name given to them today.[16] The philosopher may not realise that these algebraic equations are, in fact, a rephrasing of the principle, but the physicist is stunned by their boldness.

Why Van Schooten discouraged Huygens from publishing his treatise on collisions is a story in itself.[17] Suffice it to say that publication would have dealt Cartesianism a blow. But the treatise played an important role in the development of Huygens' thinking. Let us take a look at his later work on light, namely his *Traité de la lumière* of 1677, published in 1690. Because he regarded light as a wave effect in ether, he had to return to collision theory. The Cartesian idea was that ether was a space filled randomly by myriads

[13] OC 11, 37   [14] *Ibid.*   [15] OC 11, 81   [16] OC 16, 98   [17] OC 1, 299

of invisibly small particles. Therefore the rectilinear propagation of light, as well as its reflection and refraction, could only be explained in terms of a summation of pulses caused by all kinds of collisions between these particles. Despite his mastery of mathematics Huygens could not find satisfactory solutions based on collision theory.[18] Realising the kind of mathematics that was needed to explain the propagation of light, he silently abandoned the idea of colliding particles, and invented a new principle. This principle was yet another blow to Cartesianism. The principle proved to be correct and accurately described electromagnetic waves,[19] waves that had still to be discovered.

To complete this survey of the mountain, we return to the persistent view that history is a dialectic of conflicting grand ideas. Once upon a time, Georg Hegel tried to prove that ideas were identical to realities. He used the curious argument that reality is 'mind-like' and therefore reasonable, just as any idea must be. Physicists find such a theory difficult to accept. To physicists (and to most thinking people) the ideas in our mind are different from realities or facts of nature. Ideas may conflict with one another, but facts cannot. Hegelianism, however, is still around today – in paradigms, methodologies and research programmes. Its continuing influence has not helped Huygens' reputation, and has lowered his status in the history of science.

Alexandre Koyré was the first to misjudge Huygens' work by putting it under Hegel's microscope. According to Koyré:[20] 'Huygens paid a tremendous price for his fidelity to Cartesian rationalism à outrance.' Richard Westfall[21] used the same microscope and concluded that, if Huygens were to have pursued his ideas on dynamics, 'it is reasonable to speculate that textbooks today would refer to Huygens' two laws of motion instead of to Newton's three'. Eduard Dijksterhuis, who may have had as much affinity with physics as with history, took a broader view, but still saw an idea as

---

[18] Shapiro 208   [19] De Lang 20   [20] Koyré 116   [21] *Ibid.* 188

dominating Huygens' work, namely the idea of the mechanisation of nature.[22]

Will we ever get rid of the grand ideas? Can we not start to appreciate the subtle pragmatism of Huygens' work? Some day soon, Joella Yoder[23] will be ready to help us, by using this approach.

So far we have concentrated on Huygens, the genius. Now let us turn to Christiaan, the man. We have already seen how scrupulous and painstaking he was. This characteristic is a key to both the brilliance of his mind and anguish of his soul:[24]

> *Tristitia quodcumque agitat mens inficit aegri*
> *Nec tibi judiciis propriis tunc fidere fas est.*

> The mind infects whatever it touches with a miserable sickness
> And at such a time it is not right to trust your own judgements.

He wrote these verses during his later years at Hofwijck. Immediately we step into another emptiness, and in a way we are the first to do so.

It has become customary to claim that Christiaan's character is difficult to fathom. It is as if he were impenetrable, like a statue. A century ago Johannes Bosscha, second editor of the *Oeuvres complètes* and secretary of Holland's association of sciences, addressed a meeting commemorating Huygens' death. He began his speech about Christiaan, saying:[25] 'Paying one's last respects to a friend is one of the greater griefs of life. In our eyes he is an image of noble seriousness, undisturbed by fleeting passion, an image of clarity, hardly touched by the commotion of life.' When a man has been praised to high heaven, one wonders whether he can ever be brought back to earth.

Let us try and bring Christiaan back. We will now compare some of his letters with texts about Christiaan written by his father,

[22] Dijksterhuis (1950) 405–418; 503–509
[23] Yoder *passim*   [24] OC **10**, 719   [25] Bosscha 1

Constantijn, the redoubtable poet–diplomat. Father and son had very different characters; the son appears pale beside the father. Christiaan was reserved, tending to stand aloof from social events and ceremonies. We see this attitude in the following episode. No sooner had Christiaan finished his great work of 1659 than he was expected to attend the wedding of his sister Suzanna. Whereas Constantijn describes the party with all its sounds, smells and colours[26] – mentioning the copious dinner, the kisses over the wine, the 1600 candles at the ball, the musicians, the near-uproar outside the bride's room – Christiaan (in a letter[27]) merely regrets how much time he wasted on the 'compulsory' merriment.

We know of only one letter by Christiaan that describes some kind of merriment. He wrote it when he was twenty-six; accompanied by three young men, he went on a grand tour through France – the country where he was later to be bathed in glory. Christiaan wrote to an acquaintance in the Hague:[28]

> I wish you had a flying horse . . . so that you could be with us,
> either on our trip when we sailed down the Loire, or on the
> occasions we performed a heroic deed, for instance, when we
> decided by the *lot del fortunato dado* who should sleep alone and
> who with another, or when we needed a new horse and had to
> choose out of four, the best of which was blind . . .

The trip down the Loire was to Angers, where Christiaan had to buy a doctorate in law by order of his father.

He wrote to his father on the matter:[29] 'When we get back with the diploma I shall do my best to perceive the world as you understand it, and I think it will be possible to do that if you are kind enough to let me have the time.' Christiaan was never able to free himself from the redoubtable Constantijn. It is significant that when his father died at the age of ninety, Christiaan had a portrait painted of himself in which he was depicted as an orphan[30] – an orphan aged fifty-eight.

---

[26] OC 3, 67   [27] OC 3, 65   [28] OC 1, 353   [29] OC 1, 344   [30] Vollgraff 754

He shaped and moulded the text of his letters as if he were grinding a lens. The mastery of mathematics gave him access to the physical world, in the same way as the mastery of his passions gave him access to himself. He must have thought along these lines, as did the *virtuosi* of the Renaissance who served as an example to Constantijn. Even when angry, Christiaan was usually able to retain his composure. Nevertheless, he did write a number of angry letters; there is one about Eustachio Divini who attacked his ring hypothesis regarding Saturn,[31] and there is another about Robert Hooke who contested Christiaan's claim to have invented the balance spring for a watch.[32] His rage was boundless in a letter about François Catelan, who maintained that there was an error in *Horologium oscillatorium*. He wrote:[33]

> I am astonished at his attack on my theory for the centre of oscillation, which no one has remarked upon in the nine years since its publication. Now that I have examined his so-called refutation of my theory, I am surprised that the author has not withdrawn it in the seven months since its publication. For, to put it briefly: he finds that the sum of two line segments cannot be equal to the sum of two other line segments, if the ratios of these segments differ. Imagine that the first two are 4 and 8 feet long, and the other 3 and 9 feet, and then see how you can obtain a sum for either the one or the other that is anything other than 12 . . . It would please greatly me if this could be published, so that those not familiar with my proof [*Horologium oscillatorium*], do not think that the remarks of Catelan are of any significance. Should he still return to the subject, then you will oblige me by submitting his answer to a professional before having it published. Surely that is to the advantage of his honour. And if truth be told, I find it distasteful to be attacked by such a blockhead.

[31] OC 3, 118   [32] OC 7, 528   [33] OC 8, 349

Even when abusing others, Christiaan retains a degree of
equilibrium. So did he ever lose his temper? What did he mean when
he wrote about the sickening of the mind? In his works we do,
indeed, find traces of disintegration and darkness. Not only his
diaries and notes written on loose sheets, but also his polished
letters reveal less pleasant sides to his character: cunning, lust for
money, dirty tricks, fornication, self-pity, angry outbursts. He was
extremely rude to Isaac Thuret, who dared to apply for a patent on
the balance spring that Thuret had helped to develop.[34] He played a
dirty trick on Nicolaas Hartsoeker by presenting Hartsoeker's
microscope as his own.[35] He treated these men as inferiors, even as
servants. His behaviour may have had its roots in his social
upbringing.

He was not particularly courteous to Gilles de Roberval,
either. Roberval was the only colleague in the Academy of Science in
Paris who could act as his match, and the only one who stood up
against Christiaan's pulse theory of gravity. In clear, strong terms he
argued that the pulses in Christiaan's device for explaining gravity
did not have to be directed towards a centre.[36] It is both
reprehensible and offensive to counter such a penetrating argument
by the simple statement 'that the reason I give for the particle to be
pushed towards the centre is very clear, and cannot be disputed'.[37]
This critique, aggravated by Roberval's objections to his way of
calculating oscillation centres, prompted a crisis.

We will now say something about Christiaan's melancholy, this
striking trait in his character. Should we compare his 'melancholia
hypochondrica vera et mera' with the spleen in the poetry of
Charles Baudelaire? Perhaps, but there is an important difference;
the various depressions that Christiaan experienced fit into a
pattern. His depression of 1670, possibly the deepest, followed
directly after the debate with Roberval. Francis Vernon, a secretary

---

[34] OC 7, 408–416   [35] OC 8, 96–103   [36] OC 19, 640   [37] OC 19, 642

at the English legation in Paris, has left us a moving description of his visit to Christiaan during this illness:[38]

> His weaknesse & palenesse did sufficiently declare how great a destruction his sicknesse had wrought in his health & vigour & that though all was bad, wch I saw, yet there was something worse whch the eye could not perceive nor sense discover, which was a great dejection in his vital spirits, an incredible want of sleep, wch neither hee, not those who counceld & assisted him in his sickness knew how to remedie.

He feared that he was 'neare to the very Point of Death' and complained that the Academy was 'mixt with tinctures of Envy', since it was completely dependent on the favour of a minister. As a result of his depression, Christiaan was unable to work for almost a year and had to return to Holland to recover.

We can easily guess what prompted his depressions in 1675, 1679 (around his fiftieth birthday) and 1681. They could have been brought on by the ineffectiveness of his patent on the balance spring, by the intrigues that followed the trick he played on Hartsoeker, and by the comet debate (which was won by Ole Rømer). He took to his bed, as if paralysed, and let himself be carried by a servant.[39] Members of the family who came to see him in Paris started to speak of his guilty conscience. His older brother noted that he seemed 'to be afraid of vicars'[40] and his sister provided him with a nurse, Beguine Lacour, but she was unable to help him.

Interestingly, his illnesses, at least those of 1691 and 1693, occurred not only in France, but also in Holland. Sometimes it is difficult to distinguish them from the colds accompanied by a splitting headache, from which he suffered all his life. The first headache is mentioned in a letter of 1652, in which the genius describes how this *capitis dolor* interferes with his studies.[41]

[38] OC 7, 9   [39] OC 7, 35   [40] OC 7, 27   [41] OC 1, 184

A biographer attempting to interpret this melancholy has to venture into barren land, or into emptiness. What is the cause of this debilitating force? It is certainly not mental exhaustion after a period of activity and creativity, as the events of his life make clear. What are the properties of this force? Christiaan does not write about his suffering. Or was he, in fact, doing so when he noted in the margin of a loose sheet that 'without satisfactory business, the mind yields to casual passions which often do harm to others'?[42] If so, this again points to feelings of guilt. Not only this comment, but also other hints give the impression that his work had become a refuge from the indefinite '*tristitia*', a real abyss.

Perhaps we can attempt to understand Christiaan's melancholy by following in the footsteps of Sigmund Freud:[43]

> Melancholy is characterised by profound dejection, by the loss of interest in the outside world and the loss of one's self-respect, which is expressed in self-reproach and sometimes in the anticipation that one will be punished. We can understand this syndrome somewhat better if we bear in mind that the symptoms of mourning are almost identical, the only difference being that (in mourning) one's self-esteem is not impaired.

The next step in the Freudian approach is to identify what has been lost. Whereas mourning is the reaction to the loss of a loved one, melancholy, which impairs one's self-respect, may be a reaction to the loss of a dearly loved part of one's ego. Freud assumed that the ego is, or can be, split into several parts. But this lost part of the ego cannot be buried; it is as if it remains present in the person and can never be detached. The person in mourning, however, is detached from his or her loved one.

A major symptom of melancholy is insomnia. Insomnia testifies to a person's inability to abandon all occupations, to that abandonment which is necessary to fall asleep. We have ample

---

[42] Vollgraff 493   [43] Freud 74

evidence of Christiaan's insomnia. There is another major symptom of melancholy. Following the loss of part of one's ego, the ego that remains will regard itself as worthless and reprehensible. The melancholic rails at himself and expects to be driven out and punished. According to Christiaan's sister-in-law, who often came to see him in the last few months of his life, Christiaan presented these symptoms, and displayed this kind of behaviour.[44]

If there is some truth in this interpretation, let us now consider what part of his ego Christiaan may have lost. According to Freud's theory, it would have been the part he valued most. In Christiaan, two strands, two lives were in competition: the personal and the intellectual. Arthur Schopenhauer described such a dichotomy:[45]

> A privileged man (like a genius) leads, alongside his personal life, another life that is intellectual. It is this life that gradually becomes his only goal, and for which he comes to consider the other as merely a means towards achieving it. This intellectual life, especially, comes to preoccupy him; it acquires, through the continual growth of his insight and knowledge, a lasting coherence and intensity; it moves constantly forward towards a more self-contained perfection and fulfilment, like a work of art in genesis.

Was it, then, this life that Christiaan had lost? Was it his genius that he lost?

We can ask questions, but our answers can only be tentative because a man's soul defies analysis. If Christiaan's melancholy of 1670 was due to a 'loss of genius' as a result of being exposed to Roberval's profound intellectual critique, then he must have been preoccupied by the notion that either his ego must be intellectually superior to anyone else's or it must cease to exist. This attitude probably stemmed from his constant striving to perceive the world in the same way as his father did. Christiaan's melancholy, however,

---

[44] Huygens **25**, 472–504   [45] Schopenhauer 299 (§52)

was not just the result of a 'loss of genius', it was mingled with symptoms of true mourning.

As we have seen, in his deepest melancholy Christiaan asked to be carried around – like a child. This may have been connected with his early memories about his mother's death. He was only eight years old when his mother Suzanna lay on her deathbed, and he was the only child to be admitted to the sickroom.[46] He had to be lifted up so that he could see his mother. She kissed him goodbye, saying: 'Kom hier mijn soete mannetie, laet ick u eens kussen.'

Six months later the boy, unlike the other children in the family, was still wearing a 'long mourning skirt'.[47] Suzanna and Christiaan were said to resemble one another in that they were both of a calm and serious disposition. But several years before his mother's death, Christiaan would sometimes withdraw into himself.[48] However, he is also said to have been very obedient and helpful, though easily hurt. At a very early age he learned to take refuge in an inner world, the world of the intellect. As he grew older it became increasingly difficult for him to find any comfort in the world outside his intellectual world. Perhaps this can help us to understand the drama surrounding his death.

His older brother has left us a description.[49] Christiaan lay in a darkened room in a house in the Noordeinde, in The Hague. He was in pain, he started to cut himself with glass splinters, refused food because he thought it was poisoned, he shouted deliriously that 'people would tear him to pieces if they only knew what he thought about religion'. When he finally agreed that the vicar be summoned, it was because he felt he no longer had the strength to resist his family's wishes. But even in the presence of the vicar, he stuck to his views – although we do not know what these were. 'The Reverend Olivier addressed him for a long time,' wrote the brother, 'and prayed for him, but he is not willing to change his mind. Sadness all

---

[46] Eyffinger 107 (§75)   [47] *Ibid.* 107 (§78)
[48] *Ibid.* 106 (§68)   [49] Huygens **25**, 472–504

round.' During the night he loses consciousness. At half past three in the morning the family is informed.

When Christiaan Huygens disappeared into the emptiness on 8 July 1695, it is unlikely that his shocked family was able to grasp the fact that this scrupulous mind craved clarity right up until the very end.

I Titan

It was 14 April 1629. The mother was certain that she would give birth to a monster, because just a few days earlier, she had been frightened by the disfigured face of a street urchin. But her son was born without blemish, though extremely heavy. She brought him into the world in the dead of night, and when dawn broke it was Saturday, the day of Saturn the giant, begetter of giants. If he had been born one day later, the father thought, he would have been an Easter child. And in the name of the Father, the Son and the Holy Ghost, he had him baptised Christiaan.[1]

These are the physical beginnings of our biography of Christiaan Huygens. But if our story is to come to life, we need to know what is so special about the man. If he was truly an exceptional scholar, as many maintain, then he must be worth our exceptional scrutiny. No simple task. What was considered extraordinary three centuries ago may well seem perfectly ordinary to us today. So we must first explore the world of seventeenth-century science.

Christiaan lived in the age wherein (and now we quote):[2]

> Philosophy comes in with a Spring-tide; and the Peripateticks may
> as well hope to stop the Current of the Tide, or (with *Xerxes*) to
> fetter the Ocean, as hinder the overflowing of free Philosophy:
> Me-thinks, I see how the old Rubbish must be thrown away, and
> the rotten Buildings be overthrown, and carried away with so
> powerful an Inundation. These are the days that must lay a
> Foundation of a more magnificent Philosophy, never to be

---

[1] Eyffinger 103 (§§59 & 61)   [2] Cohen (1983) 22 (n1)

overthrown: that will Empirically and Sensibly canvass the
Phenomena of Nature, deducing the Causes of things from such
Originals in Nature, as we observe are producible by Art, and the
infallible demonstration of Mechanicks: and certainly, this is the
way, and no other, to build a true and permanent Philosophy.

This quotation dates from 1664. The otherwise insignificant
writer to whom it is attributed was describing the scientific revolution of the seventeenth century. And in laying the foundations for the
science that was both splendid and superior, and 'never to be overthrown', Christiaan Huygens played a key role.

Let us follow the masterly overview that E. J. Dijksterhuis was
able to give when all of his works had been published.[3] Despite recent
arguments that his role was rather peculiar – but what is not peculiar
in a fast transition?[4] – we think this overview is fitting, and still largely
valid.

By the time Christiaan, aged sixteen, went off to study at Leiden University, the scientific revolution was already under way. The professors were well versed in the geometry of ancient Greece, as well as
goniometry and symbolic algebra. René Descartes had made improvements in symbolic algebra that would enable it to be used as a simple
and powerful tool in mathematics. Together with Blaise Pascal, he
demonstrated how this algebra could be applied to geometry, so that
the circuitous methods of the Greeks would no longer be necessary.
However, because Archimedes was still held in such high esteem,
mathematics was mainly grounded in geometry, and was directed at
determining curve lengths, surfaces and centres of gravity of flat models with a curved circumference. Scientists were still seeking ways to
calculate the tangent to an arbitrary curve, and to estimate the highest
and lowest values. Towards the end of the century these labours led

---

[3] Dijksterhuis (1951/53) *passim*   [4] Cohen (1994) 133, 199

to the discovery of the differential calculus. Numerical mathematics evolved quite separately from geometry. At the beginning of the century, Henry Briggs had developed the logarithm, which not only simplified computation but also enabled the progress of pure mathematics. One generation later Pierre de Fermat resuscitated Diophantus' number theory and discovered, in correspondence with Pascal, several fundamentals of the probability theory.

Around 1645, Galileo played the same role in mechanics as Archimedes had done in mathematics. At the beginning of the century he had explained the motion of falling or thrown objects, using precise methods of observation and measurement, and had written a stimulating treatise on the subject. He raised the problem of the origin of motion, which would only be solved at the end of the century. He also demonstrated how principles governing inertia and relativity of motion could be applied. And in connection with these subjects, he defended the heliocentric World System that Johannes Kepler had revealed with his planetary laws. Theologians continued to reject it, but astronomers were inclined to accept this reality, though it would be a long time before they were convinced that it was the Earth that moved. The telescope, an instrument that had evolved from spectacle lenses, led to the discovery of numerous celestial phenomena which supported this heliocentric view of the universe. Kepler provided further insight into the medieval theory on the passage of light through glass (the dioptrics), but refraction, the key to this concept, would only be discovered later, by Willibrord Snel, and represented in Snel's Law. Medieval knowledge of natural history was further enriched by William Gilbert, who studied magnetism, and Evangelista Torricelli, who measured the equilibrium in fluids and gases. Pascal combined these into a single theory, reawakening interest in the molecule theory of ancient Greece, which Pierre Gassend formulated anew.

It was Descartes who invented a revolutionary theory to encompass all these innovations. When Christiaan arrived in Leiden, this theory was just beginning to find support. Descartes believed that all natural phenomena were caused by the collision of invisible tiny particles of matter. Because of their invisibility, the proof of their existence must necessarily be metaphysical, yet he was convinced of his success. All physics could be understood by means of a simple theory of collision and the use of simple mathematics. This theoretical idea held powerful sway because it fitted in so well with Gassend's molecule theory and Cartesian symbolic algebra.

But apart from such theoretical ideas, it was also strongly believed that the study of natural history must develop through reason, inquiry and experiment. Francis Bacon had insisted on this at the beginning of the seventeenth century. But his programme of experiments was not well thought out, and was virtually unworkable. It was only in the second half of the century that his plea for an empirical approach led to a logical and coherent programme, which could be carried out by English and French scientific societies. Instruments would be required, however, such as air pumps and accurate clocks, and these were yet to be developed.

Unlike in our time, in the seventeenth century the study of physics wasn't regarded as something out of the ordinary, but as something familiar – indeed, as a noble task. Descartes expected that his theory would enable people to become 'lord and master of nature', not for the sake of power, but in order 'to enjoy, with neither effort nor exertion, all the fruits of this earth', and to remain in a state of good health, for 'good health provides the basis for all further happiness'. Bacon anticipated that empirical science would enable physicians to lengthen people's lives:

> This is her new and most noble task. People will honour doctors,
> not only because they are necessary, but also because they
> dispense to man his greatest happiness on earth.

Now that we have set the stage for scientific thought in the seventeenth century, let us consider the role of Christiaan Huygens. He had been sent up to Leiden to read law, but felt drawn towards the study of mathematics. Having familiarised himself with the working methods of the Greeks, he very soon produced solutions that even Archimedes had not thought of. In 1650 he supplemented the classical work on floating bodies with *De iis quae liquido supernatant* (On Objects which Protrude above the Surface of a Liquid), and in 1654, the classical work on the circle, *De circuli magnitudine inventa* (On Finding the Circumference of a Circle). He published the second work, but not the first, as if that had been merely a finger exercise.

For Christiaan Huygens, mathematics was a means and not an end, although later on he did devote himself more to mathematics than was strictly necessary. The same applies to his book *Van Reckening in Speelen van Geluck* (Mathematics in Games of Chance), which consisted of practical rules for playing dice, while at the same time contributing to the theory of probabilities. Similarly, his work on logarithms, intended for the improvement of tuning musical instruments, also deals with their mathematical significance.

He started out, however, with pure mathematics. In 1651 he published a method to calculate the surface of a segment of an ellipse or a hyperbola, thus embarking upon a field now known as integration. He carried on with integrals for several years and finally achieved such interesting results that outstanding mathematicians of the time acknowledged him as their master. Interestingly, they only got to know about these results through his letters, as he published very little. He also revealed very little in his letters as to how he had reached these conclusions.

It was quite normal at the time to be reticent about publication. Some mathematicians actually made their living by keeping their methods secret, such as Gilles Personne de Roberval, who had to sit an exam every three years in order to maintain his professorship. But this wasn't necessary for Christiaan. The reason he didn't publish his findings was because he wasn't easily satisfied with them. Proofs still had to be demonstrated in terms of relationships in geometrical figures according to Archimedes' standard. This proved a handicap for many of the newly emerging problems; it was time-consuming and a waste of talent. The integrals of the parabola, for example, one of his unpublished masterpieces, were also demonstrated by Hendrik van Heuraet, Pierre de Fermat and James Gregory, without any of them being aware either of his work or of each other's. It became a matter of economic necessity to produce these integrals with greater speed.

The process of differentiation that would make this possible had already been set out in 1675. In fact, Christiaan had stumbled upon it when Gottfried Wilhelm Leibniz, who discovered it, had come to him two years earlier to learn mathematics. It took Christiaan twelve years to understand the new method, but even then he found it difficult to apply.

In the field of mathematics, Christiaan was clearly a virtuoso, but not an innovator. The science that he did rejuvenate was physics. In 1652, he wrote the very first physics formula (a mathematical equation for a physical phenomenon). For this alone, he may be regarded as the founder of mathematical physics.

The first phenomena that we encounter in this border area between physics and mathematics are those of equilibrium and motion. Phenomena such as heat and magnetism are more complex and were barely understood in the seventeenth century. It was only logical that Christiaan should turn his attention to the science of mechanics, which had been given new impetus by Galileo with his research into falling and thrown objects. In 1652, thanks to his own mathematical precision, Christiaan discovered the theory of collision, and in 1659, the formula for force. Yet he failed to see the causal connection between motion and force, which made his studies of motion less convincing. Nevertheless, his contribution has been exceptional. In order to understand it fully we have to see it in connection with his practical interest in the pendulum clock.

Galileo had already studied the regular motion of a pendulum. In his later years he had proposed using this to operate a clock, but had not got round to actually making one. Christiaan, who was clever with his hands and technically minded, invented a fork to connect the pendulum to the cogwheel of the clock, allowing it to swing, unimpeded. This discovery ensured the success of the pendulum clock and motivated him to extend his studies in three different directions.

First of all, he tried to adapt this clock for use in shipping, both to measure time and to determine distances at sea. Secondly, he tried to make the oscillation of the pendulum (a bob swinging on the end of a string) independent of deflection by changing the path of the bob from the circular arc. Thirdly, he tried to describe the pendulum more accurately than as just a string with a point mass on the end, meant to represent the bob, because that was incorrect.

His first attempt led to the invention of the clock with a spiral spring, a sort of wound-up pendulum. His second led to the discovery that the path of the bob that guarantees an equal time swing is a cycloid, a curved line that led to his discovery of evolutes. His third attempt yielded a theory to calculate the centre of oscillation of an extended pendulum: the first step down the long road to the dynamics of solid bodies. For this he used Torricelli's proposition that the centre of gravity of a number of interconnected objects, in equilibrium, lies as low as possible. This proposition seemed so obvious to him that he also applied it to single colliding objects. He had already discovered what is now known as the law of conservation of kinetic energy.

Christiaan did eventually publish some of these findings, though he waited a long time and had to be repeatedly exhorted to do so. When *Horologium oscillatorium* (The Pendulum Clock) finally appeared in 1673, it was regarded as a masterpiece, even by Isaac Newton, whom we shall meet again. But Christiaan's greatest innovation hadn't been included in the work – the theory of centrifugal force and the ingenious deduction of the laws of collision with the aid of a principle of relativity. By locking away this pioneering work in his study, he had far less influence on the development of physics than he might otherwise have done.

As we have seen, Descartes' knowledge of collision theory was the foundation of his revolutionary view of the universe. The philosopher had made these deductions himself and had them published. But right from the start of his own investigations, Christiaan had discovered that the Frenchman's conclusions were wrong. It was not only contrary to the laws of physics, which was forgivable for a philosopher, but also to the laws of logic, which was unforgivable. This was so awkward that Christiaan's professor of Cartesian mathematics asked him not to publish them. He complied. What was even more remarkable: Christiaan was in no way put off Descartes' idea. He simply set himself the task of meticulously rebuilding this edifice, stone by stone.

Christiaan's *De coronis et parheliis* (On Rings around and near to the Sun) took Descartes' *Les Météores* (Weather) a step further; his *Discours de la cause de la pesanteur* (Discussions on the Cause of Gravity) worked out the effect of gravity caused by scattered particles; his *Traité de la lumière* (Treatise on Light) calculated light to be caused by colliding particles. It was Huygens' collision theory on light that would have the greatest influence on the development of physics, but only a hundred years after its publication. At that time his mathematical understanding of the collision effect proved extremely useful in determining the effect of waves or vibrations. This in turn could be used to explain the interference phenomenon in light. Apart from this, his *Traité* was highly thought of by his own contemporaries, especially Leibniz.

Just as the pendulum clock constituted the practical side of his theoretical mechanics, the telescope constituted the practical side of his theoretical optics. Together with one of his brothers, he polished lenses and made telescopes. He calculated the image formation of lenses by applying Snel's law of refraction on spherical surfaces. In this way he obtained insight into the magnification of the telescopic image and the distortion of the image known as spherical aberration. With that insight, he was able to design a compound eyepiece with little colour distortion, which became known as the 'Huygens eyepiece' and was frequently copied.

This work, too, remained unpublished. In 1656, however, he published his findings on certain peculiar properties of Saturn, which he had discovered with his first home-made telescope. He reported that this planet had a moon – Titan – and a ring. His discovery of the ring, in particular, was quite unprecedented, and brought Huygens instant fame. Nevertheless, he was not the sort of astronomer to scan the skies night after night, looking for something new, like the fanatical stargazers Johannes Hevelius and Giovanni Cassini had done. He was more interested in putting these celestial phenomena, about which there was so much speculation, into the proper perspective.

In the end, this perspective led to a series of original experiments, which would make him the first to produce a reasonable estimate of the distance between stars. It wasn't until the nineteenth century that this estimate was improved upon. His *Cosmotheoros* (Contemplation of the Universe), in which he published his findings, is a fascinating cosmological work, remarkable for the stringency of his speculations, its unprecedented popularity in the eighteenth century, and the vilification that followed. Until recent times people have tried to conceal it, for fear that Huygens might be discredited.

In 1687, just as he was embarking upon this later work, Christiaan acquired a copy of Isaac Newton's *Principia mathematica*. The book made a deep impression on him. But just as he had had difficulty with Leibniz's new mathematics, he also found it hard to grasp Newton's new physics. Newton had solved the problem of motion raised by Galileo with the aid of an inertial force. But he also introduced a gravitational force, which was to work at a distance. Though how, not even Newton could explain. For Christiaan, it was a step backwards into the darkness that he had struggled to overcome. He found it absurd.

That is how our man was depicted in his own time: as an intellectual giant. Yet we search in vain in The Hague for his statue.[5] The

[5] Van Berkel 236–238

Hague was the city where he was born and died, but, more important still, it was the city where he made his greatest discoveries.
Two attempts have been made to erect a statue in his honour. The
first, in 1868, by the Haagsche Maatschappij Diligentia – a society
of mathematicians – together with the Royal Society of Sciences in
1868, soon failed, because the two parties were unable to raise enough
money among the citizens. The second attempt, in 1905, by the Hollandsche Maatschappij der Wetenschappen, after they had received a
legacy of 40 000 guilders for this purpose, actually resulted in a design.
The legacy had been left by L. Bleekrode and the design was by the
famous Dutch architect P. J. H. Cuypers. It was to be a 20-metre-high
monument, rectangular, with a niche on either side for a sculpture
of Christiaan Huygens, as well as three depictions of his life. On top
of this would be a column surrounded by the signs of the zodiac and
crowned with various female figures. The Lange Voorhout, in the heart
of The Hague, was suggested as a suitable location. The city council
turned it down.

Would we prefer a different image? We have Titan. In June 1655, Christiaan was able to confirm his discovery of a moon in the vicinity of
Saturn, for he had observed a starlet travelling four times around the
planet. He sent an anagram to astronomers in London and Prague, consisting of a line from Ovid and the letters 'uuuuuuu ccc rr h b q x'. The
line was: '*Admovere oculis distantia sidera nostris*' ('Faraway stars
move towards our eyes'). Those who had an idea as to the meaning of
the anagram were able to rearrange these letters to form the sentence:
'*Saturno luna sua circumducitur diebus sexdecim horis quator*'
('A moon revolves around Saturn in sixteen days and four hours').
This is Titan, the largest moon in our solar system.[6] It is the size of
a small planet, such as Mercury, it is two-and-a-half times as small
and forty times lighter than the Earth, and shrouded in a thick mist of
nitrogen and methane, which absorbs most of the light. At a distance

---

[6] Hunten *et al.* 671–759

of 1.2 million kilometres from Saturn, it completes its orbit in 15 days, 22 hours and 41 minutes. The fact that Titan – together with the ring that he discovered shortly afterwards – is a fitting image for Christiaan Huygens, is borne out by his own words:[7]

> *Ingenii vivent monumenta, inscriptaque coelo*
> *Nomina victuri post mea fata canent.*

> Let them remain as signs of my sagacity, and their names
> That I write across the heavens be an echo to my fame.

[7] OC 21, 315

2   Father

He was strong. His days would number 33 046, time enough for Saturn to travel three times round the Sun, and the Earth ninety times. He took life in his stride, and each day added to the store of his contentment.

To be sure, there were small discomforts. Now and again father Constantijn felt stabbing pains in his big toe and was worried it might be gout, but he was over eighty before that really troubled him. Occasionally he had long bouts of fever, perhaps from a bone inflammation; his lymph glands swelled from time to time and he would have them cauterised. He probably had tuberculosis, because he used mineral water as medicine and couldn't walk more than twenty paces at a time. For the rest, he was deaf in one ear, but this did not worry him unduly.

Actually, he had only one real disability. Severe myopia forced him, from the age of sixteen, to wear strong spectacles of hollow glass, and the vitreous membrane in his enlarged eyeballs became detached, so that everything in his line of vision seemed to be milling around.[1] Constantijn meticulously wrote down all these symptoms, in a spidery handwriting in which he recorded nearly everything about himself in thick folios or on loose pages – any scrap of paper he could lay hands on. And he never threw anything away. His physical strength was surpassed, as it were, by the strength of mind with which he created this towering mountain of paper as testimony of his vanity.

We must look upon him as a public servant. Not only does this happen to be true; he also regarded himself as such.[2] But it was as a servant to the powers-that-be that he rose to become secretary to the

[1] Van Lieburg 171–174   [2] Worp 8, 179–236 (De vita propria)

Prince and '*aulicus dicax*' (crowing courtier), as he called himself, not without self-mockery.

And that was not all; from this copious grist of official and personal writings arose 75 555 poems, or 'Cornflowers': 48 590 in Dutch, 19 962 in Latin, 6579 in French and the rest in Italian, Greek, German and Spanish. In his anthology *Cornflowers* we read the light-hearted:[3]

> Dew and darkness both descending,
> Sun and roosters gone to coop,
> All the gables, all the branches,
> All the young girls just as pretty,
> All the cheeks just as blushing,
> All the eyes just as swift,
> All the lips just as rosy,
> All the mouths just as tight.

This verse is not difficult to understand, which makes it uncharacteristic of Constantijn Huygens. He joked about his '*versus inopes rerumque nugaeque canorae*' ('verses poor in content, melodious trifles'), but that isn't to say that he simply dashed them off. He was constantly polishing the rhythm of his words. That is why his poetry appears hermetic in style and seems to stem more from the head than from the heart. A typical example:[4]

> Beauty, infirmity's veneer,
> Eye's temptress, shadow's shades,
> Trees' semblance,
> Judges not such judgement shrewd,
> Thus may your lustre fade
> With each returning winter,
> Like the linden branches
> For you an honour undeserved.

---

[3] *Ibid.* 1, 227 (*Batava tempe* 497–504)   [4] *Ibid.* 1, 233 (*Batava tempe* 713–720)

Who can understand these verses (from the same poem as the lines about the pretty girls, a poem about the Haagse Voorhout) straight off? We need time to figure out what he is saying, that it might be something like: If you, beauty, are compared to trees, do not be mistaken: your beauty fades when winter comes, but it would be too great an honour to compare you with the linden. It is then up to us to realise that bare linden branches are very beautiful in winter.

The fact that he was a poet without wanting to be, and that he is often so hard to understand, reveals to us something of the tension in his personality. Constantijn Huygens was highly gifted, and raised to be a virtuoso, in accordance with the Renaissance ideal of an educated, upright and courageous man.[5]

He was born in The Hague on 4 September 1596, the second son of Christiaan Huygens and Suzanna Hoefnagel. Christiaan senior, as we shall now call him, came from Terheijden near Breda, and Suzanna from Antwerp. Constantijn most likely inherited his artistic inclinations from his mother. She descended from a family of rich merchants and, after her father's death, moved with her mother to Amsterdam, before Antwerp fell into Spanish hands. In 1592, at the age of twenty-two, she married the forty-five-year-old Christiaan senior, who at that time was the sole secretary to the Council of State, and before that, one of four private secretaries to William of Orange. Their first son was named Maurits, after the new Regent who was godfather at his christening. For their second and last son they chose the name Constantijn, in memory of the tenacity (constancy) of Breda during the Spanish siege. Four daughters followed: Elisabeth, Geertruyd, Catherina and Constantia, of which the first and the third were to die in their teens.[6]

Thanks to his unconditional loyalty to the House of Orange and the Reformation, both inextricably linked with the Republican war of

---

[5] Bachrach 28   [6] Smit 10–20; Vollgraff 391–394

liberation, Christiaan senior 'made it' during the chaotic beginnings of the war.[7] Although not of the nobility, he enjoyed a position of high authority in the administration. As a class-conscious man he wished his position to be heritable, and as this would never be granted to a man of common birth, he drilled his sons.

The two boys received a sound education, at home, in the classics and modern languages, and were taught the etiquette of diplomacy. The latter came easily to their father, accustomed as he was to mixing with ambassadors in The Hague through his work for the Council of State, which at that time dealt with foreign affairs.

When Constantijn, the most talented of his sons, later describes his childhood, he comes across as an unbearable little genius.[8] At the age of two he sang the Ten Commandments to his mother, in French, with perfect articulation and in proper time. When he was five, he could sight-sing, after having learned the notes with the aid of the gilded buttons on the sleeve of his winter coat. And at the age of seven he played the violin to a gathering of musicians – including Jan Sweelinck – who were duly amazed at his performance. It was flawless, 'while others made many errors'. Unfortunately, in raising his eyes from the music to look at his admirers, he lost his place, burst into tears, and couldn't be persuaded to start again.

By the age of seven he could also speak French – the language of diplomacy adopted by the upper circles in The Hague – and quite impressively, too. When he was nine years old he and his brother Maurits put on a rather ponderous French play, about Abraham's sacrifice, for the benefit of several prominent individuals, including the widow of William of Orange, Louise de Coligny. He was one of her regular visitors, and she maintained that she always learned something from their conversations together. From the age of nine he could also read and write Latin.

This is how it appears in his early autobiography.[9] And it may even be true. But what this – unfinished – work expresses most of all is

---

[7] Israel (1995) 241–262   [8] Strengholt 9–15   [9] Smit 21–37

the gratitude he felt for his upbringing, and the admiration he had for his father. Christiaan senior's rigorous pedagogy, borrowed from Filips van Marnix, Lord of Sint-Aldegonde, entailed training in music, dance, horse-riding, fencing, drawing, sculpture and, of course, languages, all of which was well spent on a bright child like Constantijn, but less so on the slower Maurits.

In 1606, the Leiden University student Johan Dedel came to the house to teach Constantijn Latin, instructing him first in grammar and syntax and then moving on to read Latin writers. These works, especially Ovid's compelling poetry, inspired him to compose his own. His first poetry dates from 1607, and already he was composing in rhythmic lines.

Up until 1612 he only wrote poetry in Latin, reams of it, which Christiaan senior loved to flaunt. But once, when he took him along to Brussels to visit a Jesuit priest who wanted him as a pupil, he went too far. Constantijn was to express his refusal in Latin verse, but he couldn't say a word. 'I was unable to coax from my muse something like "My Fatherland is sweeter than any other", and was furious with myself in my childish ambition when, later of course, any number of answers sprung to my mind.' Here too, like his violin-playing in Sweelinck's presence, was an ambition fed to him by his elders, and a failure that would dog him all his life.

In 1612 his younger sister Elisabeth died, an inexplicable loss that his parents could not accept but which moved him for the first time to write poetry in French – a translation, incidentally, of a Latin epitaph. His first poetry in Dutch wasn't until 1614, another translation, this time of Guillaume du Bartas' lofty French verses, inspired by Horace and extolling the life of the peasant.[10]

Constantijn may have been clever, but he was far from handsome. He had difficulty with girls. Possibly his egocentric character stood in the

[10] Strengholt 22–24

way. When he was eighteen years old he moved with his parents to
the fashionable Haagse Voorhout.

A young lady in one of the neighbouring houses, Dorothea van
Dorp – no great beauty herself – started up a flirtation with him. She
even went so far as to declare her love for him in The Hague's Wood,
and to give him a ring. According to J. Smit, one of his biographers, his
response was bashful, gauche and overwhelmed, like a boy of fourteen
or fifteen years old. When he left for Leiden two years later, however,
her passion evaporated as quickly as it had begun. At first she wrote to
him ('How my heart throbbed with happiness when I received a letter
from her in Leiden'), but not for long. Apparently, she had her eye on
someone else. She ended up with no one at all.[11]

Constantijn was deeply hurt. This is clear from the emotional
tone of the two poems, ''Tis torment' and 'Doris, or Shepherd's lament'
in which he came to terms, rhetorically, with his unrequited love. But
twenty years later he began yet another poem with the lines:[12]

Dorothy, my first true love,
A love whose end eludes me still

only to cross them out and replace them with something more neutral.
He seems to have taken a dislike to women and to have put thoughts
of marriage quite out of his mind.

As mentioned earlier, in 1616 he went to Leiden for a year to obtain a
law degree at the university there, just as Christiaan senior had done
in Douai. And Maurits did the same. Even without fatherly supervision, they did not join in the usual brawls and binges characteristic of
student life. They even dutifully attended a few lectures.

There were no set study programmes, nor any exams. It is not
known whether or not Maurits concluded his studies with a disputation. Possibly Christiaan senior had called him back to The Hague
by then, to get used to working for the Council of State, since he

[11] Smit 38–46; Keesing (1987) 23–38   [12] Worp 4, 29 (n)

saw him as his successor. Constantijn, who had been ordered to learn English and to attend geography lectures given by an English tutor, did his disputation in 1617. This involved defending several legal theses, after which, upon payment of cash to the professors and the beadle, he could collect his degree. In Leiden it was considered improper to award a degree in exchange for cash alone; a disputation was required as well.

Why English? Christiaan senior felt that Constantijn, who could not succeed him and would therefore have to be given a different task in the diplomatic service, should familiarise himself with the power most important to the independence of the Republic. In the war against Spain this was no longer France, but England. She maintained garrisons in three fortified cities, and the English ambassador in The Hague, Sir Dudley Carleton, had a seat on the Council of State. He knew Constantijn from his musical performances in various house concerts. Christiaan senior managed to persuade him to include his son as part of his retinue for a short visit to London in the summer of 1618.[13]

Constantijn must have welcomed this break after his rather dull apprenticeship to a lawyer in Zierikzee. But at the same time it was a cultural learning experience, and had a great influence on him. After leaving Carleton's retinue, he went to stay with the Dutch ambassador in London, Noel de Caron, and came in contact with several fascinating figures of the English Renaissance. He was invited to play the lute for James I, and visited the London palaces, as well as the universities of Oxford and Cambridge. The Bodleian Library made a deep impression on him, and after his visit there he wrote:[14]

> To Bodley's Vatican! There, there's delight
> To dwell forever in, there doth thy straine
> Rise to equal to the place, and thy rich vaine,
> Frought with that learned wealth, doth seeme to bee
> In that description a new Library.

[13] Smit 58–68   [14] Worp 1, 262–267; Bachrach 33

He had never seen anything quite so impressive in Leiden. This poem also shows that he had developed quite an eye for architecture.

On his journey home, he met Anglican bishops on their way to take part in the Synod of Dordrecht.[15] At the beginning of 1619, this serious young man went to Dordrecht himself to attend one of the sittings.

In the theological dispute of that time, his father – and he, too – stood firmly on the side of the orthodox Calvinists, who refused to accept any weakening of the doctrine of divine predestination. Man could not contribute to his own salvation. Constantijn's poetry at that time is rampant with pronouncements that endorse the synod's decisions. In his later poetry, too, we see these pietistic tendencies; for example in his brief 'To God', from 1669:[16]

> When Thou knock'st upon the stony heart,
> There where no ear be granted,
> Wilt thou let that heart to hope,
> Then knock, Lord, and open Thou't.

But Constantijn wasn't one to split hairs, and hair-splitting wasn't reserved for the orthodox Calvinists. And it couldn't have been otherwise, if we look at the tension between what he professed and what he actually did; between the strict Calvinism of his church and the free humanism of his upbringing.

This explains why he sent his first published poetry to Johannes Wtenbogaert, the minister from The Hague, who had fled to the Spanish Netherlands after the condemnation of the remonstrant doctrine at the Synod of Dordrecht. In a friendly letter, Constantijn suggested that he, Wtenbogaert, had become the victim of the modernists, whereas he could have known that this friend of the family had, on the contrary, been among those who had taken their lead.[17]

As a result of this theological dispute, which had taken on social and political dimensions, many more victims fell. In May 1619, the

---

[15] Israel (1995) 450–462   [16] Worp 7, 222   [17] Strengholt 28–31

Grand Pensionary Johan van Oldenbarnevelt was beheaded in the Binnenhof of The Hague, and the solemn poet would undoubtedly have witnessed it. Joost van den Vondel protested in a fiery epigram. Constantijn remained utterly silent.

His future lay in his loyalty to the orthodox-reformed leaders who, now that the armistice with Spain was coming to an end, once again needed to be certain of foreign help. This support had to be in keeping with the pattern of coalitions between Protestant rulers against the Catholics, and The Hague was a hotbed of Protestant diplomacy. These were ideal times for a young man from The Hague who aspired to a diplomatic career.

The war effort should not be underestimated, because although the mercenary armies maintained by the Republic and the other European states were not constantly engaged in battle, they were large and therefore expensive. During the 1630s the Republic had around 50 000 soldiers in service, and in the 1670s, because the war proved to be never-ending, around 110 000. The figures for Spain were 300 000 and 70 000, England probably 40 000 and 80 000, and France 150 000 and 120 000.[18]

The efforts of the Republic were enormous for her population of 1.5 to 1.8 million people, but her riches made them possible. Yet she tried to find support wherever it was to be had. With that in mind, the wily diplomat Van Aerssen was sent to the Venetian Republic.

François van Aerssen was a neighbour of Christiaan senior on the Haagse Voorhout. He had heard Constantijn speaking Italian, and took him along. Constantijn was even more deeply impressed by his trip to Italy than by his journey to England. He wrote everything down, with a keen sense of beauty: the rapids of the Rhine, crashing down between steep rocks, with fortresses above and towns below. On the road from Eberstadt to Heidelberg, beyond the unbroken row of fruit

---

[18] Parker & Smith 4

trees, the vineyard-covered hills to their left, and to their right, green meadows as far as the eye could see. Later, rare manuscripts in the library of the palatinate of Heidelberg. The thundering waterfall of Schaffhausen, the snowy peaks of the Alps, the slippery Splügen Pass, which he crossed on horseback and where he nearly lost his footing. After the mountains, Chiavenna, abundant with fruit and flowers. The city of the Doges. There he informed the Doge, in his best Italian, on the arrival of the Dutch delegation. He heard the music of Claudio Monteverdi in the magnificent San Marco. And on returning home, there was his adventure high atop the cathedral of Strasbourg. Carried away by ambition, he climbed the outside of the spire, grasping onto the stone teeth and iron hooks. But when he looked over his shoulder on his way down, he realised he had no head for heights. He felt sick, his heart beat wildly, and only with the greatest effort was he able to reach the bottom.[19]

Van Aerssen achieved nothing, but that was not the fault of his secretary. The diplomat wrote a favourable reference. Six months later, Constantijn acted as secretary once more, this time to a special delegation to the London court of James I, who had Catholic leanings. It was an important delegation, for war was about to break out and England would once again have to provide the main support. But first there were various issues that needed resolving.

To begin with, there was the question of herring fishery, and trade with the East Indies. Then there was Elisabeth, daughter of James I, who was causing some concern. She had been Queen of Bohemia for a single winter and had set up her own miniature court in The Hague, after separating from her husband Frederick V. And throughout this time, there was the ongoing theological dispute that had been settled in the Republic by a synod, but was far from being resolved in England.

[19] Smit 75–82

For that reason, the mission, once again led by Van Aerssen, would take some time. Constantijn spent the spring of 1621 in London, was there for the whole of 1622, and in the spring of 1624, when James's Spanish policy had reached a deadlock, and a treaty could be signed with the Republic. Eighteen months in all. During this time Constantijn successfully completed his education, begun by Christiaan senior, who died in February 1624.

Before we go any further, we must first say something about Constantijn's debut as a poet.[20] This took place in 1622, when Jacob Cats published two of his longer poems. One of these was about the Voorhout (from which we have already quoted). He started writing it in the rainy summer of 1621, after a visit to Pieter C. Hooft at the Muiderslot, where he probably met the Visscher sisters, Anna and Maria, the latter known as Tesselschade, or simply, Tessel.

In October, when he was halfway through, he received a letter from Cats in Middelburg, requesting that he contribute a poem for an anthology of Dutch verse. When Constantijn had finished it and sent it in, Cats was so delighted that he immediately asked him to write another. So when he returned to London for the negotiations, Constantijn wrote his second poem – this time an ironical work about fashion fads, which he called 'Exquisite trivia'. It was a powerful debut. The two poems brought him instant fame. When Maurits complained that he found them too obscure, Constantijn couldn't help laughing.

His sudden fame coincided with his knighthood, engineered by Van Aerssen.[21] In October 1622, he went to one of the king's country homes to request a hearing. James I heard him out, beckoned to a nobleman, took his sword and knighted the half-dazed Constantijn on the spot. The favour, like his degree at Leiden five years earlier, wasn't without its price. He referred to himself, ironically of course, as 'Sir Constantine'. Still, it tells us something about his later aspirations for similar status in the Republic. But this would not be quite as forthcoming.

[20] *Ibid.* 95–100   [21] *Ibid.* 101

During his previous visit to London, and thanks to his lute-playing, he had the good fortune to meet Robert and Mary Killigrew and their wide circle of friends. These were people interested not only in the classics, but also in science and art. Robert was an enthusiastic chemist, and Mary, a fine singer, was a niece of Francis Bacon. The scholar John Donne was a regular guest at their home; Cornelis Drebbel, Salomon de Caus and many other scholars were among their acquaintances.[22]

It is no easy task to sum up their views and works in a couple of sentences, and at the same time, give some idea of their influence on Constantijn. Perhaps Donne is the easiest. The two spoke to each other as poets, and Donne honoured him with a set of copies of his poetry. His completely new, unfettered lyrics inspired Constantijn to translate a number of his poems into Dutch.

But the man who made the deepest impression on him was Francis Bacon, the remarkable chancellor. He spoke of our obligation to learn, and the dream of erudition, he was the author of *New Atlantis*, a mythical island with an official institute of research, and of the *Novum organum* with which nature should be observed. Constantijn shared his enthusiasm, but like Bacon, he was not a natural scientist.

Constantijn was amazed by Drebbel's laboratory.[23] The inventor from Alkmaar had been earning his living in London, since 1605, as 'artificer'. Drebbel polished lenses for spectacles, microscopes and a camera obscura, mixed paint in fabulous colours, constructed pumps for an underwater vessel, made wheels turn by water pressure, gunpowder combustion, and 'all by themselves'. When Constantijn saw his perpetuum mobile – which actually moved with the aid of hot currents – he wrote, *'Gloria in excelsis Deo'*.

De Caus was working on the idea that house and garden should represent the unity of man with the cosmos, and he created that cosmos by an austere geometrical plantation according to the will of man. He had designed the picturesque 'garden of symbols' for the Winter

---

[22] Bachrach 40–49   [23] Ploeg 23–28

King of Heidelberg, husband of the Winter Queen in The Hague, a garden that Constantijn had admired in 1620. Now he designed magnificent English gardens. It is as if, through his association with these people, Constantijn's own learning became replete:

> – and thy rich vaine
> Frought with that learned wealth, doth seeme to bee
> In that description a new library.

By now he was in his late twenties. In November 1624, Tessel Visscher married Captain Allard Crombach. There, in the Amsterdam *cercle*, Constantijn recited a flamboyant wedding poem, just as Hooft and Vondel had done, and flirted with a certain Machteld. Upon returning home, he wrote his 'Fire and flame', sent the poem to Tessel, and asked her to give it to her cousin Machteld – and to put in a good word for him. Tessel's reply was to ask him if he meant it. He left it at that.[24]

In the meantime he was also out of work. The coalitions had been formed and fighting had resumed (the Spanish assault on Bergen op Zoom had been beaten off, but the assault on Breda would be successful), so that for the moment, there was little point in delicate diplomacy. He stayed at home, with his mother and his brother and sisters, composed poetry, and collected all he had written into *Otia* ('Idle hours').

Then Prince Maurits died, the son of William of Orange and Anna of Saxony. He had neglected to marry Margaretha van Mechelen, with whom he had several children. It was 23 April 1625. Frederik Hendrik, younger son of William of Orange and Louise de Coligny and already forty-one years old, succeeded him as Regent. For reasons of succession, he was more or less duty-bound to marry. He chose Amalia van Solms, half his age, who spoke a charming, Frenchified German and planned to set up a glittering court in The Hague. They married on 4 April 1625.

[24] Smit 115

Naturally these events inspired Constantijn to wield his pen. This time he wrote a poem about a skipper, a man called Mouring, whose death was so lamented that the sailors quite lost heart, until Heintje (little Hendrik) spoke up:[25]

Daydreaming, spoke brave Heintje, daydreaming?
Muted measure, low the tone,
Daydreaming? Stay, I must undeceive you,
Can I be other than I am?

We can read this popular poem as a job application addressed to Frederik Hendrik. In fact, he followed it up with a formal letter of application, and the new ruler understood that, just as with his marriage, he could not do otherwise. On 17 June, he summoned the brilliant young man to his court and informed him, the following day, that he had been appointed his secretary, together with Willem Junius.[26]

Constantijn deliberately gave up his diplomatic career. Why, is not clear, for his talents were hardly appropriate to the prince's tough military duties. Did he suddenly see himself as Bacon, the learned advisor to the ruler? Did he wish to show his brother Maurits that he could make it at least as far as he had, without the advantage of being the oldest? Was it power he was after?

Whatever the case, he had a job. To begin with, he earned only 500 guilders a year, but the fringe benefits, that close to the regent, would have been considerable.[27] He could also watch the progress of the war, as Frederik Hendrik was usually with his troops in the field from May to November. In winter and spring, he could work in The Hague. Now all he had to do was get married. It sounds cynical, but that is how it was: here, too, he would steal the thunder from under his brother's nose and make off with Maurits's intended bride.

Maurits showed little initiative in such matters. First his father had arranged a job for him, and then he had more or less matched him up

---

[25] Worp **2**, 127 (*Scheepspraet 33–36*)   [26] Smit 129   [27] *Ibid.* 131–132

with a girl. She was a second cousin from Amsterdam who, in January 1623, after Maurits's clumsy advances, received a strongly worded letter from Christiaan senior:[28]

> (. . .) requesting most amiably that you do not haughtily reject him, but graciously accept him as he is. Indeed, you are acquainted with him, his disposition, his countenance, his manners and humours, together with, in my opinion, the sterling qualities of us, his parents, and not just anyone of low birth who pisses in the horse market. Dear child, if you do this, my whole life long I shall love you as my own child.

But she would have none of it, and such a recommendation from the father, even if he had been secretary to the Council of State a hundred times over, can only have intensified her feelings. Maurits's sisters had already looked her over. Geertruyd called her a 'little monkey', meaning she was 'cheeky'.[29]

Was it Constantia who drew her attention to the brother in London, with his interesting letters? Was it she who told Constantijn of her interest in him when he returned home? At any rate, in 1623 he followed in his father's footsteps by courting her, on his brother's behalf, in writing:[30]

> *Suzanne, un jour je te fis sacrifice*
> *De mon amour; et mon ame novice*
> *Se promettant de te trouver propice*
> *Me nourissait d'une ombre d'apparence.*

> Suzanna, I once offered you
> My love, and my untaught soul
> Lived in promise of your favour,
> Nourished by shadows of appearance.

---

[28] Strengholt 49   [29] Keesing (1987) 68   [30] Worp 1, 295

Constantijn had probably never even seen her. As far as we know, she did not answer.

In 1625, Hooft, who had been recently widowed, courted this very same Suzanna. In a poetical exchange with Constantijn, he spoke of this new love. She would yield to no man, he wrote, yet he had kindled her heart:[31]

> There was not one who with a loving eye
> Discerned that which love must surely see:
> The beauty of her mind.

On Hooft's insistence, she sent him a letter, as well as several poems in response to his own, but these have not been preserved. Hooft, who had seen her, described her brilliant brown eyes, graceful eyebrows and shy smile: 'It is not laughter, but laughter's dawning.'[32] When he finally got to the point, she turned him down, this bailiff of Muiden, eighteen years her senior.

Two months later, in March 1626, Constantijn saw Suzanna at the wedding of her sister Ida to Arend van Dorp. There she was, sitting next to . . . Maurits. Did he fall in love? Judging from the satirical 'Anatomy' that he wrote on 31 March, a poem that can be read as a dutiful proposal, for a marriage of convenience: no, not exactly. It certainly didn't contain any compliments for her. But he did read something that she had said she was reading: Italian work by Giambattista Marino.[33] Then he fell ill.

That summer of 1626 was no easy time for him. Junius, the Prince's first secretary, was determined to keep him under his thumb. Now that his long bouts of fever prevented him from accompanying the campaign outside Doesberg, he worried about losing his job. Frederik Hendrik might think him a coward. But acquaintances at court were able to reassure him. In the end he realised that he was

[31] Keesing (1987) 189 (*Clorinde*) [32] *Ibid.* 74 [33] *Ibid.* 59

in love, and as soon as his health had improved, he went to join the army by way of Amsterdam.[34]

In Amsterdam, he must have asked Suzanna to marry him. And she must have withheld her consent. In September he wrote his sonnets in which 'Say yes, but say it to me, that is the shortest way,' is the constant refrain. He called her 'Star'. At the beginning of November, somewhere in the vicinity of Emmerik, after having dedicated his 'Wretched awakening' to her, he wrote these ardent verses:[35]

> In stealth my vigilance I hold,
> Observe in the drowsiness of sleep,
> Through the mist of tears
> There stands my Star;
> The moisture from my eyes
> Vanishing as the dew drops
> Glisten in that morning sun
>
> Star, I say, morning star
> Who shines so far from me, and further still and further
> And still removed in time
> (As the highest light to circulate the skies),
> How can you be Star
> So constantly restraining, so haughty,
> Like a comet
> Who, far from sparkling, from assenting,
> And not know?

She remained reserved, and did not write back. Only when he came to Amsterdam on his return from the campaign did he finally obtain her consent. The lovers announced their engagement on 24 February. It was just as Geertruyd had predicted when she saw Maurits's clumsy advances: Constantijn would take care of everything.

---

[34] Smit 135–139   [35] Worp 2, 167

On the occasion of his wedding, Constantijn had his portrait painted by Thomas de Keijzer. To make it quite clear to her who he was? Let us take a look at the portrait (Plate 1).[36]

We see two male figures in a sparsely furnished room. The elder gentleman is seated, rather stiffly, at a table with a tablecloth and a number of strange objects on it; the younger man is standing at his elbow and handing him a neatly folded letter, which the other reaches out to take. Master and servant. It is difficult to say what sort of a man the master is. He looks dignified, is middle-aged, dressed in a simple black suit and a wide-brimmed hat. Light from an invisible window falls on his spurred boots. He does not look at the letter that is being handed to him; his gaze is fixed elsewhere.

This painting now hangs in the National Gallery in London. The museum catalogue describes the objects on the table: a lute, two books bound in vellum, a large pair of scissors, an inkpot and quill, and two globes, one of the Earth and one of the heavens. For the rest, we see an ornately carved mantelpiece, and above it, half a painting of a stormy sea, and half of another showing a building in flames. In the background there is an enormous tapestry depicting a religious scene and a coat of arms.

This is Sir Constantine Huygens, the catalogue informs us, poet and protector of the arts, father of the most famous Dutch scholar of the century.

---

[36] Bachrach 27–28

3 Mother

Suzanna van Baerle was born in the spring of 1599; we do not know exactly when. She had four brothers and five sisters, and her parents were Jan van Baerle and Jacomina Hoon. Jan was a rich merchant, originally from Antwerp, who had settled in Amsterdam in the 1590s. He was not related to Caspar van Baerle, the learned remonstrant Casparus Barlaeus, who had been thrown out of Leiden University after the Synod of Dordrecht and later taught logic at the Illustrious School of Amsterdam. He was, however, related to Suzanna Hoefnagel, Constantijn's mother. He died in 1605. It was left to Jacomina to look after the family and manage the property until her death in 1615 or 1616. So Suzanna was orphaned at an early age.[1]

Jacomina must have been a strong woman. Suzanna's delicate health had probably been inherited from her father. The certificate of health that she submitted at the time of her marriage speaks of 'the poor balance and harmony between the stomach and the liver', because of which 'each four or five months a severe stomach pain occurs that is accompanied by distressful feelings of fear and a tendency to vomit without effect'. She suffered regularly from migraines. For this she maintained a light diet, swallowed various remedies that were then brought up by taking a purgative, and, in summer, underwent bloodletting of her right arm, between eight and nine ounces of blood, in order to 'bring about the desired cooling effect' in her liver.[2]

Due to a lack of source material, it is hard to form an accurate picture of her youth and upbringing. Judging by the impression she made on her circle of admirers when in her twenties, it was far from unhappy. She was level-headed and self-assured. The fact that she

[1] Vollgraff 393; Smit 103   [2] Van Lieburg 174–175

read Marino is remarkable. To be sure, he was very much in fashion, with his newfangled and somewhat affected art, full of excessive and unnatural images. But even so, she must have somehow been brought in touch with his work. Had the Van Baerles hired a tutor in French and Italian who brought along the voluminous *Adone*, or some other poetical oeuvre of Marino? Did she pick it up from lessons given to her brothers Johan and David? Or did their mother, once she was on her own, give equal treatment to her sons and daughters when it came to their education?

Decidedly Suzanna was intelligent and well-read – otherwise Hooft would never have gone to all that trouble for his 'junoesque Rose' – but nearly all direct evidence is missing. It is typical of Constantijn, who kept every snippet of paper, that he didn't consider anything she had written as worth saving. Thanks to Hooft we still have two of her letters, which testify to her wit and erudition. The first, written to Hooft on 10 August 1627, regards a letter and laudatory poems written by Constantijn – at that time in the army at Grol (Groenlo) – which he requested her to forward. The letter reads as follows:[3]

> Sir, I hand on to you the best part of what I most recently received from my better half. In addition to a letter I also have something in verse, but to form a judgement I need no other Paris than my meagre understanding. I believe that re-reading your powerful writing has moved him to apply himself with zeal to write these unwarranted laudatory poems. I am amazed that there is still time to be found amidst so much pressure of work to accomplish anything worthwhile, and I reproach myself: I am surrounded by much that is good, while I see that greater repose produces less result. And so Sir, if you will kindly give me your answer, I shall request him that I may deliver this comforting news. I remain, Sir, your grateful friend at all times, Suzanna van Baerle.

[3] Keesing (1987) 80

Only a self-assured woman could have written this. She was two months pregnant, which adds an unrecognisable irony to her remark about what is produced in 'greater repose'. What is recognisable is the irony in that 'meagre understanding', which discerns Hooft's writing to be superior to that of Constantijn. She had a logical mind, not to say mathematical. In one of his poems, Hooft placed her beside a celestial globe. But for the rest?

Constantijn praises her painting skills and knowledge of medicine in his poem '*Dagwerck*' ('Daily work'), which he dedicated to her. She strikes us as rational and somewhat detached. The fact that Constantijn translated for her the more esoteric Donne need not necessarily mean that Donne, with his poignant imaginative powers, held more appeal for her than Marino's affected superficiality.

Around 1635, Suzanna and her husband had a double portrait painted, by Jacob van Campen (Plate 2). This portrait, until recently believed to have been lost, is the only likeness we have of her. We may dispute her beauty, so praised by her admirers. Dorothea van Dorp visited her in Amsterdam, long before there was any talk of marriage, and wrote to her long-spurned admirer Constantijn that, in that city, 'the sole item of beauty is little Baerle'.[4] But this comment was no more objective than Geertruyd's 'little monkey'.

In the portrait she regards us a bit warily, but we can easily imagine that shy smile. Together with her 'better half', who is as richly dressed as she, Suzanna is holding a sheet of music, the symbol of concerted enterprise in honour of the muse. But was she musical? Another question we are unable to answer, unless we have the courage to conclude that this was probably not the case, considering the little musical interest displayed by most of her offspring. On the other hand, the drawing skills of several of her children might indicate that she drew, or painted, with some distinction. When all is said and done, we still cannot be certain of her supposed artistic gifts.

[4] Smit 122

The marriage was celebrated in Amsterdam on 6 April 1627, and Hooft recited the wedding poem. Three weeks later the young couple travelled to The Hague and moved in for the time being with his mother, brother (!) and sisters on Het Voorhout, only to be separated by Constantijn's army duties. When he returned in the autumn from the successful campaign at Grol, they moved to their own home on the Lange Houtstraat, the house where Margaretha, mistress of Prince Maurits, had lived.[5]

Here, on 10 March 1628, their first child was born. It was a boy, and his father wanted to call him Christiaan, after his own father, but the mother wanted to call him after the father. Constantijn writes:[6]

> She had her first contractions between 3 and 4 o'clock in the morning, and they were at their most severe from 8 o'clock onwards, until finally she delivered, not without great distress. The child's godfather was Mr Johan van Baerle, the oldest brother of my wife, and the godmother, my mother, Mrs Hoefnaegle. She had wished the child to be named Christiaen, after the grandfather, but when she saw that my wife had not lost her fondness for him, but had only set her heart upon my name, she consented to draw lots, in which she lost.

To avoid confusion and facilitate further reading, we shall refer to the new-born as Constantijn, and to his father as 'the father'. Constantijn was a quiet, rather scrawny baby. It was four months before he began to greet his wet nurse with a smile. The father, who saw this and noted it down, was still at home in July. There were no campaigns in 1628, because the authorities in Brussels had no money for the war, so that the States-General in The Hague (which followed a defensive strategy) were spared the costs. Were these smiles withheld from Suzanna? 'They were brief bursts of laughter, and you could not look at him for long, or he would turn away his head from whoever

---

[5] *Ibid.* 144   [6] Eyffinger 89 (§17)

wished to talk to him.'[7] At least Suzanna could enjoy the hearty laughter of the father.

> Christiaan, our second child, was born in 1629 on 14 April, the
> Saturday before Easter, in the night, at two o'clock exactly, the
> beginning of the aforementioned day, in the same house and the
> same drawing room in which Constantijn was born. The morning
> of the day before, the mother began to feel that her time was
> approaching; that was why I was fetched home at nine o'clock
> from an anatomy session, at which I was present from seven
> o'clock. But it seemed to pass, and my wife joined us at table at
> midday. Only at ten o'clock in the evening did her contractions
> become more severe, and from eleven o'clock onwards she toiled
> grievously, more so than at her first delivery, because the child
> proved to be larger – it weighed nine pounds at the christening.
> But by God's mercy the mother felt reasonably well upon the first
> day, the second day she was a little feverish, which continued at
> first more severely and then became lighter, so that on the eighth
> day she was quite free of it. When the child was christened just
> before midday in the Sint Jacobskerk here in The Hague, his
> godfather was my only brother Maurits Huygens, secretary to the
> Council of State, whose christening gift was a silver, sun-shaped
> candle holder which weighed two and a half pounds. The
> godmother was the oldest sister of my wife, Lady Jacomina van
> Baerle, wife of Mr Samuel Becqer, from Amsterdam. Her gift was
> an embossed gold-plated dish that weighed three quarters of a
> pound, with in the centre the Van Bequer coat of arms.[8]

Lots were drawn for the naming of this child, too. This time the Van Baerles insisted, through Jacomina, that he be named after the other grandfather, Jan, while the Huygens, according to Suzanna Hoefnagel,

[7] *Ibid.* 89 (§20)   [8] *Ibid.* 103 (§§59 & 60)

wished him to be named after Christiaan. And 'Christiaan' it was.[9]

Let us try and look at the mothering of these two boys, and the children who came after them, through the eyes of Suzanna, for although her husband is our only written source, it was she who continually gave it shape. No doubt it was a demanding task. In November 1629 she was pregnant again, but two months later she suffered a miscarriage.[10]

Her third child, Lodewijk, born on 13 March 1631, was an easy delivery, but after three days she went down with a high temperature that continued to rise and fall until the end of the month.[11] Was this a complication of puerperal fever?

Her fourth child, Philips, was born quite a bit later, on 12 October 1633, and was at least one month overdue. It was a very difficult birth; the mother thought she would die and prayed that no daughter of hers would ever suffer as she had done.[12]

Three years later she was pregnant again, and on 13 March 1637 she gave birth to a daughter, Suzanna. During the final months of her pregnancy she continually threw up her food, and the delivery was fraught with serious complications.[13]

Perhaps this was all perfectly normal for a married woman at that time, but if we add to that her migraines and the half-yearly absence of her husband, we have to wonder whether the intellectually minded Suzanna could really accept it as such. But with only one or two exceptions, no complaint ever passed her lips. She must have greatly enjoyed the few outings she had.

When the father left for the army after Christiaan's birth (the army that would besiege and capture Den Bosch[14]), Suzanna went to Amsterdam for her sister Petronella's wedding. She could entrust the infant Christiaan to the care of his wet-nurse, 'a large, strong, lusty

[9] Smit 153   [10] *Ibid.* 156   [11] Eyffinger 109 (§79)   [12] *Ibid.* 113 (§89)
[13] *Ibid.* 115 (§97)   [14] Israel (1995) 506–508; Smit 153–156

woman, clumsier than Constantijn's had been',[15] but Constantijn had to come along. He was no trouble though; he was such a quiet, docile child.

At the end of July 1630, the following year, she accompanied her husband to Gelderland to view Zuylichem Manor, which they subsequently bought with her money.[16] Did her sister-in-law Constantia take care of the children? Christiaan was weaned on his first birthday, 'with little trouble, for within 2 days he showed no more interest in a teat'. For a whole week she ate cherries and blackcurrants from her own land, and enjoyed the village gossip.

But the following summer, after the birth of Lodewijk, she was unable to leave the house. Christiaan was still small for his age, and moody, jealous of Constantijn, who was a whole head taller than he was. And she saw that head leaning more and more to the left.

Here are a few observations on Titan:[17] Strongly determined to speak and sing, he stammers out words in imitation, imperfectly, yet with great ability to remember. (1630)

His strong character did not let him be easily compelled, and if he was alone he was quite calm and not boisterous. (1631)

Several times Suzanna heard him say the Our Father aloud at night in his sleep, although most often he got stuck at a certain point. (1631)

He often lay talking to himself and made rhymes out of all sorts of words and noises that he knew. Besides this he showed that he possessed a particularly sharp mind, though continually inflamed with jealousy towards his brother; for the rest he grew little, his limbs were weak, though he was hale and hearty. (1632)

[15] Eyffinger 105 (§63)   [16] Smit 157–158
[17] Eyffinger 105 (§§64 & 66); 106 (§§68, 70 & 71)

Meanwhile we note that it was not the older brother who was jealous of the younger, since he attracted the most attention, but the younger of the older.

Compared to Constantijn, Christiaan 'grew little, he stuttered and seemed at times to find it difficult to remember. When Constantijn had the pox, he began to show the same signs of apathy and sickness . . . , but within two days it was over and he was as lively as before.' (1633)

'In the whole of 1634 he developed so slowly that everywhere he went he was taken for Lodewijk's twin brother' and he 'was always obedient, and the first to do his tasks . . . , in character and in build as gentle as a girl, for which all mistook him.' We see that his stuttering and sudden difficulty remembering things coincided with the rapid growth of his second rival, Lodewijk, who stole much of Suzanna's attention, with his curly head of hair and comical chatter.

It is worth taking a close look at the period when Christiaan was four years old. His memories went back that far, and besides, there was plenty going on in the family at that time.

For a start, Suzanna was pregnant again. Although we lack the details, such as we had of her subsequent pregnancy, when she almost constantly felt sick, there were always her migraines. She did drawing with Constantijn. 'On her fervent insistence,'[18] the father brought along an etching by Joris Hoefnagel, which she gave to her eldest son to copy. She also gave Constantijn spelling lessons, while Christiaan eagerly copied, and Lodewijk grew more and more wild and surly (a year later he would talk of nothing but swords, fighting, hurling people to their death, and the like).

Her mother-in-law, Suzanna Hoefnagel, died at the end of April 1633, and her husband took leave from the Rhine campaign, which had only just begun. When he left once more to join the army, calling in at

[18] *Ibid.* 97 (§40)

Zuylichem along the way to settle his affairs, he fell ill, and returned home again – presumably to be nursed.[19] In August he went with Suzanna to spend a few days in Alkmaar, the home of Tessel (Maria Crombach-Visscher), to make music. The question is: did Suzanna enjoy herself there? 'She was,' according to Constantijn, 'in reasonably good health, little troubled by fainting fits, except as a result of sitting up straight for too long',[20] but she was well into her seventh month. The business purpose of this trip was to inspect a piece of land they had bought in the Heerhugowaard polder.

From mid-August onwards she was on her own once more at the Lange Houtstraat, together with the children, who were querulous and feverish with the pox, and soon she was facing her hardest delivery of all. It seemed as though she were constantly trying to put it off until her husband's return. Her notice of the birth, written from her childbed, arrived in the army camp after a week.[21] It was more of a lament. Although she never actually said so, it was clear that she never wanted to go through such an ordeal again. The father appeared on 8 November, four weeks after Philips' birth. It is very conceivable that Christiaan felt pushed aside yet again. Titan withdrew into himself, 'as gentle as a little girl'.

Did the 'little girl' remember the visits from René Descartes in those days? His father had probably got to know the intriguing yet reserved Frenchman at the court of Winter Queen Elizabeth. They were the same age, and Descartes had found temporary accommodation in either Leiden or Endegeest. Directly after making his acquaintance in 1632, the philosopher wrote to a mutual friend, 'His integrity is above all doubt . . . , despite his close ties with the court. In spite of all that I had heard about him, I could not believe that one and the same person could be engaged in so many things, and at the same time have such a thorough command of them.'[22]

[19] Smit 171   [20] Eyffinger 113 (§90)   [21] *Ibid.* 113 (§89)
[22] Smit 167; Vollgraff 408 (n28)

Descartes had more or less finished *Le Monde*, the great work that contained his revolutionary world-view, but he did not dare to publish it, even in the Republic, in accordance with his motto: *'Bene vixit bene qui latuit'* ('He who has concealed well, has lived well'). Did the father, as many believe, persuade him to write down the introductory *Discours de la méthode* (Discussion on Method), a book that would be published anonymously in 1637?

This is improbable. Their extensive correspondence was not about what should remain concealed.[23] Descartes sought in him, above all, a protector, because of his strong ties at court. It is probable that he visited the Lange Houtstraat at that time, however; but his visits would not have spanned any great number of years. Suzanna spoke with him, because he wrote that he had valued her judgement.

And Christiaan? If he did remember, then his meeting with the black-haired foreigner would have consisted of a handshake, a well-practised *'Bonjour Monsieur'*, and a pat on the head. That could have been in the spring of 1633. The deep impression Descartes would make on him is almost certainly of a later date, and based solely upon his written work.[24]

During these years the father worked methodically at his poem, 'Daily work', begun after Christiaan's birth and which would ultimately number 1971 succinct verses (not counting a conclusion added later). It was addressed to Star, 'marrow of all my joys'. It described their life together and their various occupations, and this in the form of a story about one single day. It is up to us to interpret this description:[25]

In the silence of two souls
I find my deepest wishes,
My last voyage and destin,
Thou, my sole possession

[23] Ploeg 29–41, 99–103   [24] Vollgraff 409–410
[25] Worp 3, 78 (*Daghwerck* 1049–1053)

Did they have so little to say to one another that he glorified silence? Or had Suzanna learned early on not to contradict her 'Sir'? The poem bids her listen to his words, for it is 'my – I mean *our*' plan to state how life shall be. This might be consistent with the times, but certainly for an educated woman it was humiliating not to be allowed a mind of her own.

Her Sir had no scruples about patronising the gifted Anna Maria van Schurman – or least trying to.[26] Although Suzanna's eye for quality in painting and drawing may have played a role in his appreciation for Rembrandt and Jan Lievens, he did not find her own work worth saving, in spite of the praise it earned from a number of their friends.[27] In 1634, to add to her burdens of raising a family, he plunged her into the building project of 'my – I mean *our*' little palace.

Frederik Hendrik had learnt to appreciate the father's qualities as a sound and disciplined private secretary, especially after the Maas campaign of 1632 (in which Roermond and Maastricht were taken), and he saw how he surpassed the virtuous Junius in all that he did. He had already nominated this Lord of Zuylichem as steward of his estates and raised his salary by 1000 guilders. Now he gave him a strip of land as well, on the north east side of the Binnenhof, which extended along nearly its entire breadth.[28]

Here, on Het Plein, he could build a fitting chancellor's residence, modelled on the Banqueting House in London that he had once so admired. And because he was so often away from home, Suzanna would have see to it that all progressed according to plan, stayed within the budget and followed the proper time schedule. She was a merchant's daughter, wasn't she? It is unclear whether her involvement went beyond bookkeeping, but that alone must have been a considerable job.[29]

[26] Van der Stighelen 138–139   [27] Keesing (1987) 82–83
[28] Strengholt 59–60   [29] Keesing (1987) 86; Blom *et al.* 68

Jacob van Campen designed the house in the classic style for which he had been engaged. Were its vast dimensions her choice? When the architect portrayed the couple, he painted the man looking resolutely to the left, while the woman's gaze was fixed sceptically on the painter.

The house was to measure 27 metres wide, 13 metres deep and 18 metres high, with a basement, two floors of 4 metres high, and an attic. Each floor was to consist of a 7-metre-wide hall over the entire depth, with stairways at the back, two rooms of 9 by 6 metres on either side facing onto the Binnenhof, and two more of 6 by 6 metres facing onto Het Plein. These were to be flanked by two offices of 3 by 3 metres. To make the house more distinguished, two more rooms of 4 by 8 metres were to be added to the ground floor, jutting out onto Het Plein, as a forecourt. Three man-sized statues were planned for the facade, to symbolise beauty, endurance and leisure.[30]

Building operations lasted nearly three years, longer than planned. Due to the war, stone supplies, and even wood supplies, stagnated. There was little she could do about that.

This is something Titan must certainly remember: how only a few minutes away the huge house rose where, fifty years later, he would determine the measure of the cosmos.

Meanwhile Suzanna taught Christiaan how to spell. He seemed rather slow, but by the end of 1634 he could write quite well and even wrote a few 'charming pieces'. But Constantijn, 'with little practice (and frequent interruption) was good at reading all kinds of books, and longed each time again with great impatience and curiosity for new stories'. She began to wonder whether he ought not to be operated on; he had a shortened neck muscle, which made his head lop-sided. She commissioned the architect to paint his portrait. She was having a lot

[30] Blom *et al. passim*

of trouble with Lodewijk, who at times was lively and full of beans, at other times sulky and stubborn and restless in his sleep. Philips learned to walk early, 'but answered only with gestures'.[31] In 1635, she became concerned.

When the father went to join the army in Brabant, she also left The Hague, 'where many were dying'. Apparently she was more concerned about her toddler, Philips. She brought him, together with Christiaan and a maid, to her sister Jacomina, who was married to Samuel Becquer and living in Arnhem. She kept the other two children with her. After a short visit to Amsterdam, she returned to The Hague.[32]

Why did Christiaan have to go with Philips to Arnhem? If Philips needed a playmate, then why not Lodewijk, who was closer in age? Why did the two elder children have to be separated? Christiaan never managed to obtain a satisfactory answer to this important question. He must have been angry. For when they returned, the father noted that this otherwise so obedient child could not 'easily endure that people treated him unfairly'. Aunt Jacomina or Uncle Samuel would have informed him of his son's feelings.[33]

In October, after three months, his mother arrived, together with Constantijn and Lodewijk, to fetch him and Philips and the maid. They sailed down the Rhine, but just past Schoonhoven they were caught in a freak tide and stranded, while a storm broke out. All ended well.[34]

This incident brings us to the more general question about the parents' ideas on education. Christiaan was now seven years old, and although his intellectual powers had been remarked upon at an early age, the father did not yet consider the time ripe for any kind of formal education:[35]

---

[31] Eyffinger 98 (§§42 & 43), 106 (§71), 110 (§86), 114 (§94)   [32] Smit 185
[33] Eyffinger 106 (§72)   [34] *Ibid.* 98 (§44)   [35] *Ibid.* 106–107 (§73)

He filled his time in the year 1636 mainly with children's games and reading, namely a bit of Latin, which he picked up here and there from Constantijn, and he consistently remembered what he learned. But I postponed all learning until he would be a little older, and until we had settled in the new house with a tutor.

When Constantijn had been that age, the father had given him lessons, together with Suzanna, in Latin, writing and drawing. And then to consider the father's own upbringing: when he was seven, his father had taught him to play the violin and the lute. Was this opportunism? Possibly, but it is more likely that he underestimated Christiaan.

His namesake was his first concern:[36]

In the year 1636 I had him especially read and write and learn a few basic principles of Latin grammar, such as the declensions. In the meantime I grew greatly concerned about the growing abnormality of his neck, and I requested . . . that Doctor C. van der Gall in Utrecht examine him when he was next in The Hague. When this took place not long after, I learned with relief that he judged the abnormality to be easily cured. And although I was wary of some kind of large incision that he might wish to carry out for this purpose, he would explain to me his method of treatment in a letter, before I should take a decision in the matter.

After their correspondence with the surgeon the parents decided to send Constantijn to Utrecht for the operation on his neck. He would stay with relations there.

The mother, pregnant again, seldom had a day 'without throwing up her food after a meal'.[37] She probably suffered from dehydration. Life was too much for her. There is no evidence that she received any medical care, not even a rest cure.

---

[36] *Ibid.* 99 (§46)   [37] *Ibid.* 115 (§97)

In July 1636, the only time that she had a moment to spare for a visit, with the poet of her choice, to the poet of 'Dawn rose', 'Arbele' or 'Clorinde' (all names by which she was known), she spoke to Leonora Hellemans, the woman who had said 'Yes' to Hooft. If judiciousness was Suzanna's most striking characteristic, then she must have spoken to Leonora of marital duties and imagined to herself the restful social life in the otherwise rather chilly and draughty Muiderslot. She might have stood in Leonora's shoes, and Leonora in hers. It was Leonora who had managed to avoid being kissed by her husband, turning her head away as quick as lightening.[38]

Tessel, who had lost her husband Captain Allard Crombach, was also present on this occasion. After the musical gathering – father Constantijn with his lute and risqué poetry, and then great admiration for the new house, to which they were all invited to return the following year – the guests sailed under a strong wind to Amsterdam. The learned gentlemen Casparus Barlaeus and Gerardus Vossius did not dare go aboard when they saw the swell of the grey rollers. He joked about it. She just climbed aboard. And we may safely assume that, from that moment on, she was seasick.[39]

We may also assume that Suzanna had plenty of help in the house, with at least one maid. She is mentioned on the occasion of their visit to Arnhem. Two years later we hear of a Maartje Jacobs, and possibly they are one and the same. Her husband almost certainly had a manservant, or to put it more fashionably, a valet; in any case this is how he is portrayed by Thomas de Keijzer. Apart from the various wet-nurses that came and went, she had two, possibly three servants to do the cleaning, the cooking and the shopping.

They would also have helped with the children, and that went beyond just tidying up their rooms before they could be cleaned. According to Schama it was usual that the servants played with them and shared in their upbringing.[40] This explains the father's note on

[38] Keesing (1987) 193   [39] Strengholt 65–66   [40] Schama 495

Lodewijk's sulks, 'which people tried to keep in check with good man-ners'.[41] By 'people', one does not mean one's wife. But to return to the mother, help did not mean that she could opt out. In the beginning of 1637, she must have been fully occupied with the completion and furnishing of the house on Het Plein.

In February, the house on the Lange Houtstraat was sold. It had to be vacated by 1 May. It was a tight schedule – too tight. With so much on his mind, the father also had to begin doing something about his children's education. He wrote the rudimenta (summary of Latin grammar), taught this to Constantijn, and then promised him money if he, in turn, were to pass it on to Christiaan. He also taught him musical notes, but Christiaan, sitting at his side, was the only one who could sing them straight off, 'to my surprise . . . , while, moreover, I perceived his ambition to progress'.[42]

On 4 March, Constantijn was taken to Utrecht for his operation, which meant that from then on, at home, Christiaan was the oldest. Perhaps his mother heard him singing on the day she gave birth? That was on 13 March, in the evening: 'it seemed to be about to freeze, with a clear sky full of stars'.

Her daughter arrived suddenly, three hours after the midwife had been sent away, her task unaccomplished. Only Constantia was there to help her. At first all seemed to go well, but two weeks later, on 30 March, she complained of a headache and the next day of a sharp pain in the neck and in her right side. Then she got a high fever.[43]

Much has been written about Suzanna's deathbed: first by her husband, and in our time, by J. Smit and M. J. van Lieburg. According to the latter, who has sifted through all the evidence once again, she contracted sprue, a debilitating illness, at the beginning of April, then began to hiccup violently and bring up a malodorous phlegm. At the

<hr/>

[41] Eyffinger 110 (§86)   [42] Ibid. 107 (§74)   [43] Ibid. 115–117 (§100)

end of April her high fever was temporarily accompanied by violent tremors and sweating, causing an 'insupportable dryness'.[44]

On 28 April, since the house on the Lange Houtstraat had to be vacated and the new residence on Het Plein was not yet ready, she was moved to the house of her sister Ida, wife of Arend van Dorp. On 3 May, bearing in mind the poor prognosis of her sickness, a will was drawn up in her presence, 'while she was still in full possession of her faculties'. On 10 May, fifty-eight days after giving birth, she died, 'without making any movement'.[45]

In all likelihood Suzanna didn't die of ordinary puerperal fever, nor of sprue, but of an infection. She would have suffered from phlebitis of the pelvis, complicated by blood clotting, which gave rise to painful embolism, leading to abscesses in the lungs and causing fluctuating fever.

Of the two boys, only Christiaan was allowed in to see her, possibly at her own request. She did not wish for Constantijn, whose operation had been a success, to be brought back from Utrecht prematurely. She said to the father, 'simply tell him that his little mama wished dearly to see him one more time, and that I know that he would dearly have wished to see her too, but I ask of him, now that this is not to be, that he behave well, fear God, and obey his father'.[46]

Shortly before 10 May, Christiaan came to her bedside. Still too small to climb up on his own, he was lifted up by his father: 'Come here, my dear little man, and let me kiss you.'[47] Not a word about her illness, no solemn message, just affection.

Her funeral was on 16 May. It was held in the Jacobskerk, amidst enormous public interest. 18 May: in the new house on Het Plein. 20 May: father left for the army in Breda. He had no words for his grief. Only on his return from the campaign did he turn his attention to her death:[48]

---

44 Van Lieburg 176–177   45 Eyffinger 120 (§106)   46 *Ibid.* 101 (§52)
47 *Ibid.* 107 (§75)   48 Worp 3, 46

Flimsy protector, stay close, in gratitude I thee forgive.
Come death, deliver me swiftly from these fevers:
In light eternal I long to see ascend forever
My haven my love my God my Star and me.

Of that of all his children, only Christiaan's grief attracted the father's notice, because he 'would not come out of the long mourning-skirt', as the father later wrote. In Christiaan he saw Suzanna, 'the living image of the mother's gentle character and earnest virtues'.[49]

We have already observed a pattern of loss of concentration and withdrawal, of which his conspicuous mourning is the end, not the beginning. Yet we shall never be able to prove that this is the key to the hidden Titan.

[49] Eyffinger 107 (§78)

4   Family portrait

Somewhere in the family portrait painted by Adriaen Hanneman in 1639 (Plate 3) is Christiaan. But where?

The six portraits, grouped into a harmonious whole, have no names, nor do they bear the signature of the artist. We only know that Hanneman is the painter because of his correspondence with the father, portrayed unmistakably in the centre. We can also easily pick out little Suzanna. But there is some doubt about the identity of the four boys; or rather, some doubt has arisen. It appears as though the two older boys figure at the top of the painting, with the two younger boys below. According to custom, the oldest is portrayed next to the father, on his right. That would mean that the boy on the left is Constantijn, with Christiaan on the right. Following that order we have Lodewijk on the bottom left and Philips on the bottom right. And this is how they are identified by P. H. L. van der Meulen, who made an engraving of the painting at the beginning of the nineteenth century.[1]

But confusion has arisen with the more recent, late nineteenth-century engraving made for the publication of Christiaan's *Oeuvres complètes*. Here, Christiaan is said to figure on the top left, on the authority of a catalogue written by Victor de Stuers of the Haagse Mauritshuis, which is where the painting hangs today.[2]

De Stuers gives us no reason for this remarkable deviation from the rule. Perhaps he just thought it was so obvious, since it is the liveliest face, and wasn't Christiaan the liveliest child? In hindsight: yes, and then only superficially, for outward appearances do not necessarily tell us all about inner life. But at the time that Hanneman painted

---

[1] Vollgraff 395   [2] OC 1, 1

him, he might very well not have been the liveliest child. Constantijn received greater attention from his parents, and he was by no means stupid. His gift for drawing developed early, and one of his efforts, a self-portrait from 1685, can help us to identify the portrait painted by Hanneman in the upper left-hand corner.[3]

We need to turn Constantijn's head, which is facing left in his self-portrait (which was most likely drawn in front of a mirror), to face right. If we do this, we have an almost perfect likeness to the face on the left in the family portrait. We see the same high eyebrows, shallow-set eyes, straight nose, small mouth, rather pointy chin, and the outline of the cheeks. Comparisons with other, later portraits of him and Christiaan are less convincing. But the adult Christiaan certainly looked very different. His eyes were more deep-set, the nose longer, the mouth broader, the chin heavier.

Without any doubt then, it is Christiaan who figures above and to the right in the Hanneman portrait. The rule that De Stuers, and later C. A. Crommelin, attempted to ignore applies after all. Needless to say, Lodewijk's curly head figures on the lower left; and the face on the bottom right obviously belongs to the youngest of the four. Due to the great authority of the *Oeuvres complètes*, however, we have been left with an erroneous impression of the ten-year-old Christiaan.

Did he resemble his mother? Perhaps in the expression of the eyes, but for the rest, no. Not outwardly. The likeness to which his father referred lay in his 'complexion' (nature) and 'virtues', by which he can have only meant Suzanna's judiciousness and equanimity. If we take his 'complexion' to refer to his physical and mental constitution, we might also read into this a prediction of his migraines. But up until his tenth year he was seldom ill, only an eye infection and 'a slightly bloodstained diarrhoea' at the age of two. At that time he was not delicate, just rather small for his age. He had grown more on the inside, so to speak, than on the outside.

[3] Frankfourt & Frenk 172

Naturally he also took after his father, in his looks, and in his thirst for knowledge. It was the egocentric father, so grateful for his own upbringing and education, who was to blame for not noticing Christiaan's eagerness to learn until the boy was nearly eight years old, and even then he was surprised. What he did about this later on will be discussed shortly. But first, we quote from the beginning of a letter that he wrote to Christiaan on 24 September 1638. He celebrated him in verse, thereby allowing him to share in his glory as a poet:[4]

> Te quoque, mel meum,
> Te, mei Amores,
> Te, pietatis
> Vivae character;
> Te, venerandae
> Matris imago –
> Te quoque musas
> Atque Heliconis
> Ardua castra
> Aggredior; te
> Non alienis
> Viribus usum?

After a year of lessons the nine-year-old was expected to understand these Latin verses, which meant being able to translate them according to their content and rhythm. It must have seemed a daunting task. We can barely manage it ourselves:

> For you too, my dearest,
> You sweetest one to me,
> Happiest temper
> Looks to me with longing,
> Image of your
> Honoured mother –

[4] OC 1, 2

> For you, too, I praise
> The muses, I clamber
> Helicon's steep
> Mountains, for who can
> Accomplish such things
> By one's own efforts?

He would certainly not have entirely succeeded. But he must have understood the affectionate opening lines, which called to mind the 'dear little man' he had been to his mother. The formidable father as educator and mother. When Suzanna died, he had to try and fill both roles.

In May 1637 he had taken on Catherina Zuerius (Sweerts) as a housekeeper. She was the unmarried daughter of his mother's sister, thirty-nine years of age. It was perhaps an ill-considered, impulsive decision, for as it turned out, she found it hard to adapt to his family regime, and he took a hearty dislike to her. Even on her death in 1680, he could find nothing better to say about her than that she 'hawked and harped and meddled and died'. He was extremely rude to her. One Christmas dinner, when she asked him if he had had enough to eat, he answered her bluntly, 'Yes, and I've had enough of you, too.' But he didn't send her away.[5]

Whether he had ever considered it, is another matter. In 1641, he was nearly ensnared by another cousin, Helena Liefferts, who used to send him dolls for Suzanna and whom he bid farewell – with tears in his eyes, he wrote, because she was so clearly infatuated. Every reason for him not to hire her as a housekeeper, the position she was clearly after.[6]

In spite of being treated with contempt, Catherina's status in the house was certainly higher than that of a serving maid. Her foremost task was to take care of the baby. It may well have been she who sent away the first wet-nurse, who didn't have enough milk, and replaced

[5] Keesing (1983) 19; Strengholt 67–68   [6] Keesing (1987) 97–101

her with a better one. She certainly devoted much of her time to little Suzanna, whom she raised almost single-handedly and took along regularly on family visits to Brabant. She was also a bit of a mother to the boys, if only because of her firm grip on household matters. Nothing ever happened without Cousin's approval.

If this was his improvisation on motherhood, then Abraham Mirkinius was his improvisation on education.[7] He had probably engaged him through Johan Dedel, formally his own home tutor, and still a good friend in The Hague. Mirkinius studied theology in Leiden, but was out of pocket and, for the sum of 200 guilders, was quite happy to come and teach for a year.

He moved into the house on Het Plein on 17 July, after receiving his instructions from Dedel.[8] We must remember that the father could, under no circumstances, leave the army, which was camped outside Breda. It was hard. The city, lost in 1625, could only be recaptured in the autumn of 1637, and all that time he mourned alone. This meant that he sent written directions to Mirkinius, whom he couldn't yet have met, and longed for written proof of the fruits of his education.

This arrived promptly. Christiaan wrote several letters that, unfortunately, we do not have (they must have been his first), 'and this he did in such a manner that my Lord the Prince on the battlefield, and others, were greatly astonished, when they heard, which is true, that before our departure he had never held a pen in his hand'.[9] Was this admiration for their content, or for the fact that this secretary's son hadn't been sat down to write until he was eight years old? He soaked it up like a dry sponge.

Abraham Mirkinius was undoubtedly a happy choice. He was able to offer more than just the Latin for which he had been engaged. We know, too, that he was conscientious, from the very fact that after the

---

[7] Eyffinger 101 (§54)   [8] *Ibid.* 101 (§55)   [9] *Ibid.* 107 (§77)

agreed year, he left to prepare for his exam or disputation; then came back with a fellow theology student, whom he recommended to take his place. The two older boys learnt a great deal from him.

'And he had such a sound command of the basic principles [of grammar] that not only did the children enjoy his lessons, they also made excellent progress.' The father testified further:[10]

> Both Constantijn and Christiaan sent me well-written letters, in Latin, while I was in the field. And when I came home on 7 November it appeared that, in accordance with my instructions, they had learned many Latin words, more even than in my own written rudimenta. They knew these so well that they soon lacked lesson material, and therefore had gone beyond the text and began work on the syntax. This meant that I had to prepare a sizeable piece of etymology for them. For I wished to guide them myself, at least in the Latin language, if not further, and in such a manner that the others will be eager to learn as well.

Here, again, we hear the father who liked to keep his finger in the pie. Actually, it had been Mirkinius who went beyond the lessons in rudimenta; he saw only too well that the children were ready.[11] He taught them for hours each day – Latin, Latin, and more Latin. Now and then they must have asked for something else, so now and then he would oblige. History, bible stories? Unfortunately, we know little else about this young man, except that he was a minister's son and wanted to become a minister himself. In the father's notes he is perhaps even more invisible than Cousin Catherina, the housekeeper.

But it would be hard to overestimate the influence he had on the serious-minded Christiaan. He was, after all, the first to introduce him to syntactic structure, to use Noam Chomsky's term, and possibly even in a systematic – orderly, at any rate – manner. To say that language is just for girls and mathematics just for boys is one of the

[10] *Ibid.* 101 (§55)   [11] *Ibid.* 102 (§55)

most superficial statements of our time. Linguistic structures do not emerge merely by coincidence; they have a logical, not to say mathematical character, and a meaning that one must discover. In other words, Mirkinius taught him the art of science.

By the end of 1637, Christiaan had surpassed his brother Constantijn.[12] He knew at least as much Latin, and he was the better writer. Was the father surprised yet again? He observed it, but made no comment. His remark about Christiaan not coming out of mourning for his mother must date from this time, because in May, when he had last seen him, it would not have been noticeable in any way. We recall here Bos's character sketch, 'Work must have been a source of strength for him, a fixed point from which he could combat the fits of depression, and the basis for his attitude to life.'[13] Could that be pensiveness we see in his eyes, in Hanneman's portrait?

Christiaan wanted to learn arithmetic. This was a subject that Abraham Mirkinius knew nothing about, and so the father took over. After all, it was he who had such high expectations for his sons' education. In December he began by explaining the elementary calculation of numbers: addition, subtraction, multiplication and division. Constantijn joined in as well, 'But most especially it was wonderful to see how well Christiaan understood everything, and remembered it, and even succeeded each day to discover different manner of proofs and good arguments, so that I frequently made use of his services for the education of Constantijn.'[14] Within a year it was Christiaan who received money for teaching Constantijn, not Constantijn who was rewarded for teaching his younger brother.

To try out (test), discover (invent) and prove with valid arguments (sound reasoning): we have entered that territory upon which Christiaan would build his career. That is why the time has come for us to try and see just how much of this his father was able to teach him.

---

[12] *Ibid*. 102 (§57)   [13] Bos (1980) 16   [14] Eyffinger 123 (§109)

Not much – if only for the simple reason that school arithmetic, in those days, amounted to very little. Christiaan mastered elementary calculations within two weeks, and even that seems to us a bit long. His discoveries would have been made in division, the most taxing of the four and the key to the concept of ratio. The father didn't know much more than that, though he did have books in his possession that offered more.

One clue here is his letter to Christiaan of 3 September 1640.[15] Apparently, Christiaan had asked him to explain a geometry theorem which he had found in one of these books. Translated from the Latin, we read, 'Christiaan chases me like a terrier. The fact is, if the theorem [sententia] of the great Pythagoras be true, then the foul breath [spiritum putem] of a dog must have entered you.' Even if it had been intended as a joke, it is no less discouraging. He didn't give Christiaan the explanation he had asked for; the eleven-year-old was simply being a nuisance.

Even so, we cannot leave it at that. Father Huygens was far too gifted a man for us to deny him any insight into mathematical and physical sciences. Let us recall his acquaintance with Descartes, and the testimony of the philosopher. The Frenchman had later sent him a text on mechanical apparatus that was simple enough for him to understand. He wrote back on 23 November 1637 that he was 'pleased to have a few words at my disposal with which to thank you, in a worthy manner, for the special favour (with which you have been kind enough to oblige me, by straying from the illustrious path you are accustomed to travel), of humouring my ignorance with so thorough an explanation.' In short: he thanked him for the explanation for the interested layman that he was.[16]

He was interested primarily in instruments, things that could achieve a tangible goal, and he wanted to know all about them. If there was a theoretical concept involved, then he was interested in that as well. He must have looked upon it with the fascination of a

<hr />

[15] OC 1, 3   [16] Bots 159 (n14)

man like Bacon, convinced of its greater meaning. But it was here that he learned his limitations.

In his extensive correspondence with Descartes, neither physics nor philosophy figures greatly, for he had little of interest to say on either of these topics. He considered himself a layman in all areas of science, although he was familiar with many of the arts, and understood their backgrounds. In a letter written on 5 February 1644 to Johann Gronovius, who had asked him to comment on a work of literature, he flatly refused to assume the role of judge or patron (*'judicem aut patronem plane fugio et recuso'*).[17] He preferred to act as a witness (*'patior me testem advocari'*). No judgement, merely an opinion, and he would be quite happy to help make the work better known. The father's significance to science in his time lies in just this kind of propagandising.[18]

Thanks to his wide range of interests, father Huygens was not only secretary to the established power of his day, he was also a secretary to the learned who sought the protection of that power. A considerable portion of his correspondence (over a hundred letters a month, spanning half a century) must be attributed to the seventeenth-century intellectual news circuit.

His ties with England have already been mentioned. In the 1630s there were also important ties with France, particularly after the treaty of 1635. (In this treaty it was agreed that the army of the Republic would divert the Spanish troops quartered in the north, in Brabant and Limburg, to facilitate the French campaign in the south, in Artois and Western Flanders.) There he was, like a spider in its web. He became a knight in the Order of Saint Michel; he came to know highly placed public servants in the government of Cardinal-Minister Richelieu, and later on, the Franciscan friar Marin Mersenne.[19]

---

[17] *Ibid*. 150   [18] *Ibid*. 158   [19] Vollgraff 423–425

Mersenne was the driving force behind the publications of Descartes from 1632 to 1637, and the posthumous distribution of Galileo's writings from 1634 to 1639. He wrote thousands of letters on what he considered people ought to know. This included the work of Roberval, Fermat, Torricelli, Pierre Petit . . . He also corresponded with Isaac Beeckman (on vibrating strings, in 1628), while for the rest of the world, the light of this rector from Dordrecht remained hidden under a bushel.[20] No wonder Huygens was so suddenly well informed. He, the father, saw to it that he remained well informed by writing back with reports on new discoveries.

Thanks to his mediation and status, the father acquired, without much difficulty, a large collection of manuscripts, brochures and books on the new science – a library that was impressive for its time. At his death he had 2868 titles, more than three hundred of which were on mathematics and physics.[21] And in 1640, even if there had been half as many, they would have still provided Christiaan with a tremendous store of knowledge. Just within arm's reach, in one of those rooms in the house on Het Plein, lay an abundance of knowledge, ready for the taking. He merely had to wander in, translate the Latin, then stop and think. Herein lies the significance of the father for the scientific career of his son.

Latin was still the *lingua franca* of the literati, though they had also begun to publish in national languages. Latin was the means of communication among scientists, and every hour spent on Latin was a good investment, certainly for Christiaan.

The fact that, after several years, he balked at writing Latin verse, had nothing to do with the language, but with the verse form imposed upon him by his father, the constraints of Ramtamtam ramtam Ramtamtam ramtamtam ramtam – '*Contra Maert Jacobs noli contendere verbis*' ('With Maartje Jacobs you're ne'er to bicker').[22] There is an anecdote about how, after a row with the kitchen-maid, he had gone into the hall and thought this up, and then, just to prove

[20] Costabel (1986) 6   [21] Kubbinga 165   [22] Eyffinger 124 (§113)

how clever he was, had gone into Mirkinius' room and chanted, *'No'li conten'dere ver'bis!'* But we shall let lie his father's obsession for verse, and return to the language itself; he studied this for six years. Naturally, more needs to be said about the solid, classical education of the future scientist.

First of all, Christiaan read the easier writers Cicero and Caesar. Perhaps it was Mirkinius' task to make a start on the interpretation, but in any case Henrick Bruno took this over in November 1638.[23] In 1639 he read Virgil (*Georgics*, three times!), Ovid (*Tristia*) and Justinus (*Trogus*). In 1640 Curtius, Florus, Caesar once again – just to catch his breath – then Seneca (*Troades*) and Virgil (*Aeneid*). He had to learn a scene from the *Troades* and the first half of the fourth book of the *Aeneid* by heart. In 1641, and at the beginning of 1642, he read Silicus Italicus, Ovid (*Metamorphoses*), Livy and Horace. In September and October 1642, Sallust. And finally, in the first half of 1643, Plautus and Suetonius.[24]

All this reading and memorising was supplemented with grammar studies, and writing essays and verses, which the father, in 1639, 'viewed with astonishment', and in 1640 saw reduced to prose ('there was no versified vindication, which I found strange'), in 1641 found 'fine and well thought out', and from 1642 onwards had to go without ('From Christiaan I could not obtain a single piece of work, though he was in all things wise and subtle in his judgement'). He remonstrated with him about this, but the by now adolescent Christiaan had had enough: '1643 . . . from Christiaan I could obtain nothing, either by commands, or with promises or reproaches.'[25]

Had Henrick Bruno not done his job properly? We know for certain that his teaching skills were weak. Appointed at seventeen, he was hardly more than a child himself. By December 1638, he could no longer hide his scabies and passed it straight on to Constantijn (who had so many boils on his head that, in April, he had to be shaved),

[23] *Ibid.* 126 (§118)   [24] *Ibid.* 128–136
[25] *Ibid.* 128–135 (§§126, 138, 143, 146 & 148)

and later, probably in March, to Christiaan.[26] He had studied some theology and thought about doing his disputation, but his true ambition was to become a poet.

When Mirkinius first introduced him to the father, he had several of his own published poems to show; this would have been a decisive factor in his appointment. In the summer of 1639 he was able to give a satisfactory report on the children's progress in reading Virgil's *Georgics*, and to explain to the father that his two pupils were not quite ready to tackle Ovid's *Metamorphoses*. But a few months later he came a cropper. 'To put it bluntly, my authority over them is abysmal, or at least has been doubtful for some months. They do not bother about any of my assignments; they are cheeky in their defiance and do exactly as they like. Bruno does not exist, is merely air.'[27]

The father quashed the rebellion when he returned home in November. He does not tell us how. Obviously he would have taken Bruno's side. It may be going too far to assume that Christiaan's refusal to continue writing verses had anything to do with this incident, but the coincidence is striking. It brings to mind: 'he cannot easily endure that people treated him unfairly', the remark that had been made after his visit to Arnhem in 1635. Lodewijk (who was greatly upset when Mirkinius left) was almost certainly the ringleader, or perhaps it had been Constantijn (who at twelve was approaching adolescence), but the father would have dealt with all the boys in the same way.

After this we hear no more complaints from Bruno. But this does not necessarily mean that he was in full control for the rest of their schooling. The extensive list of texts to be read consisted of considerably more than was customary at the Latin school in those days.[28] So we cannot help wondering whether Bruno was always able to adhere to that list, with two such intelligent but disruptive pupils. But even if he did not, it was still more than adequate for them to master the language. He had less success with the not-so-diligent Lodewijk, and none whatsoever with the slow-learning Philips, who

---

[26] *Ibid.* 127 (§123); OC 1, 537   [27] OC 22, 26   [28] OC 1, 548; OC 22, 24

in 1640 could still hardly read or write. In 1647, the year Bruno was finally dismissed, Philips had to be brought up to standard by someone else. Judging from his 'quick progress', it was soon clear 'whose fault it had been that he'd been held back'. Bruno's lack of appreciation for the boys' gifts is clear from his indignant report on 17 June 1643 on Christiaan's precocious attempts at handiwork.[29] They hadn't even finished their reading when

> Christiaan, who we consider so bright, almost (!) a child prodigy who I rejoice to call your delight, Sir, relapses into messing about with self-made toys, little building constructions, and machines. Clever to be sure, but utterly out of place. [Here Bruno inserts the Greek word *aprosdionousa* in his Latin, so as to emphasise 'out of place'.] Sir, you surely do not wish him to become a craftsman? The Republic, which vests such high hopes in his birth, surely expects that he go into business, following the example of his father?

Let us take a look at how this so-called regression came about. In 1640, Christiaan stopped producing verses, and in 1643 he began to produce machines instead. 'Little mills and other Models, even a lathe' is how his father describes them.[30] Just what we are to make of this is unclear.

We do know, at any rate, that Isaac Newton, whose youth is far less well documented than Christiaan's, made little windmills with a complete interior mechanism when he was fifteen. These could turn with the wind, or with the aid of a mouse chasing after a grain of corn.[31] But the father, who had gaped in admiration at Drebbel's constructions, and had so heartily thanked Descartes for his layman's description of mechanical constructions, denies us all the details with which Newton's witnesses so abundantly provide us.

Was he not really interested? When he later speaks of his 'room with a lathe', suggesting that he worked there himself, he is probably referring to Christiaan's, somewhere up on the first floor of the house

---

[29] OC 1, 552   [30] Eyffinger 135 (§151)   [31] Westfall 60

on Het Plein, in one of the four offices. So even if he did not entirely share Bruno's opinion, he would have had mixed feelings.[32]

At the beginning of 1640, a hard frost set in. Christiaan, still covered with scabies, was sent out onto the ice with Constantijn. After only a few days they skated 'gracefully, steadily, and very fast' over the Hofvijver. As well as skating, their father provided them with another treat. A garrison soldier – who was no great singer, but good enough – came to 'loosen their tongues with tremolos' and taught them to sing Italian songs 'of my own composition'.[33] He also had his former music teacher brought in to teach them to play the lute. Then something strange happened.

It is not exactly clear when Christiaan fell ill. Nor do we know what he had. It wasn't the scabies and boils, which didn't disappear until summer, after the boys' frequent bathing in the sea. In any case, the father, who never mentions it, received the following letter from his brother-in-law David le Leu de Wilhem, after he had left for the army:[34]

> Christiaan is unwell. Doctors Rumpff and Valck will come and examine him and I shall be present. The other day I asked Doctor Valck what was wrong with him. Is it some weakness or deficiency of the liver? (A deficiency, he said.) When I asked what the deficiency consisted of [both doctors said that this] stemmed from the weakness. When questioned thus, they simply waved it aside. If the children were given enough exercise, they would get better by themselves, and then we would not need to call on such folk for help.

From the date on this letter, 19 May 1640, and the date of the father's departure (8 May), who apparently knew of the illness, we gather it concerned something of a chronic nature. Was it Christiaan's first migraine?

---

[32] Keesing (1983) 20–21   [33] Eyffinger 129 (§§133 & 131)   [34] OC 1, 549

Our only clues for this are the discussions about his liver, which indicated a disturbance in his digestive system. The same was probably true of his mother, for – as we read in her medical reports – the liver would be nourished with food that was 'most refined, delicate and warm', and in cases of disturbance the head would be filled 'with manifold vapours'. According to these medical notions, it was assumed that these vapours caused pain. But we do not know if Christiaan actually suffered from headaches. Just let the lad romp around out in the fresh air, Uncle David seemed to be saying, and the problem would clear up by itself. We suspect that it is around this time that he discovered the library, and from then on, nothing could tear him away. A few months later he wrote to his father about a theorem of Pythagoras, and the foul breath of the dog had entered him.[35]

He browsed in his father's books by Simon Stevin. *In Beghinselen der weeghconst* (Principles of Weighing) he saw a description of the moments of force (as we say nowadays) in mills.[36] Mills were fashionable as toys, and were even used by children to fight with, by fixing them on the end of a long stick. Schama describes this popular misuse and its relationship to the pain on the cross, of which the mill is the visual metaphor.[37] Christiaan undoubtedly played with such mills and repaired them. He would have seen that mills were fascinating pieces of equipment. With Stevin's help, it can only have been a small step to construct the wheel mechanism. But he did not – at least not straightaway.

First he watched Hofwijck being built, with an eye for those moments of force, and for the hoists. This was a new project of the father's, this time in Voorburg, on a piece of land on the River Vliet that he had bought in 1640. The secretary of state had a modest cube-shaped house built with a surrounding moat and a long, austere, classic-style garden (Plate 4). His main interest was the garden which, according to Salomon de Caus, was a triumph of the human spirit over nature. It

---

[35] OC 1, 3   [36] OC 1, 7; OC 11, 34–36   [37] Schama 487

was a place to which he could escape when the carryings-on at court became too much for him.[38]

Hofwijck was ready in the spring of 1642. A garden party was held for family and friends, among the newly planted fruit trees. At that time Christiaan seemed slightly long-sighted, because he kept peering at the inscriptions on maps and pictures 'in spite of his sharper eyes', compared to those of the other children.[39] Perhaps this put him off working with his hands. Later on it was obviously no longer a problem for him, because he never needed glasses.

There is one other thing that might be significant. That autumn, Hendrik Bruno worked through a book with the boys written by Burgersdijck. We may consider this a concession, on the part of the father, to Christiaan's waywardness. According to the father the book was called *Logica*, according to Bruno *Collegium physicum*. It was the father's choice, in preference to Bruno's suggestion of something by Descartes.[40] How on earth Bruno came by such a suggestion is unclear, unless we assume that his wayward pupil whispered it in his ear. We are not familiar with the book, but if the father recommended it and Bruno explained it, then it cannot have amounted to very much. But this concession did act as something of an encouragement.

Our reconstruction of the 'regression' is nearly complete. But we still do not know how mathematics fits into the picture. The father considered mathematics on a par with mechanical apparatus, like the 'lathe'. Only in 1643 does he make any mention of Christiaan's quick grasp of mathematics, 'understood with exceptional aptitude all that concerned [mechanics or any other aspect of] mathematics'. It could be that he was more than usually impressed by Christiaan's knowledge of mathematics on his return from the campaign in 1642, five years after his first lesson in arithmetic. If that is the case, then this extra spurt of knowledge cannot be attributed to Burgersdijck. So then which book was it?

[38] Van Strien & Van der Leer 77–82; Worp **4**, 266–338
[39] Eyffinger 133 (§144)   [40] *Ibid*. 133 (§147); Kubbinga 164

He must have read it somewhere, says the father. Interestingly, he gives two versions of this matter: (1) 'immediately obtains a solution by making a model or other piece of handwork, making use of what he has read somewhere or from what he has understood from myself'; (2) 'immediately obtains a solution by making a model or other piece of handwork from what he has read somewhere or from what he has heard from others'.[41]

If the more modest second version is the truth, then the question arises: who put him onto it? Were there others in his vicinity who told him to go and look in François Vieta, Thomas Harriot, Marino Ghetaldi, Pierre de la Ramée or John Napier? Did the home library already contain works by these early-seventeenth-century mathematicians? We doubt it. If we take the first, less modest version, then we are left with the question of why the father did not recommend one of these books as lesson material. But how did he react to the 'exceptional aptitude' of his son? By instructing Bruno, during the summer months, to go through Burgersdijck all over again.[42]

Despite the high praise that father Huygens deserves for his children's classical training, this act of negligence must be mentioned. The fact that he also devoted great attention to their musical training and knowledge of modern languages, which was exceptional for children at that time, makes his omission all the more striking.

In 1639 Christiaan had already been given a viola da gamba worth 300 guilders. A certain Steven Eyck gave him lessons on it and he learnt to play madrigals by Luca Marenzio and other similar pieces. In the same period he also had lessons from Pieter de Vooys on the harpsichord belonging to his father. In both cases this was because he was the only son who turned out to be musical, and who showed an 'aptitude' that appealed to the father.[43]

In the summer of 1643, the father had yet another plan, 'I had Constantijn, Christiaan and Lodewijk study French seriously, under guidance of a retired clergyman . . . who came to practice with them

---

[41] Eyffinger 135 (§148)   [42] *Ibid.* 135 (§151)   [43] *Ibid.* 127–128 (§§124 & 125)

daily, so that they made considerable progress.' Unfortunately the clergyman in question concealed the fact that smallpox had broken out in his home, and he soon infected the Huygens children.[44]

This was especially tough on Christiaan, who had never had it before. In September, after spending a week in Zuylichem, he got a fever, and his face and neck quickly broke out in a rash. He suffered most from the pustules on the soles of his feet, but remained 'very sweetly behaved, resigned and patient'. He was given emetics and infusions, and on 24 October the doctor came and took seven or eight ounces of blood off him. By November, when it was Philips' and Suzanna's turn, he was rid of the pox, but was left with scars on his face that he 'will have caused by scratching off the scabs too soon'.[45] This was also the time when the father, peering over his shoulder through his thick spectacles, tested his long-suffering son on Burgersdijck: 'At the end of this year, Constantijn and Christiaan carried out exercises in argumentation on various chosen subjects, as training in logic.'[46] That is when he must have realised his own oversight.

Jan Stampioen (born in 1610) who had apparently received his degree from Frans van Schooten senior in Leiden, moved in with the family on an unknown date in the spring of 1644, as special tutor to Christiaan.[47] Possibly the father came by him through Van Schooten, who was known in the army because of his contribution to the construction of the fortifications. Like Mirkinius, Stampioen stayed in the family's service for over a year. Probably in the beginning of that year, he drew up a list of books that he wished to go through. The sixteen most important were, in alphabetical order according to author:[48]

Tycho Brahe, *Astronomiae instauratae progymnamata*
Christoffel Clavius, *Problematum astronomicum*
René Descartes, *Discours de la méthode* (the three additional essays: *Dioptrique, Météores* and *Géometrie*)
Sybrand Hansz, *De Hondert geometrische Quaestien*

[44] *Ibid.* 136–137 (§156)   [45] *Ibid.* 137 (§158)   [46] *Ibid.* 140 (§161)
[47] *Ibid.* 140 (§164)   [48] OC 1, 5–10

Johannes Kepler, *Mathematici dioptrice*
Nicolas Kopernicus, *De revolutionibus orbium coelestium*
Philips van Lansbergen, *Omnia opera (uranometria)*
Samuel Marlois, *Perspective*
Apollonius Pergae, *De elementa conica*
Ptolemaeus, *Coelestium motuum*
Christoffel Rudolf, *Coss* (on the use of letters in arithmetic)
Antoni Smiters, *Rekenconste*
Simon Stevin, *Weeghconst* and *Wisconstige Gedachtenissen*
Vitellius, *Optica*
Vredeman de Vries, *Perspectiva theoretica ac practica*

Apart from these, there were one or two books on fortifications and on astrology. Whether or not he kept to this list, we do not know. However, Descartes' *Principia philosophiae*, published in July 1644, are missing, although we know for sure that Christiaan read them straightaway. The *Opera mathematica* by Vieta doesn't figure either. This important work sets out the proofs of Archimedes' principle, which Christiaan must certainly have seen.

It is a formidable list. Pedagogue Stampioen portrayed himself as follows:

> Do not think that the books I have just named are sufficient to be able to penetrate into the heart of all sciences. In the first place you need a keen mind, and in the second place, constant training. When these conditions have been met, knowledge will not come at once, but bit by bit, only after extensive study. And it is to one's advantage to find things out for oneself. In this way one reaps greater benefit than from constant, repetitive drilling. I strongly urge that study be undertaken in this manner.[49]

This document of 23 June, intended for Christiaan's father, was not written by Stampioen himself, but by Bruno, on his behalf, so we are

[49] OC 1, 556

not sure whether the pedantry is all his own. But pedantic or not, Stampioen managed to get a tremendous amount out of his pupil.

We shall not end this chapter, however, with an attempt to assess the kind of impression Archimedes and Descartes – as expounded by Stampioen – made on the fifteen-year-old Christiaan. That will come later. Instead, just as we began, we shall close with a portrayal of the family in which he had grown up so far.

Suzanna, bright and catty, especially to Bruno, was sent to school on the Poten Street in The Hague at the age of eight, to learn French. On the whole, she grew up separately from the boys, with Catherina, who taught her to read and write and worked hard on her sewing and embroidery skills. Philips, slow and good-natured, tried to learn a bit of Latin, and was 'always the brothers' messenger'. Christiaan had a soft spot for him, though, and enjoyed his music.[50]

The three older boys clearly formed a group. Lodewijk, an average learner and fairly unpredictable, was no real match for his two older brothers. Bruno gave him simpler texts to work on, but even so, he spent year in, year out in their company. How else do we explain the fact that at the end of 1640 Constantijn and Christiaan had to learn a piece from the *Aeneid* by heart, and Lodewijk a piece from the *Eclogues*, also from Virgil? Relations between the three brothers later on were close, with Lodewijk providing the fun. Apart from a short period during their adolescence, Lodewijk was always taller than Christiaan.

Constantijn was always the tallest of the three. Quiet as a baby, he was always the quietest of the children 'and more melancholic'.[51] This intelligent, introverted boy could draw beautifully; in 1643 he was 'frequently occupied with devising his own compositions, of which he made a considerable number, to earn money that he then cheerfully spent on the Maliebaan'.[52] This apple of his father's eye was destined to become secretary to the Prince (he was introduced

[50] Eyffinger 136 (§154), 151–152   [51] *Ibid.* 133 (§145)   [52] *Ibid.* 135 (§150)

at barely eighteen years of age to Frederik Hendrik, and sent along to accompany the Flemish campaign). Although he was intellectually streets ahead of him, Christiaan was in awe of his brother. Their later relations were characterised by a deep mutual respect.

Finally, if the father needs any further describing, we shall do so with a poem that he wrote in the spring of 1645, on the occasion of a house party given for Tessel. Let us imagine the sound of music on the harpsichord in the hall, the singing of poetry, and the boys as witness to their father's gallantry. Later, Tessel sleeps in the room above him:[53]

> The widow lies close by, but I defile her not.
> See, I know her only, O strange grief.
> Or do I after all defile her? O marvel,
> The widow lies above, the widower beneath.
> Barlee, do you understand?
> What parts us thus?
> My cold garret, her cool honour.

He sent this poem to Barlaeus, who let the cat out of the bag, so that Maria (Tessel), Crombach's widow, got wind of it. The story goes that she sent him back a poem of her own, in which she rebuked him only for the innuendo in the word 'know'.[54] But long before he had the chance to chuckle over this, he had set himself the task of writing a letter, in Italian, to Pietro Paravicino.

---

[53] Worp 4, 48   [54] Smit 221

5   Student

Paravicino had to be made to understand that father Huygens was against it: no unnecessary rules were to be imposed upon his sons, (*'caricargli d'un giogo nuovo senza necessita'*).[1] And that is what he wrote him, in his best Italian. Since 11 May 1645, Constantijn and Christiaan had been Pietro Paravicino's guests in his student boarding-house on the Steenschuur in Leiden. The idea was for him to keep them on the straight and narrow and give them a bit of extra coaching in Italian, alongside their studies at the university.

He was strict and expected them, for example, to request his permission to go out in the evening after supper. They cannot have taken this rule too much to heart, because after a couple of weeks Paravicino raised the matter with the father. 'Most laudable', he wrote back condescendingly, 'but such a strict hand is unnecessary with my boys. When they were still with me at home, I never needed to be severe with them, thank goodness; just a friendly word and a fatherly admonition (*sempre con dolcezza et ammonitioni paterne*).'

We have an astonishing example of these fatherly admonitions, written at the time the two boys began their studies. It is called 'Rules for study and for life thereafter, written for Constantijn and Christiaan Huygens, students at Leiden Academy'. We cannot help wondering if they didn't take these too much to heart, either:[2]

> They shall rise at five [!]. When dressed they shall read a chapter of the New Testament in Greek and then, on their knees, they shall recite their daily prayers in Dutch. At 6 o'clock they shall apply themselves to study law according to professor Vinnius, part of

---

[1] Vollgraff 404   [2] OC 1, 4

which has been written down, and part of which is yet to come.
This exercise shall continue until nine o'clock. Then breakfast
and relaxation of the mind. From nine until ten o'clock they shall
attend a lecture by Vinnius, who at present is well-versed in the
rules of law. From ten until eleven they shall do exercises in
mathematics for Van Schooten. From eleven until midday,
drawing or music including Parerga, for which they need to bear in
mind that there is but one organ in the house; indeed they must
ensure that this is at their disposal. From the midday meal until
two o'clock, or two thirty if they think they do not need so much
time to prepare themselves for Vinnius' lecture, they may laze a
bit. At three o'clock they are to attend that lecture. From four
until six they are permitted to be free to divert themselves,
to take a walk or to exercise body and mind in an appropriate
manner. What free time remains before the evening meal should
be spent on the arts. This applies as well for those other hours that
remain, for not all free time is to be frittered away in diversion and
going for strolls. In summertime after the evening meal they may
go out for a stroll, in wintertime they may amuse themselves with
parlour games, with music-making or in doing as they please. At
ten o'clock, before they retire to bed, they shall once more read a
chapter from the Greek Testament and recite the home prayers,
just as they do upon rising.

As far as the father was concerned, the lads were allowed out after
supper, at least in the summertime. These precise instructions are
followed by a moralistic admonishment:[3]

On Sundays they shall attend the one church service in Dutch, the
other in French. They are also to spend an hour or so in reading
the Holy Scripture, the catechism and their clarification. They
should always attend church and the sermon together; only as a
rare exception may they walk along the street alone or apart. They

[3] OC 1, 5

are to approach city fathers and professors with deference, and when addressed by them, at least to regard them in upright fashion without appearing to be proud. They are not to become friends with drunkards, but only with virtuous and clever young men, or at least with those who are honest and interested in their studies. They should avoid idlers and swindlers like poison; however much they may be carried away by such fellows with their smooth talk and friendly banter; such behaviour will lose them an official post. They shall write to their father as often as possible on matters of their health and progress in their studies, especially when he is camped with the army. Finally I beg the Almighty God, (who sees all, does He not?), that my best friends may be charged with these instructions and counsels.

We would be quite mistaken if we were to imagine that Christiaan immediately buried himself in mathematics, even though he did arrive still filled with Jan Stampioen's lessons, and hounded by an enthusiastic Frans van Schooten junior. The first letter he wrote home shows that he had run through his pocket money, visited professors and, following Constantijn's example, took time off to paint pastels. As this is the first of his letters to have been conserved, we reproduce it in full. It is dated 29 July 1645, written in Latin and addressed to Lodewijk, the father being stationed somewhere in Flanders; there is also a message for Catherina, though he never addresses her directly:[4]

I am greatly surprised that Cousin does not send the money that I am in need of. If she still wishes to discuss the matter, then let her do so. According to me, the money may be disbursed. At any rate it is the same sum that father had agreed to apportion me. This is why I ask her to make haste, otherwise I shall no longer be able to manage. Jacob's letter on the death of Piccolomini is being talked of all over town; it has gone to Spanheim, to rector De Bondt and to many others whose acquaintance I have not yet made. Some

[4] OC 1, 12

> recommended me to make a copy of it. We are doing pastels using
> the tortillon technique. If you could see what I drew yesterday,
> you would never again use Spanish pencil. I have copied
> Rembrandt's portrait in oils of an old man. I have done it so well
> you can hardly tell the difference. I shall send you an example of
> the technique, but it takes much time and effort. Greetings.

How lively they are, these first lines of his! This sixteen-year-old
knows how to manipulate, displays both anger and enthusiasm, and
is boastful, yet still maintains his equilibrium. No sign of inner conflict, which might have come upon hearing of the death of Joseph
Piccolomini, a young adventurer who was killed in a battle between
Poland and Sweden on 6 March. There are signs of sensitivity, though –
he must have spent hours working on the deeply furrowed face of
Rembrandt's old man. Clearly he was preoccupied by the closeness of
death; a year later he would copy figures from Hans Holbein's *Dance
of Death*.[5]

It is doubtful whether Christiaan learnt much from the lectures
of the fifty-seven-year-old lawyer Arnold Vinnius, or whether he rose
at the crack of dawn to swot for them. In any case, they didn't do
him much good. He had a far more inspiring teacher in the person of
the thirty-year-old mathematician Frans van Schooten. For one thing,
he excelled in his field, and he was a freethinker, remonstrant and,
above all, Cartesian. The fact that he was tolerated at the university in Leiden can only mean that the governing committee considered his subjects – mathematics and physics – to be innocuous. When
he was appointed professor in 1646 (as successor to his father, Frans
van Schooten senior), he bluntly declared himself to be 'not of the
reformed religion'.[6]

Among his students were Johan de Witt, born in 1625, Johannes
Hudde, born in 1628, and Hendrik van Heuraet, born in 1633,
who would all three make a name for themselves in the field of

mathematics. And, however innocuous their field of study may have been, the first two got appointed to high administrative posts. The foreign students who studied under Van Schooten just had to fend for themselves in learning Dutch, for this was the language he used in his lectures. There were, after all, always more mathematics students who understood Dutch better than Latin, he reasoned, 'because many more students attend Dutch than Latin lessons in this subject'.

In 1637, Frans van Schooten met Descartes in person. Even before the meeting he had studied a printer's proof of the third and final essay, the *Géometrie*, which served as a clarification to the *Discours de la méthode*. His boundless admiration for this work meant that he could not abide any criticism of the maker, the more so after the 'horrible pamphlet' against this perspicacious mind that had been written by the orthodox Calvinist Jacobus Revius – his Leiden colleague, if you please![7] But we shall return to this in the next chapter. After his appointment he began on a Latin version of the *Géometrie*, which he would complete and publish in 1649. Perhaps he lectured on it when Christiaan turned up in his class, because this was certainly the kind of mathematics that appealed to him most, but it is more likely that he presented this material only to a select group of his more gifted students.

There is no doubt that his lecture was mainly introductory, and therefore traditional. Without doubt he would have referred to the classical edition that he published on his appointment, namely: a version of Vieta's aforementioned *Opera mathematica*. This contained Archimedes' proofs, which were frequently incomplete or not properly written down and which would have been further worked out and discussed.

Van Schooten gave direction to the enormous potential that Stampioen had released in Christiaan. We know nothing of how this all began, how Christiaan attracted his attention, which assignments he

[7] Verbeek 57–61

gave him and whether perhaps, strolling along the Rapenburg, they may have discussed the recently published *Principia philosophiae* by Descartes. The letters they exchanged, when they were no longer able to speak together, are collegial and full of mutual respect, which certainly tells us something about Christiaan's celebrated professor, fourteen years his senior.

Perhaps his role as Christiaan's inspirer has been just as underestimated as Mersenne's (to which we shall return later) has been overestimated. But nothing is to be deduced from correspondence, only from Christiaan's booklet of exercises from 1646.[8] Here we find the rough notes on calculations on a lever (*Mechanica elementa*), the free fall (*De motu naturaliter accelerato*) and the catenary curve, which is the form assumed by a chain hanging freely from two points (*De catena pendente*). These exercises preceded his correspondence with Mersenne and already referred to physical topics raised by Stevin and Galileo. If Christiaan raised these with Van Schooten – which is conceivable, considering his lessons from Stampioen – the teacher would gratefully have seized upon these to use with his student as exercises in analytical geometry. But we do not know, just as we do not know whether Van Schooten looked through, and discussed with him, the final draft.

The exercises on the catenary curve deserve extra attention, as they contain the axiom that will prove to be at the heart of all his later work on mechanics.[9] First the result: the catenary curve is not a parabola, as asserted by Stevin (and indeed by Galileo as well). What form was it then? Christiaan was only to return to this at the end of his life.

How did he come by this result? Firstly, by representing the chain as a succession of heavy balls of identical weight, attached to one another by light cords of the same length. Secondly, by presenting the combined downward force (*actio*) of two successive balls (following the method of Stevin) in one parallelogram from the forces of the

[8] OC 11, 34–36   [9] OC 11, 37–44

balls separately. Thirdly, by stating (this is the axiom referred to) that in equilibrium the point of gravity of two (or more) successive balls lies as low as possible.

He never doubted the accuracy of this axiom that Torricelli was the first to articulate, and for which various equivalent formulas existed. Eight years later he wrote, '*Nisi principium ponatur nihil demonstrari potest.*' ('If you do not take this as point of departure, you can prove nothing').[10] In fact this is true, and the catenary curve is not a parabola.

Imagine achieving such results at the age of seventeen, which challenged the authority of Stevin and Galileo. But his proof was, mathematically, not entirely flawless – something he realised twenty-two years later, when he noted the error with the following harsh judgement: '*Non sequitur neque est verum*' ('It does not follow and it is not true').

Just before the summer of 1646, Constantijn was called home; his father judged one year of Arnold Vinnius' lectures to be sufficient, and Van Schooten's efforts would be wasted on him. Whether or not Christiaan had to plead to be able to stay on, we do not know, because the letters he undoubtedly wrote to his father on his studies have not been preserved.

(As to the whereabouts of the father: in the latter part of the campaign of 1645 the army, after a period of unsuccessful fighting, finally managed to take Hulst by surprise. This meant that he was away from home for a long period of time. But in the campaign of 1646, the army was inactive, due to the deterioration of Frederik Hendrik, so he was able to be at home much more.)

On 3 September Christiaan wrote a teasing letter to Constantijn about calculations on parabolic conic sections, which weren't really all that difficult to understand, and about the free fall.[11] We recognise the problems in his book. But teasing his brother was not his real

---

[10] OC 16, 214   [11] OC 1, 18

aim. This was hidden in the postscript: 'You can give this to Father to read.'

Well! Now that he had been presented with his son's erudition, he promptly addressed himself to his Paris correspondent, the friar Marin Mersenne: 'I have two young lads, my oldest and the one who follows, who greatly wish to see your quadrature of the hyperbole and your [calculations of] the centres of impact. And just to show that they are able to form an opinion on them, I shall have a copy made of a letter written by the youngest to the oldest . . .'[12]

Mersenne appeared to be amused, for even before he received the copy he sent the boys a mathematical problem: is the difference of the quadrates of two figures, of which the one is the sum and the other the difference of two quadrates, also a quadrate? After he had read the copy he was no longer simply amused, but quite serious, when he responded on 13 September. He, the great intermediary of science, addressed himself directly to Christiaan.[13]

His letter, in a nutshell, informed Christiaan that: 'the proof of your argument on the free fall is missing, send it up to me; but take care, the calculations of Galileo are not quite accurate, you must improve on them'.

Christiaan answered on 28 October:[14]

Sir, I rejoice in my good fortune that the letter that was intended for my father's eyes only has fallen into your hands, and I further gratefully acknowledge your benevolence in deigning to find me worthy of so kind and courteous a letter. But now I shall answer your arguments. First you say that no heavy body mass can move as fast as a stone that falls downwards from a height of a few miles. I am afraid I must disagree. I refer you to the philosophy of Descartes: it is only the friction of air that prevents other body masses from moving just as fast.

[12] OC 2, 547   [13] Beaulieu 23–29; Vollgraff 423–430   [14] OC 1, 24

STUDENT 77

Let us sum up the rest of his answer. The proportions $1:3:5:7$, which Galileo has proposed for the distances covered in equal time spans by a falling body mass, result from mathematical arguments, which also show that $1:2:4:8$ which you, Mersenne, propose, are 'geometrically absurd'.

Explanation of the proportions according to Galileo: for figure number 1, the 1 is calculated: $(1 \times 1) = 1$; for figure number 2, the 3: $(2 \times 2) = 4 = (1 + 3)$; for figure number 3, the 5: $(3 \times 3) = 9 = (1 + 3 + 5)$; for figure number 4, the 7: $(4 \times 4) = 16 = (1 + 3 + 5 + 7)$; and so forth.

Christiaan concluded with a remark on the parabolic path of canon balls and the promise to write something on the catenary curve. This quite floored Mersenne, who immediately decided to broach his most difficult problem.[15]

This is the reason for his letter to Christiaan, of 8 December 1646, on the *'centre d'agitation'* of a pendulum; not a simple one that can be made by hanging a small, heavy mass from a long thin wire, but a body mass with a specific form (such as a triangular sheet) that swings on a specified fixed axis (with the triangular sheet, for example, perpendicularly through a vertex). His question was: what was the conceivable distance between the axis and the weight of that body, if that weight were contracted to a point, so that it swings just like a simple pendulum, with this 'point weight' as pendulum and this distance as length of wire? That was the *'centre d'agitation'* (*centrum oscillationis*, centre of oscillation). With this letter he had given Christiaan something new, something difficult, which would keep him occupied for a while and put his gifts to the test. For a brief time, at any rate, it had him stumped.

On 26 November he answered rather evasively that he enjoyed solving problems, that he could discover solutions and that he only wrote these down when they were quite complete (*'je ne les mets pas*

---

[15] OC 1, 45

ecrit qu'en tout dernier lieu') and that they were often on points of gravity, tangents, curves . . .[16] For the rest, he would send separately the promised work on the catenary curve. What he sent has been lost, so we do not know exactly what Mersenne actually got to see, in spite of the efforts of the editors of the *Oeuvres complètes* and the pains they took to reconstruct it. Also, it is not certain whether the friar had already seen the piece when he wrote to the father on 3 January 1647, 'If he carries on like this, he will surpass Archimedes himself', but even so, his judgement was no less felicitous.

From then on he was 'My Archimedes'. It was a nickname father Huygens relished, although he could not yet see the consequences it was to have. But Mersenne spread the name of this Archimedes far and wide, which undoubtedly contributed to the fact that, after four or five years, it was thoroughly established. Herein lies his greatest significance. Christiaan would eventually have come across the problem of the centre of oscillation on his own, for many were preoccupied with it at the time. And we have more proof of Mersenne's stimulating influence.

Looking back on this episode, ten years later, Christiaan wrote to Pierre de Carcavy, 'Father Mersenne honoured me with his letters, which spurred me on in my mathematics, and sent me writings of [French] renowned scholars, and especially from Monsieur de Fermat, which I began to understand as I took the trouble to go into the subject more deeply.'[17] As if he needed spurring on in his mathematics, and as if Van Schooten had not already gone deeply into the work of Pierre de Fermat, as well as that of Descartes. It is highly improbable that the professor had less to offer in this respect than the friar. So let us consider this rather as a sign of Christiaan's politeness, and an indication that he had little to learn from Mersenne.

Nevertheless, their correspondence is interesting, because it reveals what Christiaan was working on in 1647 and 1648: the

[16] OC 1, 34   [17] OC 1, 427

quadrature of the circle according to Gregorius van Sint-Vincent, the impact centre in a sector of a circle, the ratio of the circumference of a circle and an ellipse, or of the surface of a ball and an ellipsoid, the body of revolution of a cycloid, falling bodies and paths of a projectile according to Galileo (he knew the *Discorsi*), and vibrating strings (he knew Mersenne's *Cogitata*, and together they reflected on why it was necessary to tighten a lute string four times over to make it sound one octave higher).[18]

This may not be all, because we seem to be missing the letters between February 1647 and March 1648, which may have dealt with other matters. Or were they never written?

In 1647 the father lost the secure position that he had enjoyed under Frederik Hendrik. The pliable Stadholder, his Heintje (Little Henry), died on 14 March. He was succeeded by Willem II, born in May 1626, the only son of Amalia van Solms. A tempestuous character, who, for dynastic reasons, had been married off at fifteen to the ten-year-old English Princess Henrietta Maria Stuart, now took over the reins of power.[19]

There was no telling what would remain of father Huygens' position at court, now that he was confronted with this new generation of go-getters. By careful manoeuvring between the self-assured Stadholder, his capricious but powerful mother and the conceited child-queen who was his wife, he managed not only to maintain his position but even to consolidate it. Willem II trusted him, even with his mother in the background, and in May he granted him Zeelhem Manor at Hasselt, in Brabant. (This explains the higher status attached to the title of Lord of Zeelhem, a royal gift, rather than the title of Lord of Zuylichem, which had been purchased with 'my – I mean *our*' money. Not to mention cheaper titles, like Lord of Heerhugowaard and Lord of Monnickendam).

[18] OC 1, 98 & 101   [19] Israel (1995) 546, 595–596

It is not entirely clear what the father had to do in order to obtain this trust. We do know that, shortly beforehand, he had lent his assistance to the foundation of a school for higher public administration, the College of Orange (*Collegium Aurasiacum*) in Breda, with André Rivet, the learned tutor of Willem II, as rector. Lodewijk was among the first students to be sent there. After six months the level and status of this college (which was closed down in 1669) was still unclear – which is putting it nicely – and in the beginning of 1647 there were fewer than sixty students. May we assume that the father, who was curator of the college, played his trump card and enrolled Archimedes at Breda to please the Prince of Orange?[20]

However compliant Christiaan may have been, he can hardly have obeyed the paternal command and turned his back on Leiden, and therefore Van Schooten, without protest. We have no proof of hard feelings on his part, however, unless this be in the 'missing' correspondence with Mersenne, and in the dearth of letters to his family in 1647. We learn not from him, but from Rivet, who told the father, that he participated in a public debate in August that year.

As to his general state of well-being, we have only the testimony of the lawyer Johann Dauber to go on. Christiaan went to board with him in March, but that was the following year. On 3 April 1648, Dauber wrote to the father:

> [Your son] who lodges with me is young in years but old in virtues. Never before have I seen so much insight and knowledge, such a quick understanding, so fine a judgement, such extraordinary circumspection, such honesty and modesty in discussion. . . . I write this of my own accord and am not afraid of being suspected of flattery.[21]

And of course this is the sort of letter that was preserved. After this we do find letters to the family, but not to the father. On

[20] Smit 225–226   [21] OC 1, 87

2 March 1648, Constantijn informed him of his father's displeasure at his not having written to him. A letter he writes back to Constantijn is of interest, because it reveals to us a new side of Christiaan:[22]

> I have no doubt that you have danced this Shrove Tuesday, for dancing has been so in fashion this winter. I fare very reasonably. Among other things we were splendidly entertained at the house of the bailiff. There were girls as well. After we had played at some cards in the afternoon at five o'clock, we ate an excellent meal with a dessert of preserves. These were quite as pleasant to the eye as they were to the palate, so prettily were they arranged by the girls, with garlands of every vegetable and flower that was to be had. Now that I regale you with such detail, I shall not be reproached for not naming the girls. Fifteen of us sat down at table: the mayor [Johann van Aerssen], the oldest Rossem sister, the Misses Ceters, Bornius, Miss Veecken, the son of the bailiff [Cornelis], Miss Stas, Miss Boxstaert, [Willem] Stas, the second daughter of the bailiff [Amarantha], Captain Despon, Becker, the oldest daughter of the bailiff [Maria], myself and the second Miss Rossem. The bailiff was in Den Bosch. After we had eaten, we danced the courante (with three violins), until three in the morning. Now you can't say that I have deprived you of details. If I come to The Hague at Eastertide, you will see how I have learned to fence. *Adieu.*

Apparently thinking back to this time in Breda, Christiaan referred to *'les belles qui me charmoient si fort'* ('the pretty girls who charmed me so').[23] But we are never to know whether he fell in love with one of these girls around the time of his nineteenth birthday. Even in this very detailed letter, we cannot tell whether he was of the group that danced until deep into the night. Clearly, though, Christiaan had fallen amongst convivial company from Brabant, and not into intellectually stimulating circles.

[22] OC 1, 80   [23] OC 6, 133

Perhaps he was in some way influenced by the philosopher Hendricus Bornius, who, according to the above quotation, was sitting at that table laden with candied fruit; he would later become professor at Leiden and tutor to Willem III. In his inauguration speech, this Cartesian in disguise said that he looked upon the Earth, Sun, and Moon as stars of differing greatness, and Christiaan would later send him his work on the pendulum clock and Saturn.[24]

There is no trace whatsoever of any influence of the mathematics teacher John Pell, a mediocre mathematician who attracted only a very few students; after attending a few of his lectures, Christiaan never saw him again. He corresponded frequently with Frans van Schooten, who wrote, for instance, in June 1648, that he had bought an old but undamaged copy of a Greek–Latin edition of Archimedes for seven guilders and five stivers, from an estate in Haarlem, and would Christiaan like to purchase it. It goes without saying that he did (the edition was from 1544).[25] And when he wasn't working at his mathematics, he joined half-heartedly in his brother Lodewijk's escapades.

It was the quarrelsome Lodewijk who unintentionally hastened their departure from Breda. Whether blood actually flowed, we do not know. But when frustrated adolescents are given fencing lessons and go round carrying swords, the inevitable happens. On 23 March 1649, Rector Rivet notified the father and curator:[26]

> Lodewijk did not start the fight; that was a certain ruffian De Vries, son of the rector from The Hague, who provoked him for some paltry reason, in a state of drunkenness. But he should have only responded with contempt to the provocation, and informed his tutor . . . I have always been of your opinion that students should not carry swords when at school, but only when travelling. But until now I have been unable to forbid it, for each time I was given to know that this was permitted at the other academies. To

[24] Vollgraff 414–417   [25] OC 1, 101   [26] OC 1, 104

be sure, in my time at Leiden, such gladiators were unknown. And it has always surprised me to see them enter the lecture hall not wearing a cloak, and with innocent-looking gold braid, a wide sash across the breast and a sword, to be decked out in such fashion with their hands on their swords as if poised for battle. Your sons were the first, after a young man by the name of Berk from Dordt, who sported a moustache as well. What I request of you is that you add the weight of your fatherly authority to that of the curators, so that they may cease in such conduct. As a general rule swords will be banned for all.

Instead, the father and curator wrote back angry letters to the rector, in which he accused him of weak governance and explained that when necessity arose these weapons served to defend oneself, 'but not to be brandished, as a soldier would do'. And that was the last straw. He called his sons home. So Christiaan, who had lived away for four years, returned in August to his ancestral home, without a degree.

Just to complete this 'tale of moral conduct' of the two brothers, we quote a few passages from Constantijn's journal; he wrote it partly in code, but this has recently been deciphered. For the past two years the eldest, now twenty-one, had acted as assistant at his father's secretariat, and after the war (the Peace of Munster had been signed on 30 January of the preceding year) he went on an extended trip to the south, passing through Brabant:[27]

> 24 May: Someone else of the party had a girl with him. He said she was his sister-in-law, but we took her to be a doxy. H. said he had fondled her breasts, but later we found out this was not true. She was an Aernout and came from Wynox near Berchem, her brother-in-law was a city collector. After we had eaten, I rode with W. to Berchem. We did the city. R. visited a doxy in Berchem.

[27] Heijbroek 170–171; Huygens 46, 89, 129

> That evening R. and I stood in front of the window where our
> little Miss Aernout made as if she were preparing to go to bed.
> R. wanted her to open the door so he could bid her farewell, but
> she would not. She stood there with her breasts bared.
>
> 26 May: This morning we went out walking. At noon R. and I rode
> in a coach with a whore to Berchem and there we drank brandy.
> No swiving to be had though.
>
> 6 September: Immediately after noon we went to Madame van
> Santen and found her sitting in a small room. She was young and
> pretty and fair, well-endowed and quite naked, like most of the
> girls here. At once her sister entered, who was also handsome
> indeed, but scarred by the pox. We went all together to their
> house. There we saw their father, a somewhat thin-looking fellow,
> and the mother and another sister who was so-so. We made our
> acquaintance with all these people and I accepted a glass of apple
> wine from Madame van Santen. W. said that he had asked her for a
> bit of swiving. I danced a courante with her.

These confidences were no secret to the father, for he checked the
journal for accuracy and correctness, and would have easily been able
to understand the few coded sentences. We note that Constantijn, in
the meantime, nipped down to Angers on 23 July to buy himself an
academic degree.[28]

At that time both France and England were in a state of turmoil. In France, the Fronde was challenging the monarchy, and in
England, Parliament under Oliver Cromwell had brought down the
king. During his visit to London, which preceded his long trip south,
Constantijn had witnessed, to his horror, the beheading of Charles I.
Did the father have a similar ragging in store for Christiaan and
Lodewijk? He had written to André Rivet that he was taking them
out of the Academy of Orange to go on a 'grand tour'. In other words:
to obtain their degrees at Angers.

[28] Vollgraff 446

He abandoned this plan for the time being, possibly because of the Fronde. But he was determined that Christiaan should become a highly qualified public servant. He himself had been sent by his father to be an apprentice in a lawyer's firm at Zierikzee. Now he, in turn, considered sending Christiaan and Lodewijk to the bar in The Hague.

In September Christiaan wrote to Constantijn, 'I believe that my father . . . wishes us to visit the lawyer's office, but I hope that it is not for long.'[29] Not for long! Did he actually protest? We do not know, just as we do not know whether he actually went there, or instead, made the trip to Antwerp and Leuven, as he reveals in a letter to Constantijn. All that we know is that, shortly afterwards, he was sent off to join a mission on its way to Flensburg.

Equipped with a lofty recommendation from the father ('the boy is not only learned in law . . . , but also in French, Latin, Greek, Hebrew, Syrian and Chaldean, and is moreover an outstanding mathematician, musician and painter'[30]), on 16 October he joined Hendrik van Nassau in Bentheim, who had been sent to the Danish court to plead for a moderation in the toll on passage through the Sont. He reported to Constantijn on 25 December:[31]

> Five or six days ago I returned from my journey to Denmark, where, as you know I accompanied Count Hendrik. For him Holstein was far enough, but out of curiosity I continued further, in good company, as far as Copenhagen and Helsingør, where ships must pay toll to the King. Had the season allowed, I should have crossed over to Schonen and Sweden, to see Descartes and the queen about whom he writes so wondrously. But that could not take place, and I had to return to the Count. At the beginning I spent ten or twelve days with him at the court in Flensburg. Here it was all eating and drinking and dancing and play. After dining, one proceeded 'ins frauenzimmer' where there were twelve ladies-in waiting and several frauleins, all in French dress but

[29] OC 1, 111   [30] OC 22, 57   [31] OC 1, 113

none speaking a word of French. Mostly the King [Frederik III] and
the Queen [Sophia] danced just like all the rest. The Queen is only
eighteen, the King forty-three or older. I was surprised that he
enjoyed dancing minuets with the rest of us. The count was
permanently on the dance floor, and Beer taught us English
galliards that are no longer in fashion. I could tell you much more
of what I have seen on my travels, but to keep it short I shall just
say which customs were new to me. At table, only the king has a
plate. They drink beer from large silver goblets, that can hardly be
raised. Their bread is full of cumin, as are their sausages, and
much of their meat. Everyone sleeps between two thick feather
quilts. They dine at ten and sup at five.

He fell ill on his return, with painful swollen glands in his neck and
below his ears. In January these burst open, soiling his arms. Did
he really imagine that as diplomat's attendant he could travel across
the Sont to Stockholm? And even then, the bitter winter proved too
much for Descartes, so he could not have spoken to him anyway. The
philosopher died on 11 February 1650, six months after he had arrived
on the invitation of Queen Christina.

Christiaan's letter to Constantijn on the death of Descartes is
badly stained, till well above the opening lines, probably because he
had strewn with ink instead of sand.[32] Was this absent-mindedness or
distress? In the sixteen-verse epitaph that he wrote for Descartes in
March his emotions are restrained, not to say rigidly in check. Here
follows the most elegant of his quatrains:

> Cette ame qui toujours, en sagesse feconde,
> Faisoit voire aux esprits ce qui se cache aux yeux,
> Apres avoir produit le modele du monde,
> S'informe desormais du mystere des cieux.

[32] OC 1, 124 & 125

Not a day passed, or his keen eye
Would unravel for us, what we could not.
Now and forever, the world he hath arranged
And has gone to charter heaven's mystery.

The third verse brings us back to science. But before we embark once again on Christiaan's scientific works, we must say something of the political developments that made them possible, and which maybe even furthered his career.

In 1650 Stadholder Willem II came into conflict with the States.[33] After the Peace of Munster they sought to reduce the costly army, the Stadholder's most important power basis. Willem II thought he could impose his will by locking up seven leaders of the opposition in Loevestein, and the next day, 31 July, he recklessly dispatched his army to crush the troublemakers in Amsterdam. The city closed her gates, however, and brought her guns to bear in reply, so that the coup turned into a fiasco. After this, the hot-blooded Stadholder flung himself into wild hunting parties across the Veluwe, became seriously ill, and died suddenly, one week before his Henrietta Maria gave birth to her first child on 14 November – a son, Willem III. Without a capable Stadholder (Willem III became 'Child of State') the States seized all power for themselves.

The result was that father Huygens, whose influence depended upon the House of Orange, lost his grip on the diplomatic services of the Republic in the course of that chaotic year. This, in turn, meant that he was no longer able to manoeuvre his sons into any official position.

So for the time being there would be no official post for Christiaan. The political developments that drove the mathematician Johan de Witt from his studies and propelled him into the high office of Grand Pensionary of Holland, the most powerful State of the Republic

---

[33] Israel (1995) 603–609

of The Netherlands, also drove his younger fellow student, Christiaan Huygens, from an official position to become a mathematician and, indeed, the greatest authority in the Republic of the Learned.

This development also spared him the conflict with father Huygens about his future, a conflict that continued to simmer beneath the surface, and its symptoms cannot be retraced either in March 1647, or in September 1949. It is remarkable that Christiaan never recognised this development as favourable to himself.

On 2 August, when Willem II was still considering whether to press ahead with his attack on Amsterdam – at gunpoint, if necessary – Christiaan wrote to Constantijn, 'I hope that all may turn out for the best, and patiently await the outcome of this whole business, without taking it seriously to heart.'[34] He observed these developments through his father's eyes (who stood on the side of Willem II), but was determined to see nothing. We see his aversion to politics as an act of protest against his father. We think he must have decided that he would fail in any public position, just to save himself.

And his aversion did not stop there. Father took him along to the wedding, in Amsterdam, of Christina Hooft, daughter of the late bailiff of Muiden and Leonora Hellemans. On 22 March he wrote to Constantijn, 'What I enjoyed most was watching the antics of the youth of Amsterdam, who strike me as excessive and insufferable in their behaviour.'[35] At the time Christiaan was busy writing his first scientific work.

This is how it begins:[36]

De iis quae liquido supernatant
Libri 3
Hypotheses

    I. *Si corpus sponte seu gravitate sua moveri incipiat, deorsum moveri; id est ut centrum gravitatis proprius fiat plano horizontali parallelo.*

[34] OC 1, 128   [35] OC 1, 123   [36] OC 11, 83–210

II. *Si corpora plura gravitate sua moveri incipiant, ea deorsum moveri; id est ut centrum gravitatis ex omnibus compositae proprius fiat plano horizontali parallelo.*

III. *Si liquido corpus solidum immergatur, tantam liquidi molem supra propriam superficiem ascendere, quanta est moles corporis infra eandem superficiem depressi.*

In our own language:

On parts which protrude above the surface of a liquid
3 books
Hypotheses

I. If a body begins to move either of its own accord or due to its own weight, it moves downwards; that is to say parallel to [?] the horizontal surface, if the centre of gravity is appropriate.

II. If various bodies begin to move due to their weight, they move downwards; that is to say parallel to [?] the horizontal surface, if the centre of gravity of all the bodies together is appropriate.

III. If a solid body is plunged into a liquid, the weight of the liquid that rises above its own surface will be as high as the weight of the body that has sunk beneath the surface.

(Remark: the third proposition is only true if the density of the solid body and that of the liquid are the same, otherwise one must read volume where weight (moles) is now given; furthermore, apparently we must replace the question marked 'parallel to' with 'perpendicular to' and we must consider the bodies of the second proposition as being connected.)

In the meantime he had become thoroughly familiar with the work of Archimedes on floating bodies. He had gone through it under Jan Stampioen; after this, he had studied it in Van Schooten's detailed version; then, finally, he had bought the (1544) special edition. There was ample room for further development of this renowned work, because the Greek had only considered forms that could be rendered as sections of spheres and parabolas being rotated on their axis. So Christiaan could test his powers on other simple forms.

But he had something more: a physical (not a mathematical) principle that the Greek did not know of, or at least had never admitted to in his mathematical calculations. It was the principle of the minimal height of the centre of gravity, the same as he had used in his study of the catenary curve. This enabled him to elaborate on Archimedes' work in an original manner.

This, in fact, is the aim of his first book. In the first four theorems, which follow immediately after the propositions quoted above, Christiaan deduces from this principle the level of the liquid surface, and subsequently the equilibrium of floating bodies with a density which is as great, or smaller, than that of the liquid. This latter case leads to the famous law of Archimedes, which Christiaan deduces in three variants, followed by the new theorems on equilibrium. Theorem 6 says that the difference in height between the centre of gravity of the floating body as a whole, and the centre of gravity of that part of the body that has been immersed in the liquid, is as small as possible. Theorem 7 says almost the same, but in relation to the part not immersed in the liquid. If the body is a homogeneous mass, then these two theorems are not inconsistent with one another. To prove stability, Christiaan draws an arbitrary horizontal surface through the body that must correspond to the surface of the liquid. Then he calculates the centres of gravity of the upper and lower part, through which he draws horizontal surfaces (s). Finally he works out whether the distance to the surfaces (s) from all points in the vicinity of the centre of gravity of the body as a whole is greater than the distance from that centre of gravity to the surfaces. And so he discovers, once again, the equilibrium conditions that Archimedes had found for sections of spheres and paraboloids with a vertical axis. But similarly, he also finds the conditions for a straight cone with a vertical axis, if the smallest angle of the cone is pointing downwards (theorem 14) or upwards (theorem 15).

After this highly innovative work in his first book, his second comes as a bit of a disappointment. Here, no theorems on minimal distances are used. The reason for this is that he rewrote the first book using this insight, whereas the second and third were left in their original versions. Theorem 1 of the second book, which therefore does not follow on from theorems 6 and 7 of the first book, repeats Archimedes' treatment on the weight of body mass, which works its way downwards, and the force of the liquid, which works its way upwards. (It is difficult to read this theorem without intruding our conception of force; we may speak of *actio* or effect.) Even so, the second book is profound. Christiaan attempts to find a solution that is as complete as possible for the equilibrium of a floating cuboid, in which we use cuboid as shorthand for a rectangular parallelepiped with a short side (a), a middle side (b) and a long side (c). If we assume that (c) lies constantly parallel to the liquid surface, then there are six ways in which such a cuboid can float.

The first two ways are easy to work out; these are the ways where the short or the middlemost side is parallel to the liquid surface. To work out the remaining ways we separate side (a_1) from side (a_2) that lies opposite and similarly side (b_1) from side (b_2). If we tilt the cuboid, then four possibilities emerge. Firstly, all of (a_1) remains floating above the surface of the liquid and all of (a_2) will remain below the surface, and (b_1) as well as (b_2) remain partially below. Secondly, (b_1) remains floating right above the surface and (b_2) right below, while (a_1) as well as (a_2) remain partially below. Thirdly, (a_1) remains right above the surface and all of (b_1) also emerges above, which must mean that (a_2) as well as (b_2) only remain partially below the surface of the liquid. Fourthly, not all of (a_1) remains above the surface and (b_1) does not entirely emerge above the surface either, which must mean that (a_2) as well as (b_2) both remain totally submerged.

For these various ways in which a body can float, he calculates how the location of the centres of gravity of the part above and the part below the surface of the liquid depends on the ratio of the sides (a)

and (b) and on the ratio of the densities in the solid and the liquid. He then requires, in accordance with the theorem with which this book begins, that these centres of gravity lie on a vertical line, as otherwise the beam would turn through a torque. He could also have required that theorems 6 and 7 of the first book should be carried out, but he did not get around to this elaboration. So he succeeded in deducing the conditions of equilibrium for the first four ways, but in the case of the last two he found the problem too difficult (this requires the solution of an equation of the fourth order).

The third book deals with a floating cylinder with flat extremities. It begins with four theorems on the centre of gravity of cylinder pieces and then follows, though in a different order and more concisely, the argument of the second book. Experience proves that the axis of this cylinder stands either perpendicular to the liquid surface (think of the pieces in a game of draughts), or parallel to it (think of a pencil). Which of the two is more stable depends on the relationship between the length and the diameter of the cylinder, which can be compared to the sides (a) and (b) in the second book. What is interesting about this book is not so much the proofs of these solutions, but rather the mathematical theorems with which it begins.

Christiaan wrote all this down neatly on twenty large sheets of paper (the size of four pages), drew the figures, and sent in *De iis* for his professor's approval. Van Schooten had almost nothing but praise – as his letters of 27 September and 21 November show – and was quite awestruck: '*Praestantissime Domine*' ('Most Prominent Gentleman').[37] But neither praise nor awe prevented his student from embarking straightaway on a second version. He finished the first book, but not the second or the third. Why not?

The fact that he gave priority to other mathematical works (*Exetasis*) did not mean that he was no longer interested in *De iis*, because he spoke of it years later in letters to Gregorius van Sint-Vincent, Gottfried Kinner von Loewenthurn and René François

[37] OC 1, 130 & 135

de Sluse. But the form did not appeal to him. On 23 March 1652, he wrote on the title page *'omnia mutanda'* ('all should be otherwise').[38] And later, in 1679, he added that only certain theorems (numbers 6 and 7 of the first book and the first four of the third book) should be preserved, and that the rest could be tossed onto the fire (*'reliqua vulcano tradenda'*).[39]

This explains why the work was never published, although it contains results that Daniel Bernoulli, Pierre Bouguer, Leonhard Euler and Michel Guyou (in that order) would rediscover. Bernoulli's work is from 1738 and Guyou's from 1879. This latter date alone clearly shows that *De iis* was something extraordinary. It was, and is, proof of his genius.

[38] OC 11, 91   [39] OC 11, 92

6 Collisions

'It pleases me exceedingly . . .' This was how Professor Frans van Schooten began his commentary on *Exetasis* (Research) on 20 September 1651.[1] He had just read through Christiaan's latest mathematical work.

Not even a year had passed since his remarkable thesis and the respectful *'Praestantissime Domine'*. Titan had just turned twenty-two. He spent all his time on mathematics, and was interested in nothing else. We imagine him closeted with his calculations in his room in the vast house on Het Plein, not even noticing the food that Cousin placed in front of him.

Constantijn, now out of work, chased after the girls and described them so provocatively to his friend, the young Van Aerssen, that Van Aerssen could feel it 'in a certain bodily part'.[2] But the studious Titan appeared to deny his sexuality entirely and, instead, lit a candle when darkness fell, so that he could carry on working, 'for candlelight doubles the day'.[3]

The father, now nearly out of work (he had only his property to manage and had handed over the income and title of Zeelhem to Constantijn) completed his poem on Hofwijck all in one go: more than 2800 verses.[4] But Titan, even more possessed, was deaf to all; he formulated theorem after hypothesis and calculated anew the vast work of Gregorius van Sint-Vincent (the Jesuit professor at Gent) on the quadrature of the circle . . . making heads spin when he discovered its well-concealed flaw.[5]

This is how it must have been in 1651, because otherwise it is hard to understand his new achievement. Even so, we have precious

---

[1] OC 1, 145   [2] Keesing (1983) 44   [3] OC 21, 727 (OC 4, 104)
[4] Van Strien & Van der Leer 11–18   [5] OC 11, 280

little to go on. Things might have been quite different from what we imagine. At Hofwijck, the source of the father's inspiration, he might just as easily have been tempted to while away his time with a game of billiards or bowls, with a glass of beer in the sunshine. And could he really have ignored all the usual humdrum of family life or, at his age, that 'certain bodily part'? Whatever the answer may be, we know of no other personal document than a request to the father (in Latin) which must date from the end of that year:[6]

> I offer no more than this small token, but I thought to have filled this small volume with something great on my part. I set out the essentials only, with the main substance of the proof, and I shall add the remaining proofs if your appreciative and discerning mind so wishes. But why should that be necessary – unless of course it be asked of me, for your smallest request shall be granted – as we encounter one another daily, you and I who am the smallest bird in your beloved nest of sons? To this gift I add one request: that the son who bears your name, who is less heedful than I, as you are good-naturedly wont to say, and who has clearly been shown also the sum of my own patience, be tolerated by you, with love. My regards [vale].

We can say this is a moving document, and leave it at that. If we look at it more closely, however, we see a striking balance between the small and the great, the requests and demands, the banal (Constantijn's drunkenness?) and the sublime (the forty-three quarto pages of *Exetasis*).[7] You do not write such a measured request for nothing. There must have been words between them now and then at Het Plein. If the atmosphere at home was so acrimonious, then Christiaan, who tried to handle conflict 'without taking it to heart', might have found it unbearable. Like everyone else in the house, he must have suffered, and tried to put an end to it. In this case he staked all he had: his first publication.

[6] OC 1, 163   [7] OC 11, 270–337

Van Schooten had written that he was exceedingly pleased, then continued:[8]

> Indeed, to the extent that it (in my opinion) may be placed versus any work in geometry, due to the subtlety of the findings, the conciseness and clarity of the proofs, and also the meticulous care in style, and may be compared with the works of Archimedes himself. I would like you to publish it as soon as possible, so that it be clear to all, that up till now no new ways for the quadratures have been found, and that you are the first to have proved that of the hyperbole, because the efforts of Pater Vincentius have failed. You could use this dissertation for other matters that you, as forerunner, have published, and because this is short, have it printed with large lettering in a small format, as in 8vo.

So it was Van Schooten who advised him to make his results public, straightaway – though he had more up his sleeve, for he was deeply critical of the *Opus geometricum*, published in 1647 by Gregorius van Sint-Vincent.

To begin with the most spectacular of Christiaan's discoveries: Gregorius believed that he had proven that the quadrature of the circle was possible, that is to say, that the surface and the diameter of a circle were in measurable proportion, or to put it in yet another way, that the number π (3.1415926535897932384 . . .) could, as it were, be constructed with a compass and ruler.

(Because this is impossible, contrary to Gregorius' assertion, we shall call π an irrational number; in the course of that century John Wallis discovered that $\pi/4$ could be calculated by dividing the product $3 \times 3 \times 5 \times 5 \times 7 \times 7 \times 9$ . . . by the product $2 \times 4 \times 4 \times 6 \times 6 \times 8 \times 8$ . . . , and subsequently Leibniz discovered that $\pi/4$ could be calculated by $1 - 1/3 + 1/5 - 1/7 + 1/9$ . . .)

Christiaan had discovered the flaw in this extraordinarily lengthy and obscure *Opus geometricum*. In fact, Gregorius had not

[8] OC 1, 146

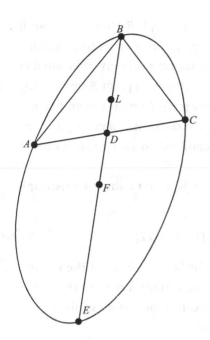

FIGURE I

applied the so-called method of indivisible line elements, proposed by Francesco Bonaventura Cavalieri, to line elements, but to the sum of their ratios. If this was not a blunder, it came very close, and by using a mathematical example, Christiaan could clearly show this application to be flawed. However, 'the fact that Gregorius had been in error in his quadrature of the circle' wasn't worth more than an appendix of *Exetasis*.[9]

His quadrature of the hyperbole (once again, that is the surface, and in this case, a portion of a surface) to which Van Schooten refers was not even his most significant result. That was, undoubtedly, his discovery concerning a proportion for a surface section of an ellipse, which we shall now attempt to demonstrate.[10]

Draw an ellipse and choose two arbitrary points A and C on the circumference (Figure 1). B lies at the top of the segment determined

[9] OC 11, 280 & 317   [10] Bos (1980) 129–130

by A and C. These points determine a section $e(A, B, C)$ of the ellipse. If they are joined by straight lines, these points also determine a section $t(A, B, C)$ of a triangle, which we can see immediately to be smaller than the section $e(A, B, C)$. Now draw a line from point B through the midpoint F of the ellipse to the opposite point E on the circumference. This middle line passes not only through the centre of gravity L of the ellipse segment, but also cuts the cord AC in two equal pieces at point D.

When we make this drawing, it is easy to show Christiaan's discovery:

$$e(A, B, C) : t(A, B, C) = (2 \times ED) : (3 \times FL)$$

Because we can easily work out the surface (A, B, C) of the triangle and we already know the length ED, the surface section of the ellipse follows from FL, the distance of the centre of gravity of the ellipse segment to the midpoint of the ellipse.

We shall not attempt to present his precise and flawless proof. This result, which he accomplished by classical means, was not only new; it was also compelling. He applied it immediately in *Exetasis* to calculate the surface of a circle (which, after all, is an ellipse whose axes are both the same length) or, as we now understand it, to find a more accurate value of π. Ten years later he would use it to estimate logarithms.[11]

Exetasis established his name. It was his scientific debut, alongside his father's literary debut at twenty-six years old – four years older than Christiaan was when *Exetasis* first appeared. It was Van Schooten, a scholar of European renown, who encouraged the son, and Cats, a poet from Zeeland, who inspired the father. For the son this meant his 'negotium', for the father his 'otium', to define, in classical terms, the contrast between exertion and ease, labour and leisure, or slavery and freedom.

[11] OC 14, 451–459

As early as 6 October, Christiaan set out to inform Gregorius van Sint-Vincent, politely, of his blunder, '*Theorematum quorundam causa quae conscripseram intelligens non parum mihi obstare posse si tua pro veris haberentur*' ('The theorem that I submit cannot, in my opinion, be applied in the manner you hold to be correct').[12] The professor from Gent did not give in straightaway, and a spirited correspondence followed. In the end Christiaan actually visited him; it was worth his while to persuade the other of his error.

After the publication of *Exetasis*, correspondences followed with other scholars of the Spanish Netherlands: Daniel Seghers in Antwerp, Andreas Tacquet and Gerard van Gutschoven in Leuven, William Brereton in London, Gottfried Kinner von Loewenthurn in Prague, Daniel Lipstorp in Rostock, but temporarily in Leiden, and various mathematicians in his own country. Even without studying these letters in great detail, it is clear that there is much more under discussion than just the proofs in *Exetasis*.[13]

His *negotium* weighed upon him heavily. The vivaciousness of his letters to his family (of which we shall give two examples later in this chapter) and the polish of his letters on science conceal his anguish. How could it be otherwise?

He had refuted assertions on the form of the catenary curve and the surface of the circle; there is nothing particularly shocking about this. What is astounding is that he undertook the painful task of refuting the assertions of the man he had only recently praised, unreservedly, in his verse 'Now and forever, the world he hath arranged'. He had stumbled upon serious flaws in Descartes' collision theory, the physical foundation of his world vision! What should have been sublime was now shattered to pieces.

On 17 January 1652 he voiced his doubts to Van Gutschoven, but on 29 October he expressed his certainty to Van Schooten.[14] The renowned philosopher, like his mother fifteen years before, had

---

[12] OC 1, 147   [13] Vollgraff 437–443   [14] OC 1, 166 & 185

abandoned him to that inconceivable emptiness. And he dared to go forward:[15]

$$bx + ay = ac$$
$$axx + byy = bcc$$

In all probability these formulas, which he wrote down, are the first physical formulas ever to be written; a and b are body quantities (we would now call this: mass), while x and y and c are speeds. The second formula shows that energy, as we now call it, is conserved. How did he work this out?

He noted down other things not yet heard of, in a jumble of calculations, bits of text, and geometric results. The chaotic impulse to forge ahead? 'No wonder I suffer from headaches,' he wrote to Van Schooten on 13 August (it is his first mention of a migraine), 'not continuously, but at most inconvenient moments, for I want nothing more than to continue in my work, especially mathematics. Up until now, however, I must deprive myself of study, unless my strength of will can combat the pain.'[16]

Knowing this, it is surprising to read his letter to Lodewijk, written very shortly before on 24 May. His brother was in London, and just one week later would be there to see the cannonades outside Portsmouth that heralded the first of the sea wars between England and the Republic:[17]

> I am obliged to admit that your trip sounds most pleasurable and that you spare no effort in inducing me to regret not having visited England in your stead. I hope that Mr [Diderik] van Leeuwen will speedily acquaint us with all the details, for you shall not be permitted to return before the ambassadors. We arrived home from Cleves on the twelfth of this month, after a visit of one week. That week must have cost the Elector [Frederik Willem of Brandenburg] a pretty penny, for it is certain he paid for

[15] OC **16**, 98   [16] OC **1**, 184   [17] OC **1**, 180

everything. The city was so full of people it was nearly impossible to obtain a roof over one's head. I was obliged to change lodgings three times in order to find one that was the least filthy and noisy. The wedding [of Count Willem Frederik of Nassau, Stadholder of Friesland, with Albertina Agnes, sister of the deceased Willem II of Orange] took place on the second. It was splendid, with ceremonies that are quite customary in Germany. If you wish to imagine it, just think back to all you saw in Kassel, and then many times more splendid. On the sixth I saw the procession, which quite outdid the wedding of Mr van Brederode. The one half, under the leadership of the Elector, came dressed as Romans, the other half as Moors with the Count of Waldeck. They marched in perfect time, and the three leaders brought up the rear, I think you can guess how. The next day they went in procession around the town, but without their costumes. But I have forgotten to tell you about the magnificent fireworks on the fourth, when often three to four hundred rockets would be let off into the air, all in one go. On the eighth there was a ball, but that was nothing special for those who have attended such balls here in The Hague. As well as this, there was entertainment to be had almost each afternoon at the Comedie, where the same French orchestra was that you saw play here for His late Majesty, and every evening the usual ball. But here it was only princes and counts who danced (indeed, there were so many of these), so that the nobles and their ladies must be satisfied with the spectacle alone. Sometimes I ate with the nobles, at three or four tables, or else with the ladies – there were quite a number of these, and now and again with my father and Count Maurits, who always dined at his leisure in his hotel, with just a few servants in attendance. Mr van Brederode took us one afternoon in his carriage to Xanten, where it is said that Caesar ruled when it was still called Castra Vetera. This gentleman provided us with food the whole day through, and in good company, for his family was there too. The day after he showed us the plantations of the Count Maurits, with gardens and grottoes finely laid out and at very moderate cost, a most agreeable setting. I have no time to tell you more. I should like to return there, for I greatly enjoyed this short spell spent away from home.

One month later Christiaan left home once more and made a round trip as far as the Frisian capital, where he spent eight days looking around. This time he accompanied his father to the ceremonial entry of the newly wed Willem Frederik and Albertina Agnes into their residence. His father's presence at this occasion was, of course, imperative. For had he not been behind every arrangement to join the Houses of Orange and Nassau?[18] Afterwards they made a detour to see other small towns in the north. On 5 July Christiaan wrote to Lodewijk:[19]

> Never have I been in a country as agreeable as our own [Holland]. I have just returned from Friesland. I have to tell you that it greatly resembles Denmark, certainly regarding the customs and manners of the inhabitants. As you cross the sea and arrive in Staveren, you find yourself, in no time, in a most peculiar village called Molkwerum. The houses stand practically one on top of the other and all higgledy-piggledy, without there really being a street at all, so you can barely find your way out . . .

To discover energy conservation and then lose his way in Molkwerum! What is surprising about these letters describing his travels is that we see Christiaan behaving as a nobleman (indeed he writes: *gentilhomme*) who is taking time to enjoy himself. He makes a point of dining in the company of charming women, and takes in the wide-open Frisian countryside as if it really means something to him. Most probably his father (who belonged to the court of The Hague) demanded that he accompany him to the east, so that he could show him off, but it is unlikely that he then expected him to go north as well.

Christiaan appears to assume two very different guises. At least one of them is a mask, but perhaps the other is too. So who is the real

[18] Smit 243   [19] OC 1, 182

Christiaan? He is the gentleman of leisure (*otium*), while at the same time, judging by his notes of 1652, he works feverishly at his scientific studies (*negotium*). Picking up one thread after another, forever starting something new but never seeing it through. Hardly anyone ever finds out. Does it keep him awake at night?

Let us try to imagine something of this feverishness. He is tremendously preoccupied with the rules of collision. René Descartes had already set these out in his *Principia philosophiae* of 1644, but in January Christiaan already knows that, except for the first one, they are incorrect.[20] He goes in search of the correct laws and discovers that the principle of relativity, described in Galileo's *Discorsi* of 1638, can be of use to him. To make it easier he works with physical formulas, deduces the most important laws and notes them down on a large sheet of paper.

Then he becomes fascinated by luminous rings around the Sun (which are caused by the refraction of light in high-floating ice crystals). Descartes had described these aureoles too, and in May, he finds errors here as well. He sets out to discover the correct rules for the refraction of a beam of light, and, perhaps sometime in July, he works out the theorem that when a bundle of rays falls on a spherical surface, it contracts into exactly one point. Is this before or after the discovery of the rules of collision?

Then it is lenses that attract his attention. Probably in October, he measures the refractive index of his glass and discovers it to be 600 : 397, but then he already knows that this index is not the same for all glasses, not even for all water.[21]

And all the while, he is engrossed in pure mathematics. After his findings in *Exetasis* he devotes himself to describing a conchoid. He remembers that Nicomedes, who at some point had applied himself to the construction of just such a curved line, must have thought of an equation of the fourth power: $(x + a)(xxx - 4acc) = 0$.[22] However

---

[20] Vollgraff 453   [21] OC 1, 192   [22] OC 12, 13

did he come by it? It might even be true, too. Once again, he picks up *De iis*, writes down that everything should be otherwise, and possibly considers a somewhat freer algebraic application. Yet he begins – in April? as an exercise? – on problems of Pappus that amount to solving equations of the third and sometimes the fourth order . . .[23]

The mathematical groundwork produces simultaneously and indisputably a harvest of kinetic and optical discovery. If we postpone the discussion of Christiaan's work on lenses until a later chapter, we can restrict ourselves here to his work on collisions.

Some time in the summer of 1652, we get *De motu corporum ex percussione* (On the Motion of Bodies as a Result of an Impact). Christiaan notes down all essential theorems for the concept of collisions on a large single sheet (most likely there are more, but these have not been preserved).[24]

This condensed form makes *De motu* what it is: a rare masterpiece. Or should we say: in this very form? Indeed, he boldly writes that not only the impulse or momentum of the motion is conserved, but also conservation of the energy of the motion (as we call it) may be assumed, which he later preferred to see as an effect rather than a cause.

It is rather unfair to view these pages with our present-day knowledge of physics, but we can hardly do otherwise, and it does allow us insight into his achievement. If we then attempt to view the work with the knowledge of Galileo or Descartes, then perhaps this sin could be forgiven.

While Christiaan, four years after this decisive first step, in 1656, formally deduced the laws of collision from three assumptions, two laws now suffice: that of impulse conservation (IC) and energy conservation (EC). His first assumption comes close to (IC), and (EC), which he certainly saw in 1652, seems to follow from a universal principle that he does not even include in his assumptions. We shall return to this later. But first: how do we derive the laws of collision?

---

[23] OC 11, 213   [24] OC 16, 92–99

Consider two balls (or bodies, for their form is not important) with a mass of M and m (hereby indicating that the first is greater than, or just generally different from, the second). Let the balls move along a line and collide with one another, and the speed of the first will change from B to A and the second from b to a (B and b are thus speeds before collision, and A and a are speeds after collision). (IC) and (EC) now mean that

(IC) $MB + mb = MA + ma$

(EC) $MBB + mbb = MAA + maa$

It follows from (IC) that $M(B - A) = m(a - b)$, so that the ratio of the ball masses becomes

$$(m : M) = (B - A) : (a - b)$$

Consider now that it is difficult to measure these four speeds. Their 'true' magnitude cannot be determined, as it depends on the speed of the person who measures. What is B for one person may be B^1 for another. When both persons move parallel to the line along which the balls are moving, and when the second person is moving faster than the first with a velocity of w, then $B^1 = (B + w)$. Such a transformation also applies to the speeds A, b and a. According to the second person, the ratio of the ball masses will be

$$(m : M)' = (B' - A') : (a' - b')$$
$$= [(B + w) - (A + w)] : [(a + w) - (b + w)]$$
$$= (B - A) : (a - b)$$

This is the same as for the first person. So it doesn't actually matter how the speed is measured. And this means that each transformation of the type $B^1 = (B + w)$, which does not affect the validity of (IC), is permissible.

It is the relativity principle of Galileo, who had discovered it. By applying this principle the solution of the equations for (IC) and (EC) given above is simple, although it does not appear so. The speeds A and a, which exist after collision, are expressed in the speeds B and

b, which existed before collision, and read:

$$(M + m)A = (M - m)B + 2mb$$
$$(M + m)a = 2MB + (m - M)b$$

The proof proceeds as follows. Consider the velocity w with which the centre of gravity of the balls moves, and the corresponding impulse $I = (M + m)w$. Then write $MB = p$, $mb = q$, $MA = r$ and $ma = s$, so that the conservation equations are simplified to

$$(IC) \quad p + q = I = r + s$$
$$(EC) \quad mpp + Mqq = mrr + Mss$$

The latter is obtained by multiplying on either side of the equals sign by Mm. Now insert in (EC) the q and the s from (IC), so that

$$\begin{aligned} m(pp - rr) &= M(ss - qq) \\ &= M[(I - r)(I - r) - (I - p)(I - p)] \\ &= M(II - 2Ir + rr - II + 2Ip - pp) \\ &= 2MI(p - r) - M(pp - rr) \end{aligned}$$

Resolve the difference of squares $(pp - rr)$ and insert I, so that

$$\begin{aligned} 2MI(p - r) &= (M + m)(pp - rr) \\ 2MI(p - r) &= (M + m)(p - r)(p + r) \\ 2MI &= (M + m)(p + r) \\ 2M(M + m)w &= (M + m)(p + r) \\ 2Mw &= p + r \end{aligned}$$

This is the solution in its most simple form. The first-mentioned formula follows after substituting the abbreviations:

$$\begin{aligned} 2Mw &= MB + MA \\ 2w &= B + A \\ 2I : (M + m) &= B + A \\ 2I &= (M + m)(B + A) \end{aligned}$$

$$2(MB + mb) = (M + m)B + (M + m)A$$
$$(M + m)A = 2(MB + mb) - (M + m)B$$
$$= 2MB + 2mb - MB - mB$$
$$= MB - mB + 2mb$$
$$= (M - m)B + 2mb$$

The formula for a follows by substituting MA with $MB + mb - ma$, as is required by the conservation of impulse.

To appreciate fully the significance of the mathematics, we can look at the simple case of two balls of identical mass $(m = M)$. The formula for A then becomes $(M + M)A = (M - M)B + 2Mb$, so that $A = b$; and the formula for a becomes $(M + M)a = 2MB + (M - M)b$, so that $a = B$. This means that the velocity of the first ball after collision becomes, whatever its magnitude was, as great as the velocity of the second ball before collision; and that the velocity of the second ball after collision, whatever its magnitude, as great as the first ball before collision. In this case the balls have adopted each other's velocity.

This is perhaps even clearer when numbers are used, for instance, metres per second. Let velocities to the right be positive and those to the left negative. Take $B = +1$ and $b = -1$, then $A = -1$ and $a = +1$. Or take $B = +2$ and $b = 0$, then $A = 0$ and $a = +2$. In both cases the balls approach each other with the velocity 2. We can write this symbolically as

$$(1, -1) > (-1, 1)$$
$$(2, 0) > (0, 2)$$

Now, Descartes got the first one right, but the second one wrong, because he thought that $(2, 0) > (-3/2, 1/2)$! Christiaan had both right.

And this brings us back to *De motu*. How could Christiaan find the correct answer without taking (EC) as point of departure? The first part of the answer is that (EC) follows from (IC) if the masses are equal. In this case the equation for (IC) is indeed

$$B + b = A + a$$
$$(B + b)(B + b) = (A + a)(A + a)$$
$$BB + 2Bb + bb = AA + 2Aa + aa$$
$$BB + 2Bb + bb = AA + 2bB + aa$$
$$BB + bb = AA + aa$$

The correctness of the second line with the squares follows from the correctness of the first line; the third line follows from the calculation of the squares; the fourth line gives the result of the insertion of the equations $A = b$ and $a = B$ proved above; from here follows the accuracy of the last line, which gives the equation for (EC) if the masses are equal. So Christiaan did not even need (EC) for the simple case of equal masses. If he assumed (IC), then the logic of the relativity principle was sufficient, which we shall shorten to (RP).

Let us describe (IC) and (RP) as they stand in *De motu*, mentioning, however, that they appear in the worked-out text of 1656 and not on the sheet of 1652.[25] It concerns the first and the last of the three theorems with which *De motu* opens (the middle one requires complete elasticity of the colliding balls).

> (IC): If a body is put into motion, it will continue that motion at the same speed and in a straight line, unless it is obstructed.
> (RP): The motion of bodies, whether or not their speed changes, should be seen in relation to that of other bodies, which are thought of as at rest even though they participate in a further common movement with the (first-mentioned) bodies. Therefore if two bodies collide, but otherwise have in common an unchanged speed, they give impulses to one another that, seen

[25] OC **16**, 29–91

by someone who participates in the unchanging common movement, are as great as when the common movement did not exist.

To illustrate this, Christiaan used Galileo's idea of a collision experiment on a boat sailing along a bank. One man stands on the boat with his arms outstretched in the direction in which the boat is sailing. In each hand he holds the end of a piece of string that is attached to a ball at the other end, at knee height. Another man stands on the bank looking directly at him and can, if he so wishes, take hold of the top ends of the string with his hands, which are raised likewise. The balls can make an oscillation in the direction of the boat's course, and collide.

This provides a clear illustration for what Christiaan proposes next:

> If a man standing on a boat travelling at a steady speed lets two
> balls of equal weight approach one another with a speed that is
> seen by him or from the boat as being the same, then we say that,
> from his viewpoint, the balls must rebound just as fast as when he
> does the same standing on a stationary boat, or when he stands on
> the bank.

Galileo had already defined both (IC) and (RP). The first had already been around for some time and formed a ready starting-point for Descartes when he discovered his collision theory. But Christiaan was the first to discover the consequences of the truth of (IC) and (RP), and it looked as though he did not need any formulas to do so. An argumentation was sufficient, but then one in which the kinetic aspect of collision was so clearly distinguished from the relative aspect that his results were always correct. Was it instinct that kept him from going astray?

Let us take as an example his argumentation for the case of collision $(1, -1) > (-1, 1)$ which is only valid if the balls are equal. It was perfectly clear to him that the movement to the right and to the left, before and after collision, was the same, which complied with (IC). Then he imagined this collision taking place on the boat that was travelling along the bank at an arbitrary speed of *w*. According to the man on the bank, the speeds before collision would be *w* greater than for the man in the boat, but because the balls 'do not know' who is holding them, the speeds must also be *w* greater after collision for the man on the bank than for the man in the boat. This enabled him to understand that (RP) was complied with if speeds for the man on the bank were $(1 + w, -1 + w) > (-1 + w, 1 + w)$. If he chose an arbitrary speed *w* for the boat equal to 1, then he got $(2, 0) > (0, 2)$.

For those who consider this far too simple to waste their breath on, it is worth remembering that Descartes, one of the greatest minds of all time, did not understand it. And that is not all.

Christiaan also saw that this argumentation broke down if the balls were unequal. He may have carried out his own experiments, up in the attic, which proved this, but he may also have delved into Galileo's *Discorsi*, in which case it would have been obvious to him. This brings us to the second part of the answer to the question of why he was able to obtain the correct answers even without (EC).

Apparently there was something in his General Principle (GP) that could replace (EC), and to understand this we shall look once again at how he worked out ideas in the *Discorsi*. Galileo had written about a falling body that acquired a speed whose square was proportional to the height from which it had fallen. He had also written that the speed didn't alter if the body was diverted from its vertical path through a sloping surface, or by some other cause.

In Day 3 of the *Discorsi* we read a highly suggestive dissertation on the oscillation of a ball on a string: from its lowest point, the horizontal impulse of the ball will be just sufficient to cause it to rise to the same height from which it fell. Is it mere coincidence that Christiaan provided the man on the boat, who carried out the test, with two strings? It seems unlikely. Just like Galileo, he would have made the connection between the speed of the balls in the experiment and the height of their fall.

And now we come to his master move, which revealed the size (mass) of the balls.[26] He linked the height of their fall with the idea of their centre of gravity, and posited that this would not change as a result of collision. For, he argued, there is a General Principle that determines that the centre of gravity of the two balls lies as low as possible, and that this – because it is a General Principle – applies both before and after collision.

If G is the constant height of the centre of gravity and M and m are the ball masses, then the product $G(M + m)$ is also constant with a value of C. C is determined by the sum of the ball masses, multiplied by their heights. If H and h are the heights of fall of these ball masses before collision, and H' and h' their heights of rise after collision, then the General Principle leads to

$$(\text{GP})\ HM + hm = C = H'M + h'm$$

And according to Galileo, because the squares of the speeds B and b before collision are proportional to the heights of the fall H and h, the proportionality constant being g, and likewise AA and aa to the heights H' and h', we get

$$HM + hm = H'm + h'm$$
$$(g\,BB)M + (gbb)m = (gAA)M + (gaa)m$$
$$MBB + mbb = MAA + maa$$

This clearly shows that the application of (GP) to movements is equivalent to (EC). It cost him quite a bit of trouble (he needed three lemmas to prove it!), but Christiaan could now write in proposition XI of *De motu*:

[26] OC 16, 72

If two bodies collide, what one shall obtain by taking the sum of their size, multiplied by the square of their speeds, shall be as great before collision as after it. For this, one must know the ratio of sizes and speeds in numbers or in line segments.

Why did he consider it advisable to deduce (EC), which was his point of departure, from a (GP) extended to motions? Because it would be less shocking? But in doing so, he concealed his brand new discovery in his clever, but controversial generalisation of the General Principle which, earlier on, he had used for the catenary curve in *De iis*.

And to think that he would never have heard the roar of the canons on 10 August 1653, outside Ter Heijde, which sent Maarten Tromp to his grave and brought Johan de Witt to power, if he had dared to postulate energy conservation![27]

The problem of collision occupied his mind long after his *annus mirabilis* of 1652. In fact, it preoccupied him for the rest of his life. On 29 October 1652 he sent his first results to Van Schooten, who, on 4 November defended Descartes' collision theory against him.[28] On 16 December 1653 he informed Gottfried Kinner von Loewenthurn of his solutions, which we have given above as $(2, 0) > (0, 2)$, and let him guess the solution for a collision of a ball against a stationary ball that was twice as large (heavy).[29]

By 1654 he was apparently so much further that he was no longer interested in the solutions, but in the background of his assumptions on the perfect elasticity of collision. 'No instrument surpasses the hammer,' he noted in his studies of the impulse transfer.[30] He came incredibly close to the concept of force, for which we have Newton to thank, and which we see evolving in his notes between 1664 and 1666.[31]

[27] Israel (1995) 720–723   [28] OC 1, 186   [29] OC 1, 260
[30] OC 16, 104–105   [31] Westfall 140–175

Van Schooten wrote to him in some despair on 25 October 1654, in reply to a letter we do not have. He could hardly believe that such a great and visionary genius as Descartes would have published something that was untrue; and that he, Christiaan, would do well to let the matter rest, for it was a waste of his time and his talent (*'quam tempus atque industriam tuam inutiliter impendas'*).[32] He didn't mince words.

Christiaan replied that he would surely change his mind when he had seen the manuscript, but added that, 'if all Descartes' rules – apart from the first – are not wrong – then I am obviously no longer able to tell the difference between what is right and what is wrong'. For Van Schooten, a reputation was at stake, a reputation not entirely divorced from his own, and that is why he tried to dissuade Christiaan from publishing *De motu*. And with evident success.

It will always remain a mystery whether or not Van Schooten's influence was a decisive factor. At the end of his life Christiaan merely said that, in 1654, he still understood too little about the nature of collision to be able to actually publish. Still, he went to the trouble of rewriting the manuscript according to the rigorous convention of three *hypotheses* (assumptions) and thirteen *propositiones* (proposals), which was then circulated within a very small circle. It was completed in this form between 6 and 20 July 1656.[33]

Between 1656 and 1661 he revealed his discoveries to René François de Sluse in Liege, and to various scientists in Paris. In January 1668 he defended his collision theory at a sitting of the French Academy of Sciences, and his most significant propositions were discussed in 1669, in London, at a sitting of the Royal Society. Only then did he take the step to publish his most important finding.[34]

On 18 March 1669, seven propositions (without proofs) appeared in the French *Journal des Scavans* (Journal of Scholars), including the General Principle. Its Latin translation was published in

[32] OC 1, 301   [33] OC 16, 11–14; 99–155   [34] Gabbey (1980) 169–170

April 1669, in the English *Philosophical Transactions*, several months after publications of similar findings by Christopher Wren and John Wallis.

Why did he not yet reveal all? From time to time he revised the text, improving on it and dabbling in questions of philosophy, but never adding anything new to his calculations.

Nevertheless, in 1692, after the publication of Newton's *Principia mathematica*, he thought he had found something new, because on 11 July of that year he wrote to Leibniz, 'Talking of motion, there is much that I can demonstrate in the way of new paradoxes and other matters, as anyone shall see if I publish my proofs of laws of collision that could previously be read in journals in Paris and London.'[35] But by the time *De motu* was finally published in 1703, in *Opuscula posthuma* (posthumous works), it was outdated.

'Tomorrow we travel,' Christiaan wrote to Constantijn on 20 August 1654.[36] He was in Spa staying with his father, who was suffering from tuberculosis and needed the fresh air.

> Our cases are packed, and we have only to bid farewell to our friends. We have been here exactly one month. I have taken the waters only a few times, however much people tried to persuade me of their great benefit to my health, and thank goodness I feel no worse than the medicinal bathers. Brother Lodewijk took medicines yesterday, as one usually does on leaving the springs, then threw up the moment he came outside. He had the greatest difficulty reaching his lodgings and thought, to quote his own words, that he would surely die. We are certain that the apothecary made a mistake, or that it was a fake medicine of some sort. Mr von Steyn Callenfels will accompany us as far as Vianden or perhaps even further, so we need have no fear of bandits [who roam these parts], in spite of the Prince of Condé's troops in

[35] OC **10**, 302   [36] OC **1**, 293

Vianden. The guard will consist of eight men on horseback with two on the coach, who are to be given protection by the cavalry. Our own carriage goes empty from Maasdricht to The Hague, after bringing Mr van West-rhenen, along with his wife and sister, Miss Westerbeek, to the garrison. As you know, this Mr van West-rhenen lodged in the same house as we did and [medically] treated my father and his following on two occasions, so he richly deserves such courtesy. Undoubtedly Aunt Sint Annaland [Geertruyd] will be relieved to see her horses back, for she writes no letter to [coachman] Joncker without recommendation of some kind concerning the animals. I have made two drawings of Spa that I shall show to you when I come home. One is made from the mountains and the other from my cousin's [Philips Doublet's] window. If Flip [the youngest brother] makes fun of my drawings, he should know that they are vastly better than those of aforementioned cousin, who chose the same vantage points. There are so many people around us that I have difficulty not listening to them, and thinking of what to write. In any case I wish to add that you have my sympathy now that you have fallen into disfavour, and that amidst all this entertainment I am continually angered at the injustice done to you. [Constantijn had requested to join the delegation to Switzerland and been given a curt refusal by Amalia van Solms.] I put it down to the beneficial effects of the Spa water that my father is not visibly dispirited by such disdain, for undoubtedly it is he who should feel it most deeply. Adieu. Hug the aunts for me if you should see them. And tell our little sister that my father has purchased some jewellery for her. The royal princess [Henrietta Maria of England] also departs tomorrow, so as to escape the pox, and she leaves behind Miss Killigrew [Maria], who is covered with it. Tomorrow we shall dine in Botschenbach, which lies four miles from here, but you will not find it on the map. My warm regards. In spite of all allegations on the part of Mr Ruijs, brother Lodewijk does not have E.V.A. engraved upon his Spa bracelets.

'God bless you,' Constantijn answered, moved by his brother's sympathy:[37]

> While you were in Spa, a little fellow came to see me, someone
> with whom you are acquainted. Whatever is the damned chap's
> name again? He lives in Breda and manufactures lenses for
> telescopes. He regaled me with all the many things he could do
> and swore he had made a telescope with which you could tell the
> time in Dordt, all the way from Breda. He fetched from his pocket
> a pocket-sized telescope that looked useful enough, for you could
> read a letter from a distance of ten or twelve paces, but the trouble
> was it was too large to secrete in your hand. When I studied the
> lens, I observed that it was not polished in any masterly fashion.
> He had more of these in his bag, a veritable repository for all sorts
> of bits and pieces, also a microscope such as the one I made but
> then rather inelegant and with a set of worthless lenses. I told him
> of your invention with metal mirrors, but did not disclose to him
> what he was dying to know about it . . .

So Titan was not only writing his *De motu;* he was engaged
in all sorts of other things as well. He was corresponding extensively
with Gerard van Gutschoven, mainly about practical chemistry in the
laboratory, and that most probably would have provided an answer for
the peddler from Breda. In October 1653 he noted down a formula to
plate copper with silver:[38]

> Take a half a measure of silver and dissolve this in nitric acid, let
> this settle with salt or salt water, then wash it down with
> rainwater until the calcium of the silver is quite smooth. Then
> take a measure of ammoniac solution, a measure of silica or
> silicon, a measure of salicornia and a pinch of purified mercury.
> Mix these together with a little rainwater into a paste, and paint
> or spread this with a brush onto well-cleaned copper. Then hold

[37] OC 1, 295   [38] Vollgraff 435

this in the fire, but do not let it become too hot, just until it glows, until it has ceased to smoke, then cool it off in water of tartar. Repeat this four or five times and then it is well-silvered.

What exactly happened was as much a mystery to him as it was to his contemporaries. So he left it at that, when he wrote to Gottfried Kinner von Loewenthurn in 1655, 'If you have a mind for chemistry you must simply begin, then we shall exchange ideas on this noble science.'[39] Later on he felt somewhat differently about that 'noble-ness'. In 1692 he informed Leibniz of the death of Robert Boyle: 'I find it very odd that, with all the experiments one finds in his book, he was able to establish nothing. But it is, of course, such a difficult field.'[40] Chemistry was not yet regarded as a science. It was more of a recipe. We imagine him up in his attic, hovering over the hot copper, amazed, tears in his eyes from the smoke, and a nagging headache.

Undoubtedly, plating copper with silver was part of his work on mirrors and optical instruments. In fact, it must have taken up a great deal of his time, even if much of the polishing and grinding of the lenses also became a pastime for the out-of-work Constantijn. 'Optics absorb me completely,' he wrote in the above-mentioned letter to Van Schooten of 29 October 1652.[41] The following gives some impression of what he was up to in the 'room with a lathe'.

In the beginning of 1653 he gave the measurements for a telescope tube to the instrument maker Paulus van Aernhem. On 20 September of the same year, he informed Frans van Schooten that Van Berckel (possibly either the employer or assistant of Van Aernhem) had made a new telescope according to his own design, 'which I would like you to look through as well'.[42] In the meantime he obtained instructions from Van Gutschoven on how to polish lenses, and the prescriptions used by Sirturus and Hevelius.

He not only got Constantijn to do his polishing for him, but also professionals such as De Wijck in Delft and Kalthof in Dordt. These

[39] OC 1, 366   [40] OC 10, 263   [41] OC 1, 185   [42] OC 1, 242

were not just lenses for 'far-seeing instruments' (telescopes), but also for 'near-seeing instruments' (microscopes), as appears in Constantijn's letter of August 1654, part of which we have already cited.

What was Titan himself doing all this time? Probably making accessories, such as the vital grinding disc and the lens mountings, carrying out the polishing work using chemicals, attaching the lenses in their mountings (sometimes using cardboard, according to a later letter), and examining the images of his instruments to see whether they tallied with his calculations.

His calculations became more practical. The optical arithmetic, which belongs to the field of applied mathematics and on which he began work, together with *De motu*, in 1652, will be discussed in the following chapter and again in the final chapter, in conjunction with the late edition of his *Dioptrica*. Here we shall mention just two more results in his practical mathematics, which he presented in 1654.

In his *Illustriam quorundam problematum constructiones* (Solutions to Several Famous Problems)[43] he summarised the fruits of three years of hard work, supplying the answer to the problem of two proportional averages of two given quantities, of perpendicular lines on a parabola and of the greatest and the smallest values. It was good for his name, and also for other mathematicians who then had to seek the missing proofs.

His calculation of the figure π was particularly clever. Archimedes had once determined π at a value between 3.140 and 3.143, and although the number had great practical significance, no one had yet found a more accurate value. In 1596, Ludolph van Ceulen had determined 20 decimals by approximating the curve of a circle by a large number of connected straight-line segments, and expressing their length in the diameter. Titan showed that such a result could be

[43] OC **12**, 182–237

achieved more quickly by approximating the circle: not by connected straight lines, but by segments of parabolas. For those segments that touched the inside of the circle, he found 3.1415926533, and for those that touched the outside, 3.1415926537. He published this elegant calculation in *De circuli magnitudine inventa* (On Finding out the Circumference of the Circle).[44]

'It pleases me exceedingly . . .' And yet, is there something missing? A portrait perhaps? None exists from this period. Some sign of excitement or disappointment then? There is only restraint. What we see of him is smooth, as if a patient hand has polished and silvered him over in a mirror in which, if we do not take care, we may see ourselves. We have already filled in a great deal, more than is really justifiable.

The world in which he lived speaks for itself. There is life, depth and colour. A few examples, drawn, at random, from the year 1654: Constantijn publishes his father's great poem on Hofwijck. The father describes his travels through Luxembourg, Trier, Koblenz, Dusseldorf and Moers after his cure in Spa. And then the bitterly cold winter, when the Waal freezes over at Zuylichem.[45] The Peace of Westminster, signed on 15 April 1654, ends the first of the Anglo-Dutch wars, sealed with the tears of fifty-five-year-old Oliver Cromwell and a Psalm 133 for four voices. Not to mention the seventeen-year-old Louis XIV's first and last visit to the Parliament of Paris, and his renowned words, '*L'état, c'est moi.*'

A reflection:[46]

> First-rate minds are distinguished by their straightforward judgements. All that they produce comes as the fruit of their own reflection and proclaims itself everywhere as such, even as it is uttered. This is why, just as monarchs, they hold the kingdom – the kingdom of minds – in their immediate grasp.

---

[44] OC **12**, 113–181   [45] Strengholt 88   [46] Schopenhauer 581 (§265)

Another reflection:[47]

> Decidedly the special relativity theory requires the concept of an inert system, for it cannot be proved epistemologically. The contradiction in the concept of such a system was clearly revealed by Mach, but it was already vaguely recognised by Huygens.

[47] Albert Einstein in a letter to Michele Besso from 1954; French 267–268

7 Saturn

It had already been recognised by Huygens . . .

Not because he could reason so logically (for that in itself yields no new knowledge), but because he could see something unfamiliar that others had failed to notice. If there is some sort of logic in seeing the unfamiliar, then it is the logos of a poet who joins fragmented ideas to form indivisible knowledge. Now that we have come as far as Christiaan's first contribution to empirical science, we shall strive to discover the poetic in him, or, to paraphrase his own words, to see the world as he understood it. He who knew the classics directed his gaze to the heavens, and we ask:[1]

Titan, son of Kronos, is the logos to be found in the myth of creation? The Greek twilight of the gods: Life burst forth in the heaven Uranus, and he coupled again and again with Gaia, mother Earth. He concealed his offspring in earthly caverns. Finally Gaia could bear her burden no longer. Sighing, she gave to Kronos – well-hidden – a sickle. And he hewed off Uranus' member as it stiffened again in lust. Thus was Kronos assured of power, and he devoured his offspring, born of his sister Rhea. She, in abhorrence, found resort in deception, gave to Gaia Zeus, the last-born, and to Kronos, who knew no better, a stone. The resolution that brought this creation to fulfilment: Zeus seized power and Kronos belched forth once more that which he had begotten – Titan among them.

Zeus is Jupiter to the Romans and Kronos is Saturn, the god with the sickle, the god who sows, also commemorated in December amidst merriment and not seldom in debauchery.

[1] Kirk 113–116

Pale giants roaming the skies, high above the earth: Zeus and Kronos with Uranus, the castrated, and far from Gaia, fertilised to abundance. This is how it was for Ptolemy, for Ibn al'Hasan (Alhazen), Nicolas Oresme, and Koppernigk (Copernicus). This is how it appears in Kepler's *Mysterium cosmographicum* (The Mystery of the Cosmos) in 1597, in which the sphere of Saturn determines the sphere of Jupiter with an inscribed regular hexahedron and thereby orders the whole solar system into a harmonious logos. Saturn was the farthest away. And there was something about that planet . . .

We know what it was, or rather, what it is.[2] This giant planet, rotating at great speed on its own axis, trails a system of dust rings and more than ten moons in its equatorial plane. Five of these moons, in order of their distance to the planet, are known as Tethys, Dione, Rhea, Titan and Japetus. Titan is by far the clearest. It would have been simple to spot with the telescopes, modelled after Galileo, which were available in the seventeenth century, if it had not been outshone by the ring system.

Saturn's equatorial plane makes a large angle (of 28.5 degrees) with the plane in which this planet and most of the other planets rotate around the sun. It is no coincidence that this last plane, the ecliptic, is also the equatorial plane of the Sun. The Earth's equatorial plane also makes a large angle (of 23.5 degrees) with the ecliptic, but for the rest of the planets, and for Jupiter especially, the angle is small. The intersecting line of the equatorial plane of a planet and the ecliptic plane has a fixed direction, because its rotational and orbital momentum cannot change.

This explains why we, here on Earth, have a side view of the system of Saturn only twice within a brief period in its rotation around the Sun. The system of Jupiter, on the other hand, in which Galileo had already discovered four distinct moons in 1610, can be seen continuously, from the side. After Galileo's discovery that there was something unusual about Saturn – it had *anses* (ears), which we shall

[2] Van Helden (1980) 149–153

call sickles – these brief periods fell in the years 1642, 1656, 1670 and 1684. These were the years in which the ring system did not shine brightly, but was reduced to a feebly glowing line.

Under the circumstances, the discovery of Saturn's moons was a prize for any astronomer with a slightly better telescope than Galileo's, and as long as he remained unbiased towards the first moon. In 1643, just after the most favourable moment had passed, Antonius de Rheita claimed to have seen no fewer than six Saturn moons, and five new Jupiter moons as well. This last was impossible. Pierre Gassend and Johannes Hevelius, who were suspicious of his findings, had a good look but saw nothing special about the sickles, which were illuminated again.

Not only they, but also astronomers such as Johannes Wiesel, Richard Reeves, Francesco Fontana and Eustachio Divini became convinced that they would not discover any moons around Saturn. They simply stopped looking. Still, their telescopes were just as accurate as the 12-foot one that Christiaan used to discover Titan in 1655. But Christiaan knew nothing about the discussion being waged at the time amongst the astronomers. He was unbiased. And he just happened to look when the moment was right.

Let us carry on awhile with our consideration of those 'right moments' for the discovery of the Saturn moons. Because the other moons were less clearly visible than Titan, larger objective lenses were needed to gather their light than those used by Christiaan and others around 1655. Giovanni Cassini had them polished, and waited until 1670. In 1671 he discovered Japetus, and in 1672, Rhea, outer moons just like Titan. He waited until 1684, searching more closely to the planet itself, and subsequently carried off the double trophy of Tethys and Dione. Cassini was a hunter; Christiaan captured by chance.

Forty years after Galileo had first spotted the sickles, they were still a mystery. A telescope to view the ring accurately didn't yet exist. One could just make out vague forms that seemed to change shape. By 1656, there was still no suitable instrument available. In that year Johannes Hevelius published the results of fourteen years patient observation, covering the entire cycle of 1642 to 1656, with drawings of the various forms, or phases of the sickles.

Hevelius saw the phases as being more complicated than they actually were, and thought up complicated geometric explanations for them. Divini, and others, also had their ideas about geometry. When Christiaan, after his discovery of Titan, claimed that the sickles simply represented a flat ring, it was not based on any observations of his own. What he had observed was no more than a thin line. But what he also saw was – breathtakingly – the ring. It wouldn't be until 1673 that Cassini could observe the first details. A piece of good luck? No – recognition:

> Let them remain as signs of my sagacity, and their names
> That I write across the heavens be an echo to my fame.

If we wish to know how all this actually happened, we need to return to the beginning of 1655.

Early that year, Johan de Witt sent Christiaan – who was busy building his 12-foot-long telescope – a document describing a way to determine geographical longitude at sea. Johannes Placentius of Frankfurt, the author of the document, claimed that you could do so by making use of the moon, and applied for a patent from the States-General. On 4 March Christiaan wrote back that Placentius' request was based on a deception, and contained flagrant mathematical errors.[3] We must assume that his answer, with its implicit advice to the States-General not to grant a patent, marked the beginning of his own ideas on a way to determine longitude.

De Witt had been Grand Pensionary, the highest official post of the Republic, since 1653. Even in this capacity, he continued with his mathematical studies. Van Schooten added De Witt's *Elementa curvarum linearum* (The Elements of Curved Lines) to his second edition of Descartes' *Géometrie*, which was published in 1659. This meant that his work appeared alongside that of Christiaan, who

[3] OC 1, 318

also contributed. In 1671, his *Waerdye van lyfrenten naar propor-tie van losrenten* (Valuation of Annuities according to Current Rates of Interest) was published, an original piece of work on insurance mathematics.[4] Political differences, and perhaps social differences as well ('We all know how he dresses,'[5] Constantijn wrote in 1666), made his relations with the Huygens family difficult, but he and Christiaan got on well as colleagues.

De Witt attempted to build up the Republic as an independent state, one which could live from her free sea trade, not subordinate to maritime England and, as a true republic, without the monarchi-cal ambitions of the House of Orange. This was the unique product of the Peace of Munster, and the premature death of Willem II. 'With incomparable lucidity De Witt succeeded for many years in maintain-ing his domestic and foreign system, but eventually his policies broke down,'[6] writes E. H. Kossmann:

> In spite of De Witt's superior handling of the problems that faced him, the odds were far too great. Under the earlier stadholders the country had served to break Spanish hegemony. Under William III it served as England's ally to break French hegemony. Under De Witt it tried to act as a great power with the sole ambition of serving its own interests. Perhaps one must call this pragmatism, realism, conservatism, or whatever term one wishes to use to indicate that De Witt was a practical man, not given to daydreams about Dutch greatness or the ideal state. Yet, whatever he may have allowed himself to think, his government was untraditional, unconventional; it was an adventure.

His life ended in a hideous murder. They must have spoken together on quite a number of occasions, Titan and Johan. We have evidence of this in Christiaan's letter of 22 May 1659 to Pierre de Carcavy, in which he requested a work of Fermat on behalf of De Witt.[7] It also appears in the fragmentary character of their notes on mathematical

---

[4] Rowen 60–62   [5] OC 5, 436   [6] Kossmann (1987) 192   [7] OC 2, 411

method, and the publication of their work by Frans van Schooten. Similarly, on 18 June 1655, in May 1658 and on 23 February 1663 Christiaan sent pages of calculation, with no further explanation, to the Grand Pensionary.[8] He in turn, busy as he was, wrote a request to Christiaan on 9 March and once again on 9 April 1659 'to read once more the first part (of the to-be-published *Elementa*) more meticulously, so that you can assure me that it contains no more faults or errors'.[9] After 1665 they would correspond via Johannes Hudde, who was not only a mathematician, but later also mayor of Amsterdam and administrator to the United East Indies Company.

It was in March 1655 that Christiaan notified the States-General of the application for a patent by Placentius.[10] In the same month he wrote to Andreas Colvius that, in the near future, he wished to publish *Dioptrica*, on his calculations for lenses. This friend of his father's was a minister in Dordrecht. He had sent him a small microscope, but because the minister did not realise that Christiaan, together with Constantijn, had polished the lenses himself, it seemed at first that he would be paid in money rather than praise.

A good two years before this, in December 1653, Christiaan had finished a manuscript of 108 pages on the theory of lenses, which he had been working on since the summer of 1652. It was very precisely set out in three chapters and with twenty-one propositions, but in spite of its being so complete, he hadn't done anything further with it. On 25 October 1654 Van Schooten, who had read it, wrote to him saying that he was unsure whether his criticism of Descartes was justified. He had suggested adapting it into an appendix to a new edition of Descartes' *Dioptrique*, to be published by Elsevier.[11] Unlike his verdict on *De motu*, Van Schooten did not oppose publication of *Dioptrica*, yet he did have his reservations.[12] Christiaan, however, was adamantly opposed to any adaptation.

At the beginning of 1655, with his remarks to Colvius in mind, he decided to have the work printed exactly as it was. We don't know

[8] OC 4, 311   [9] OC 2, 371, 388   [10] OC 1, 320
[11] OC 13, 1–153   [12] OC 1, 301

for certain why he didn't succeed in doing so, but we may assume that Van Schooten was behind it. For although Christiaan attributed the law of refraction to Descartes – erroneously, as he later realised – he made it all too clear that he considered that the great man had applied it carelessly. He calculated the focal length and focal plane for a large number of lens forms, finally arriving at a general formula in proposition XX. His proposition XIII on a convex or cylinder lens, which he used to measure the refraction index, is particularly remarkable.

The work was to lie fallow in his office for the next ten years – until he had realised the possible extent of the flaw in his paraxial approximation. In 1665 he would apply himself to the highly complex mathematical problem of spherical aberration, a master stroke that enriched the work still further.[13] Once again, it just stayed where it was.

The 12-foot telescope, which he and Constantijn had been working on for more than a year, was ready in March 1655. The instrument has not been preserved, so we can only guess what sort of telescope it was. This is important if we wish to know the circumstances under which he discovered Titan.

No doubt it was a telescope in the Dutch tradition.[14] This means that the large frontal lens (the objective) was convex and the small rear lens (the eyepiece) concave, with the vanishing point of the eyepiece in the centre point of the objective. In keeping with the times, the eyepiece probably consisted of a number of lenses placed one behind the other, so as to lessen the effect of blurring and colouring (chromatic aberration) in the image. Seven years later Christiaan would invent a special lens combination in which this defect was reduced to a minimum: the 'Huygens eyepiece'. The vanishing point of this combined eyepiece most likely lay at a distance of 3 inches, or around 7.5 centimetres. We know with certainty that the objective had a focal length of 12 (Rhineland) feet, which comes to 12 times 31.39 centimetres, or 377 centimetres. We know this because the lens, by good fortune, has been preserved. It is made of ordinary grey-green

---

[13] OC 13, 272–388   [14] Van Helden (1980) 148; OC 15, 10–15

plate glass, 0.32 centimetres thick and 5.7 centimetres in diameter, with '*Admovere oculis distantia sidera nostris*' engraved on the edge, as well as the dates 3 and 16 February 1655. The telescope has a magnification of (377 : 7.5), or approximately 50 times.

One winter's evening, we don't know when, he opened the attic window of the house on Het Plein and, for the very first time, slid out the nearly 4-metre-long telescope. He balanced the long tube, held the eyepiece with both hands, and directed it towards the moon. It can hardly have been otherwise, for we always direct our gaze at what is most conspicuous. His hands must have nearly frozen. He gazed at any planet he could focus on: Venus and Mars, then later in the night, Jupiter and Saturn. Afterwards, back in the comfort of the warm house, he made sketches of what he had seen. The sketches weren't done by Constantijn, who was probably there from time to time as well, and who was certainly more proficient at drawing, but by Christiaan himself. And these drawings cover not only one night. We see the date 25 March on the first detailed sketch of Saturn, which shows a vague outline of the sickles.[15] Even if it wasn't he who had drawn it, he must have already been aware of Titan, because in April, May and June, as the nights grew less dark, he continued to follow Saturn.

The far-off movement he saw enthralled him. And of course, he had a magic lantern.[16] This, too, he had made in the meantime, to gratify his father, just as he had gratified his father's friend by giving him a microscope. Although the *laterna magica* had been invented long before, possibly by Giambattista della Porta, and it was a long while since his father had so enjoyed the shadow shows Cornelis Drebbel used it for, it lacked a proper projection lens. We shall not venture to give an opinion on whether it was Athanasius Kircher or Christiaan Huygens who first saw the necessity for such an extra lens to make the shadows stand out more clearly.

The magic lantern served the dual purpose of satisfying his own desire to understand optical images, and his father's desire to impress. In 1662 his father would write to him from Paris requesting him to

[15] OC **15**, 239   [16] OC **4**, 269

send the magic lantern, so that he could entertain the French court. Christiaan then wrote to his brother Lodewijk, who had accompanied their father:[17]

> As I have promised to send the lantern, then this will have to be done. I have not been able to think up a good excuse not to do so. But when it arrives, you could, if you so chose, easily put it out of order. You must take out one of the three lenses standing together. I shall act as though I have no idea what is wrong, and the ensuing explanation shall cause just the necessary waste of time. This is all for his own good for, in my opinion, it does not befit my father to put on such puppetry in the Louvre, and I am sure you would not wish to help him to do so.

But what he saw moving with the aid of his 12-footer, every 16 days and 4 hours in the spring of 1655, was no puppetry. And he knew, when he addressed his anagram in June to Gottfried Kinner von Loewenthurn in Prague and to John Wallis in London – whom he appealed to more or less on the off-chance – that he was a lucky man, and the first, since Galileo, to discover something new in the heavens.[18]

This was his happiest time.

He set forth. After Constantijn, it was his turn for the 'grand tour', an event that had been postponed for six years, due to the general unrest in France, and which was to provide him with that piece of paper from Angers that his father so desired. With him went Lodewijk, Philips Doublet, the son of his father's sister Geertruyd, and a fourth young man from the Hague, Gijsbert Eickbergh. We quote from Lodewijk's journal:[19]

> 6 July. We departed around ten o'clock from The Hague in Uncle Doublet's coach-and-four. After we had arrived in Maaslandsluis between two and three o'clock, we dined, sent back the coach, and crossed over to Brielle.

[17] OC 4, 111   [18] OC 1, 334, 331   [19] OC 22, 461–491

7 July. At four in the morning Captain Kerckhoven came to
inform us that the wind was favourable and that we must set forth
as soon as possible.

9 July. At daylight we found ourselves between Calais and
Dover, both easily distinguishable. High coastlines both to the left
and to the right, especially the coast of England, which looked
completely white. Because the rising tide, which is more
turbulent in this narrow channel, was against us and the wind
constantly changing, we made little headway that day. But
during the night the wind became more favourable and brought us
to into the roadstead of Dieppe at about five in the morning.
Straightaway a jolly boat came up alongside to take us to land,
which was finally accomplished, after much bargaining and
swearing, for the sum of three louis and fifty cents each. Once we
were on land they demanded as much for each chest, which never
happens. Because we would not comply with their demand, they
went and complained to the governor, who immediately sent his
secretary. When he was informed of what had happened, it
appeared that we had paid more than twice what they were
owed . . .

11 July. From Dieppe to Rouen the distance is put at twelve
miles, but these are rather long, as in some faraway province, for
the closer one gets to Paris, the shorter the miles become. In
Rouen we put up in the Rue Herbière with a man called
Bougeonnier, who had a charming wife and a learned sister-in-law.
The latter did not neglect to visit us the following morning. We
had heard about her in Dieppe.

13 July. They brought us to one of the lovely islands [at
Rouen] in the river. Our little barge lay moored in the shadow of
some trees, and we were served with seven or eight tasty meat
dishes, and after this as many dishes of fruit, with excellent wine
throughout.

15 July. It did little else than rain from Argenteuil onwards,
so that we saw little of Paris, due to the dark weather, until we

were nearly there. We finally arrived, at nearly six o'clock in the evening, passing through the suburbs and the port of Saint-Honoré, and went to lodge in Saint-Germain on the Rue de la Serne, with Montglas.

23 July. We have been in this city for nine days [writes Christiaan to Constantijn]. If only you knew how we have passed the time, you would call Gargantua a sluggard. Tassin, our guide, calls for us each morning. He is just the braggart you describe, and you hit the mark about his voice, which I should dearly love to set to music. I have not yet even visited poets or musicians, but just wandered with my companions through the streets. (These are muddy and stink dreadfully, for people simply tip their chamber pots from the upper windows, with the warning: 'Watch out for the water!') We have visited the very aged ambassador [Willem] Boreel though . . . Everything is expensive here; we shall spend an immense sum on hired coaches alone. I have a room for myself, almost entirely carpeted. But there are rats and mice in the attic and they frequently pay me a visit. And then there are the bedbugs, which get me during the night, so that I have bites on my hands and forehead.

2 August. At the castle of Fontainebleau we saw a lake teeming with enormous fat carp and we greatly enjoyed tossing them bread.

12 August. Tomorrow we leave for Orléans [writes Christiaan to his father]. After a quick glance at the city we go on board to speed up our journey along the Loire [to Angers]. We have quite had our fill of viewing buildings and other silly things. When we return with the doctor's degree, I shall do my best to see the world as you understand it. And I think it is possible to do so, if you are willing to grant me time.

In Angers, on 1 September, they paid out just over fifty guilders to become doctor *utriusque juris*. 'One of the great abuses of the University of Angers is that the professors of law sell degrees for five years of study to people who have not studied with them; they make their acquaintance only through the money that they receive.' This is according to a report, sent thirty years later, from Charles Colbert to the king.[20]

The degree? One paid more for a wig. If Father could just send some more money, for they were spending it like water! They carried swords, called themselves knights. The letter that Christiaan wrote at the end of September to his friend the diplomat, Diderik van Leeuwen, who lived in The Hague, is thoroughly light-hearted:[21]

> To that most noble knight who is without fear of the Lion. Up until this day, two of your knights have failed to write to you any word of thanks or recognition for the honour of your lordship's favourable reception prior to their departure. Rather like Our Lord when he cured the ten sick men and only a sole Samaritan came to thank Him, you too have every right to say: Did I not favour four knights with my courtesy and beneficence? Where then are the other two? This similarity entered my mind while hearing the sermon that a poor friar delivered to our ambassador last Sunday. So it comes that the undersigned confesses to owe you one thousand five hundred thanks, to be paid out in lofty words. But I beg you most humbly not to claim this debt, for you are only too familiar with my inability to repay it. I would be powerless to avail myself of the elevated and florid style suited to your lordship, whose letters overflow with eloquence and wit and merit, to be printed alongside those of Balzac and Malherbe.
>
> But let us cease to be flippant, and instead, speak the truth. I may boast in always having been the first to drink to your health and to that of your beloved, and to often wish for the good fortune of your presence. Also, it is my opinion that there needs be a flying horse to bring you to us in just one second, whether it be for a trip

---

[20] Brugmans 30–31 (n3); Vollgraff 491   [21] OC 1, 353

along the Loire, or for heroic action, such as when Dame Fortune
should decide who must sleep alone and who with another. Or to
choose between four horses, the best one of which be blind, and
more of these daunting occurrences. I would have invited you to
join our distinguished debate that took place in full counsel,
when we had just arrived in Paris. There you would have heard
each one of us venture his reason for coming to France: the one
professed to have come to learn how to behave in genteel society,
another to be presented to the celebrities, yet a third wished to
view fine architecture and the latest fashion, and a fourth just to
be away from home. After much lengthy and heated discussion, it
was decided almost unanimously that it is not worth the trouble
of travelling such great distance for all that is to be had here. After
this, you might have witnessed us plunged into a debate on
sovereign possession, in which there was even greater divergence
of opinion. I recall that there was one who considered that he
would possess absolute sovereignty if only he were allowed to
add a coach-and-four to all that he already possessed, so that he
could ride to The Hague whenever he wished. You are the best to
judge whether or not your presence in all this was urgently
required.

Just this once do we hear this sort of tone from him. The twenty-
six-year-old is rather more serious-minded when he quotes a verse of
Seneca:[22]

Latrunculis ludimus, in supervacuis subtilis teritur.
We play at soldiers, squander delicacy in trivia.

As if to exonerate himself, he writes, '*Potest dici in eos qui difficilia
problemata sectantur sine delectu*' ('This can be said of people who
take no pleasure in tackling difficult problems').[23]

What did Christiaan discover from the musicians his father
insisted he visit? The first one he met was the organist La Barre,

---

[22] OC **22**, 171   [23] Vollgraff 493

undoubtedly a man of rousing psalms, but he had a theorbo-playing son so he could make a bit of music. Also he met the bandmaster Thomas Gobert, and a certain Michel Lambert, whose daughter-in-law sang beautifully. Finally the harpsichordist, Jacques de Chambonnières, who had set up the music society called 'The Curious at Heart', warmly welcomed him. He used to go along quite often, but we may well ask if this is really what he was looking for.

Tassin, steward of the equerry De Behringhen who, in turn, was a friend of father Huygens, may have been a braggart, but he did manage at least one sought-after introduction, namely the master of research, Habert de Montmor.[24] It was at his house on the Rue Sainte-Avoy that Christiaan met the elderly Pierre Gassend, the man who had breathed new life into the atom theory of the Greeks. According to Jean Chapelain, the physicist was favourably impressed by the discussion they had together. He died the following 24 October, which meant that his highly coveted chair at the Royal College was now empty.

It went to Gilles Personne.[25] This son of a farmer, more often known as Roberval, the name of the village where he was born in 1602, had originally travelled around the country working as a home tutor, and studied mathematics at various universities along the way. In 1628 he settled in Paris, where he met up with Marin Mersenne and became a school coach. Equipped with all this knowledge, he managed to pass the competitive examination for the Ramus Chair in mathematics at the Royal College. He had to be renominated for the chair every three years, but he found this so easy that he carried on even after he was nominated for Gassend's permanent chair. Van Schooten, visiting him in 1646, reported that his lectures were drawing in no fewer than a hundred students at a time. It was then that Christiaan met him, in the circle of Claude Mylon.

It was the astronomer Ismael Boulliau[26] who brought him into contact with Mylon; Christiaan had visited Boulliau on his own initiative above the Royal Library, at that time in Rue de la Harpe.

---

[24] Mesnard 37; Roger 43–45   [25] Costabel (1986) 21–22   [26] Hatch 106–116

Naturally, they discussed Saturn. Boulliau introduced him to his neighbour, the conservator Jacques Dupuy, who showed him the well-stocked personal library in his office. This was also the meeting place for an academy formerly led by Mersenne. Le Pailleur succeeded him on his death in 1648, and when Le Pailleur died in 1654, Mylon in turn succeeded him. There were several dilettantes among them, but the members of the circle addressed one another as 'we mathematicians'. Roberval, certainly no dilettante, was a member, as were Adrien Auzout, Pierre de Carcavy and another dozen virtuosi.

Christiaan attended a couple of their meetings. 'Here, the great meet daily', he wrote to his father.[27] Almost certainly it was here that he heard the frivolous problem put to Pascal by the knight Antoine de Méré, a formidable card and dice player: 'Could winning be predicted?' There was more frivolity attached to the academy. Although Marie Perriquet did not actually attend the meetings, she was a member. And naturally enough the proud chairman introduced Christiaan to this learned beauty.

Then he walked straight into a hornets' nest.[28]

As it happened, others were also involved in his affair with Marie Perriquet: Mylon, Roberval, and the not-yet-mentioned Valentin Conrart. Marie was the daughter of a royal advisor and lived with her sister Geneviève very close to Conrart in the Rue Saint-Martin. Conrart considered himself her protector and found it improper that Christiaan, on being told where she lived, should visit her on his own. Marie was obviously impressed with Christiaan's 'qualities', because just before he left for The Hague on 30 November, she discussed them with Conrart for four hours straight. How are we to understand these qualities? And was that the only topic of the conversation? Christiaan had his misgivings, and as soon as he arrived in The Hague he wrote to Roberval:[29]

[27] OC 1, 348   [28] Gabbey (1982) 71–73   [29] OC 1, 369

It would be a great embarrassment to me if you have divulged what I know about P. [Marie Perriquet]. To him I referred to her as being a saint. Also to M. [Mylon], to whom I said that, in any case, you had not spoken to me confidentially, which he had assumed. Whatever comes of this, I would be greatly obliged to you if you would ensure that this shall in no way be used against me.

Roberval replies to this letter of 30 December 1655 on 16 January 1656:[30]

I have revealed nothing of those things we spoke of in private. But I fear that they have made a fool of you. I am to be reproached for this, for I should have told you that C. [Conrart] is a past master in such matters. He has a full-blown relationship with P., just as a number of other worthy elderly gentlemen who are respectably married. I have discovered him to be neither worthy nor respectable. As you and I enjoy confidentiality together, I ask you once more to tell me what has passed that I ought to know of. Then henceforth we shall be able to correspond [upon scientific matters], without the hindrance of such trivial matters. And as far as M. is concerned, he is a good-natured greenhorn, the crow in the fable. But please write so that only those familiar with the affair will understand.

What was it that Christiaan knew about Marie Perriquet, and what did he fear would be used against him? What role did Conrart and Mylon play? The crow that reminds Roberval of Mylon is probably not the crow from the fable by Jean de la Fontaine, which is a raven, but from the fable of the crow and the goddess Athene, who scorns his offers. Most probably all four of them were in love with her and were therefore treading cautiously. Christiaan, in his reply to Roberval, was careful not to give anything away:[31]

---

[30] OC 1, 374   [31] OC 1, 395

Nought has escaped me that might cause him to suspect. In his correspondence to me he regales me so thoroughly with the virtues and skills of P. that he cannot possibly know that I am perfectly informed. And M. remains reticent upon the matter, from which I presume that he has wind of it. Let us leave it at that.

However vague the affair may have been, clearly Christiaan's first acquaintance of an erotic nature was with a woman who was not free.

Quite an important figure, this Valentin Conrart: secretary of the French Academy, the exclusive association for the arts that he had set up in 1635, with Richelieu's blessing, in his own home; also the most renowned writer of his day, one whose influence reached the Republic and certainly his friend Huygens.[32] We can fully imagine Christiaan's misgivings. He and Lodewijk had visited him at his summer residence at Athis, but could not be received by him as he was sick at the time. His influence with Marshal Abraham de Fabert procured them a pass for their return journey through the Spanish Netherlands, where the French were pursuing their war of conquest. This was quicker than going by sea, the route by which they had come.

Jean Chapelain had more time for them. He was another protégé of the late Richelieu and, for some obscure reason, the recipient of a princely pension of 1000 *écus*. He marked Christiaan out for a place in the French scientific world, and encouraged him to publish his findings on Saturn. He introduced him to Paul Scarron, the celebrated writer of satire and burlesque, who lived in the Rue Neuve Saint-Louis. We do not know whether he became part of Scarron's circle and met the famous beauty Françoise d'Aubigné, Scarron's twenty-five-year-younger wife who would later become the Marquise de Maintenon.

They left before the season in Paris had begun. Conrart wrote to father Huygens, 'I would have wished that you had granted them your permission to partake of life at court during the winter, when

[32] OC 1, 350 (n8)

it is usually more lively and more splendid than at other times.'[33] Christiaan wrote to Constantijn, 'Without doubt the best time to be in Paris begins just as we are to depart.'[34]

Next time he would stay on to spend the winter there.

On arriving home he concentrated on Saturn. As soon as he saw it once more on 16 January 1656, he made a sketch; the sickles had almost disappeared. Together with his previous observations, he could now come up with something definite. On 5 March he wrote:[35]

> aaaaaaa ccccc d eeeee g h iiiiii llll mm nnnnnnnnn ooooo pp q rr s ttttt uuuuu

These were the letters of: 'Annulo cingitur tenui plano nusquam cohaerente ad eclipticam inclinato' ('It is surrounded by a thin and flat ring that is nowhere connected [with the planet] and makes an inclination with the ecliptic'). He placed these letters under the short caption De Saturni luna observatio nova (New Observations of a Moon of Saturn), in which he established his priority. But Chapelain urged him to have an announcement printed in the meantime, for an astronomical work would take at least a year to write.[36] The single sheet containing this text was printed in either April or May.

Meanwhile Wallis, a professor at Oxford, who corresponded from an address in London, had made Christiaan unsure about his priority. He had responded to Christiaan's anagram by sending him another. When he pursued the matter in a lengthy letter written on 4 February, it turned out that he had, in fact, understood Christiaan's anagram and had never thought of disputing his discovery, 'In a homely manner, I played a trick on you.'[37] 'I thought it not wholly in accordance,' came the dry reply.[38]

Saturn was no more than a footnote in their mathematical expositions. We mention only the result of Wallis's formula to calculate π from the ratio of two infinite products. According to Christiaan's

[33] OC 1, 368   [34] OC 1, 362   [35] OC 15, 299
[36] OC 1, 398   [37] OC 1, 401   [38] OC 1, 423

laborious mathematical calculations, the tenth cipher behind the comma must lie between the 3 and the 7. On 22 August, Wallis calculated for him that, for practical purposes, the lower limit could be raised from 3 to 6; to be exact, he found the limits of 5(69) and 6(96).[39] It is 5(897932384 . . .). It was not a practical joke.

In May the ring was no longer to be seen. On the 6th of that month brother Philips, who had been sent out with a message once again, wrote him a letter. He had been working as an apprentice at the embassy in Berlin, and was travelling through Brandenburg and Prussia to learn his diplomatic skills. Now he was in Danzig:[40]

Brother, Three days ago I delivered the letter to Hevelius. He is such a pleasant gentleman, that I greatly regret not having made his acquaintance upon the first day that I arrived here. He was immensely pleased with your letter and papers, all the more because he is at this moment quite engrossed in Saturn. He is busy writing a dissertation on the subject that is very soon to be published, and you shall receive the first edition of it. He showed us two or three large illustrations that will appear in it, in which the growing and shrinking of the ears, and the time at which this occurs, is beautifully rendered. He says that if the ears are not partly visible now in May, these will remain invisible until September of the year '57. He can say nothing of the comet, and he doubts whether you have properly seen it. He will look out for it. His largest telescopes measure 17 or 18 feet, with four or five lenses that he grinds in red copper dishes, and are twice the size of yours, and he asserts that the size of the dishes contributes greatly to the perfection of the lenses. His lenses are as clear, in their material as well as in their smoothness, as any I have ever seen. He is now grinding hyperbolical lenses, and had already made a couple of small ones. He considers it very possible to make them quite clear, and also that more can be done with them. I have not

[39] OC 1, 379   [40] OC 1, 419

yet looked through his telescopes, for it was cloudy weather, but according to him, they are very good. Today I shall take a thorough look through them and then I shall write to you concerning them in my very next letter. He has shown us a microscope with three lenses, which is as good as yours. He is a most dextrous fellow, cuts all his plates himself, makes also very fine instruments of copper. He is not uncommonly boastful and is most courteous. I have never seen such a tidy study as his. Farewell in great haste.

One year later this loyal servant would die, in Marienburg, on 14 May 1657. Two weeks later the news arrived in The Hague, and Christiaan informed, amongst others, his Uncle David van Baerle:[41]

People write to us that the progress of his illness was painful, coupled with a continuous fever, from which he died after seven days, his sole complaint being that his father and close friends would be unable to take their leave of him. You can imagine how dismayed we are by this news, because we were quite uninformed of his indisposition, and what great grief awaits my father, who travelled but a couple of days ago to Luik on business of His Highness. But we hope that the good God who has seen fit that such grief should befall him and ourselves, will also grant us the strength to overcome it in the end, and will preserve us from further misfortune.

Misfortune. Did he choose this rather cold word, after his declaration of 'our' devastation and sorrow, because he viewed life and death as merely the outcome of a game that could yield only fortune or misfortune? In any case, this word is admirably appropriate for his small booklet titled *Van reckening in speelen van geluck*.[42]

He had come into contact with the subject of probability calculation in games of chance in Paris. Even though it was Méré's frivolous remarks that had led to the discussion, it also had its serious side. They must have known, at Mylon's academy, that Pierre Fermat and Blaise

[41] OC 2, 32   [42] OC 14, 50–150

Pascal had been interested in the calculation of probability since the summer of 1654, but most likely they were not aware of their results – or not completely. In spite of all the pains taken by the editors of the *Oeuvres complètes* to put Christiaan's findings into their correct perspective, much still remains uncertain. How much did he pick up and how much did he work out for himself?[43]

In 1656 he brought up the subject in a letter to Claude Mylon, as if he were picking up the thread of previous discussion[44]; he did the same with Roberval on 18 April.[45] Maybe the former was not capable of making any personal contribution, but the latter certainly was, even if he maintained his customary silence about it. But why was Christiaan, contrary to his usual restraint in such matters, anxious to publish his findings so hastily?

'A few days ago I wrote on the fundamentals of the calculation of probabilities in games of chance,' he wrote to Roberval, adding emphatically, 'on the request of Van Schooten, who wishes to have them printed; this was among my suggestions to him. If I play with two dice against another, with the agreement that I win if I play a seven and he wins if he plays a six and that I give him the dice, then I ask which of the two has the advantage and how great that advantage is.'

On 6 July he wrote to another member of Mylon's academy, Carcavy:[46]

> The essay that I sent to Van Schooten two months ago also contains a theorem that serves me in all my queries concerning games of chance. It is as follows: If the number of times a person is lucky enough to throw b is represented as p, and the number of times to throw c is q, then
>
> $$(bp + cq) : (p + q)$$
>
> is its value.

Or, as we would call it: the expectation value.

---

[43] Coumet 127–129   [44] OC 1, 391   [45] OC 1, 404   [46] OC 1, 442

It was a beautiful finding, useful in its application, but it couldn't possibly serve as the foundation of the calculation of probability. This is why the subtitle of Annie Romein's biographical sketch of Christiaan, 'Inventor of Probability' is unfortunate.[47] In his introduction to *Van reckening in speelen van geluck*, the 'inventor' refers vaguely to the discovery of probability in France. Clearly what he meant here was not its applications, but the development of the fundamentals by Fermat and Pascal, which he did not understand. This makes our need to understand his sudden haste to publish all the more urgent. He seems almost apologetic when he writes in his introduction that it was 'useless and not praiseworthy, but that the foundations are laid down within for a most goodly and profound theory'. Do we see here once again the influence of Van Schooten, who had been so against *De motu*, but was now so rapturous about *De ratiociniis*?

As it happens, Van Schooten translated Christiaan's booklet himself into Latin and published it in 1657 as *De ratiociniis in ludo aleae* (On the Calculation of Games of Chance). It appeared as part of his *Exercitationum mathematicorum libri quinque* (Five Books of Mathematical Exercises), which was also published in Dutch, in 1660, with the original text by Christiaan. Because it was, for a long time, the sole publication on probability calculation, it exerted a great influence on the theory of probability.[48]

With the aid of the expectation value, most of the problems raised by Fermat and Pascal could be solved. Incidentally, it was not this value alone that enabled him to calculate luck in games of chance. It was also his astounding capacity to elucidate what happens if, for example, a game is interrupted in mid-play. The player can determine what might happen in the rest of the game, as well as what chance he has of winning, just by calculating the expectation value. In this way it was shown that the formula he so fortuitously discovered was generally valid.

---

[47] Romein 397   [48] OC **14**, 5–10

These were his 'games of chance'. But didn't his little sister describe them equally well? On 19 July 1656 Suzanna wrote to him of the delightful holiday she had spent near Den Bosch:[49]

> We have passed two days in Haenwijck, which I find to be much better than I had thought. Mrs Stanton came to eat cherries, with all her family, and all the eligible young men from Den Bosch came too. Cousin Zuerius had prepared a most attractive meal for all these people, which they all greatly enjoyed, for they partook of it most heartily. In this manner we pass the time here most pleasantly, and are most cheerful and merry.

That summer he drew her portrait, and impressed one of her girlfriends from Brabant with his talent. It is not clear which friend it was, but evidently he knew her well enough to be greatly inspired by her, and he was at his most charming when he wrote:[50]

> To me it is no matter of indifference whether or not I am on the best of footing with the kindest and most beautiful person in the entire world. And furthermore I am a man who keeps his word, though you may believe to the contrary. Therefore, before departing for Zuylichem with all the family, I wish to settle an old score by sending to you this drawing of my sister. If you should see two eyes in this portrait, you must not be misled into thinking that you have another drawing before you. Believe me, the change was gradual and has occurred in a manner quite unnoticed. Perhaps while you were waiting, she turned these eyes towards you and maybe even the lips began to move a little, for you must surely notice that their colour has become more lively. If this is not so, I hope that in your goodness you will still accept the drawing for what it is. All I ask in return is that you take back any imprecation you may have formed against me in my absence, and that I not be exposed to your reproaches if, by

[49] OC 1, 455 [50] OC 2, 53

any chance, I shall have the good fortune to see you one of these days in Den Bosch. There is nothing in the world that I hold more dear, just as there is nothing that I cherish more than your graciousness towards me and to be, dear young lady, your most humble and obedient servant Chr. Huygens van Zuylichem.

This letter must have been written in mid-August 1657, barely three months after the news of Philips's death. We have quoted from a draft. In the final version the lively passage about Suzanna's eyes has been left out. Was it inappropriate in one who mourned? Or did he not wish to pursue the matter with this girl much further?

In August and September of 1657, he spent about three weeks with the family in the dark castle of Zuylichem. Something seemed to have snapped in his father. He had become apathetic.[51] He had already been complaining for the past seven years that his state duties no longer offered him any challenge. Was he also suffering from the lament of the child who had died without his father's farewell? Nearing his sixty-first birthday, back in The Hague and short of breath, the patriarch took to his bed. He lay there quite motionless: 'All take into account that perhaps my end has come.' But as he slumbered he heard the whispers around him and could even laugh about it afterwards. Christiaan put aside his mathematics (on the 'pearls' of De Sluse?) and devoted himself to his father's work, 'the errors and faults of Rammazeijn the printer'. Sometimes the father was feverish, sometimes he suffered the hiccups. By November, however, he had got over his depression, and on Christmas he went outside again for the first time.

His collected poetical works, *Cornflowers*,[52] were published at the end of 1657. This gigantic oeuvre took two years to print; its preparation had also been a considerable task. There were printed manuscripts of earlier anthologies, but large amounts of poetry needed

---

[51] Strengholt 95   [52] Worp 6, 314–341

to be collected for the first time; they also had to be put in order, neatly copied out, then set and proofread by Rammazeijn.

And this is how Christiaan spent October and November. He enlisted the help of friends and acquaintances of his father. He also wrote to his former tutor Henrick Bruno, and we can imagine the grin on his face when he opened his letter with 'Hero!'[53] This was his own bombastic style exactly, but Bruno sensed he was being made fun of. His reply arrived from Hoorn, where he was now rector of the Latin school, saying that he did not wish to be so addressed.

Did the son carry out this momentous task out of motives of duty or affection? Perhaps it was just his turn, after Constantijn's publication of Hofwijck. But that need not exclude affection. At any rate, it was out of affection that he had drawn his father, in profile, in much the same way that we see him in the double portrait by Jacob van Campen. Not only from this, and from his sketches of Suzanna, do we see that he enjoyed drawing by way of a change, but also from his drawings of willows bordering a stream, a farmhouse at the edge of a wood, the dune at Scheveningen and – not to forget – Hofwijck. Now he came across this portrait of his father again, because there was a poem on it by Joost van den Vondel in praise of his father, which had to be printed in *Cornflowers*:[54]

> The virtuous son received from God and from his father
> His very substance and his grace that Heaven chose
> To grant as fame and honour:
> It is gratitude that now befits the son,
> Who returns via art his substance to the father.
> Thus Christiaan becomes the father of his father
> And father Constantijn, the son of his own son.
> Thus art may rival nature, whom we must honour with a crown.

As Titan puts the finishing touches to his tome, Kronos hiccups a counterpoint to the ticking of his pendulum clock.

---

[53] OC 2, 81   [54] Worp 6, 323; OC 1, 529

And there it is, as if by chance, rather like its casual mention in a letter to Van Schooten of 12 January 1657: the pendulum clock.[55]

It came as a result of one of the many ideas, either his own or adopted from others, with which he had kept instrument-makers busy for the past couple of years. Ever since Placentius' application for a patent, which he had reported, he had been trying to work out improvements in measuring time at sea. At the beginning of that year he had commissioned Salomon Coster in The Hague to couple a clock-work to a pendulum. Galileo had already recommended the pendulum to regulate the movement, and Coster, an expert clockmaker, had, amongst other things, invented a double drum spring. Christiaan had the idea of a lightweight fork coupling, and 'cheeks' to lift the pendulum slightly during wide deflection.

For the moment he begrudged himself the time to carry out the calculations for this experimental model. He did not even manage to produce a short description establishing his rights to the invention. After completing his work on *Cornflowers*, he wanted to finish his book on Saturn.[56]

If we include his preliminary announcement of April or May 1656, it took him a good three years. He would not have begun with the actual text before August 1656, because until then he was busy rewriting *De motu*. He wrote down his theory about the planet in the beginning of 1658, but in the course of that year he extended his material for the book to include the solar system; it was not until 5 July 1659 that Vlacq finally printed his *Systema Saturnium*.

While his isolation in The Hague had been ideal in enabling him to discover Titan and the rings, it now worked to his disadvantage. In order that his findings be completely sound, and before he could publish, he needed to read up on all that had already been written on astronomy with telescopes. Many of these works were not yet available in the Republic, nor was it easy to get hold of them. In order to be fully convincing, he needed more experience of observation, and

[55] OC 2, 4   [56] OC 15, 209–353; Van Helden (1980) 153–156

to prove his practical expertise. Gathering extra observational data and making improvements on his telescope also took time. When *Systema Saturnium* finally came out, it was a thoroughly polished piece of work that went far beyond the sickles of Saturn. It is, without a doubt, the most important book on telescopic astronomy since Galileo.

Let us first mention its weak point. Because of his discoveries, Christiaan considered that he possessed the best telescope of his time, which was highly unlikely; and he certainly made no attempt to hide the fact. His boastings infuriated Hevelius, as can be seen from his letter to Ismael Boulliau of 9 December 1659 (nowhere to be found in the *Oeuvres complètes*).[57] In 1660 the temperamental Eustachio Divini would even challenge him to a kind of duel, on behalf of the Florentine Accademia del' Cimento, to settle who had the better telescope. Now that he had carried off the trophy, he need not have pursued the reason for his success. But he wished for more honour than he deserved.

In a way, the same applied to the description of a micrometer, with which he could determine the angle that a planet subtends on the sky. This micrometer consisted of copper platelets in the focal plane of the objective, which could be bent far enough apart or close enough together so that the planet, viewed through the eyepiece, could fit exactly in between. Guillaume Gascoigne was not only ahead of him in this, but he had also made it easier to use. His relatively unknown micrometer could be adjusted with a screw during observation, while Christiaan had to keep bending the platelets during viewing, until they fitted as closely as possible around the planet disc. His micrometer could not be used with a Dutch telescope, as the focal plane of the objective lens lies behind the eyepiece. To do so, he had to equip his telescope with a convex eyepiece.

In his notes we read that he regularly observed planets from 13 October 1656 onwards, but he probably only began in 1658 to measure their size.[58] In November that year he even tried his luck with the

---

[57] Van Helden (1980) 154, 163   [58] OC **15**, 7

elusive Mercury, but without success. He also carried out observations on stars and nebulas. The stars all appeared to him to be pointed, which meant they must have been far away, and unchangeable (on 14 November 1659 he noted, 'I have observed no change at all in the star of the Swan, P Cygni'[59]). In the autumn of 1656 he drew the nebula of Orion.

To return to his measurements on the four easily observable planets: by measuring the distance of the copper platelets once they were in place, the angle the disc subtended on the sky could be calculated with the aid of his dioptric formula. This made him the first person able to report the distance in arc seconds from Venus, Mars, Jupiter and Saturn to planet Earth when at their nearest. And because the ratio of the distances of these planets to the Sun was known by their period of revolution, he could also show how large they would appear if all four stood at the distance of the Sun. He discovered that Jupiter was the largest planet, then came Saturn, then Venus, and finally Mars. But because he did not know how far the Sun stood from the Earth, he could not, of course, convert these angle measurements into distances. Estimations made by Boulliau and Giovanni Battista Riccioli placed the Sun at a distance of respectively 1486 and 7000 times the Earth's radius. But this seemed to him too small.

Once more Christiaan embarked upon a game of chance. Here we come up against remains of Kepler's notion on the harmony of the universe. To be sure, the planets did not appear to be regularly larger or smaller as they orbited closer or further away from the Sun. But Christiaan thought that the Earth, the third planet, 'in order to maintain the greatest possible harmony of the system', must fit in size harmoniously between Venus and Mars, the second and third planet. This led him to estimate the size of the Earth disc, if placed at the distance of the Sun, to be 111 times smaller than the solar disc, and he arrived at a distance to the sun of 25 086 times the Earth's radius. Taking an average Earth's radius of 6367 kilometres (at the time

[59] Vollgraff 520

this figure was fairly well known, among others by the Dutch physicist Willibrord Snel), this comes to approximately 160 million kilometres, while in reality this figure varies between 147 and 152 million kilometres.

This meant that his estimate of the distance was surprisingly good; in hindsight, though, he was just very lucky. Due to diffused light in his telescope, the angle of the disc of the planet was almost as overestimated as the size of the Earth was underestimated, in relation to Venus and Mars, using the harmonious rule. This does not detract from the fact that he made a tremendous contribution to astronomy with his daring distance measurement for the solar system.

We find a second remnant of Johannes Kepler's harmony vision in Christiaan's figure for the inclination that Saturn's system makes with the ecliptic. He suspected that it was a considerable angle, but was unable to determine it, from his limited observations, with any great accuracy. So he fixed it as 'harmoniously' equal to the inclination of the equatorial plane of the Earth with the ecliptic, which is 5 degrees too little, and in his drawing of the largest opening he did not have the ring shine over Saturn's poles, which, in reality, it does.

The third remnant we find in his assertion that, with the discovery of Titan, the last moon had been found. According to him, there were just as many moons in the solar system as there were planets: namely, six. Earth's moon had been known since time immemorial, Galileo had discovered the four moons of Jupiter, and he himself had discovered the remaining moon, next to Saturn. Ten years later Cassini would prove him wrong and, much against his will, he would have to reconsider his figure for the inclination of the system of Saturn.

Finally, we come to his suppositions about the ring. We must not confuse Christiaan's estimations with our present knowledge of the ring system, which we can now study close up with the aid of a satellite. He thought that the ring consisted of a solid single entity. Therefore, to be strong enough it must be sufficiently thick. He drew the thickness only 40 times smaller than the outermost radius, which comes to approximately 3000 kilometres; in reality this system of loose dust is only a few hundred metres thick. But because he had not been able to observe the moon in May 1656, he had to conclude that the side reflected little or no light. It was a claim that cut little ice with his colleagues who, apart from this, were happy to embrace his hypothesis on the ring.

Christopher Wren soon saw that, 'in any event, the thickness is insufficient to be able to be observed by inhabitants on earth, and therefore the corona [ring] must be viewed as a surface only'.[60] Eustachio Divini and his colleagues at the Accademia del' Cimento were of the same opinion. Mylon and Boulliau, too, proposed that Christiaan make the ring very thin and not worry about the problem of its consistency. They were aware of his speculations before *Systema Saturnium* was published.

The fact is: on 28 March 1658, Christiaan had written a long letter to Jean Chapelain about his theory of Saturn, after informing Boulliau, in confidence, on an earlier occasion.[61] The letter containing his declaration was read aloud to the Academy of Montmor (a scientific circle that met weekly), and its elegance delighted the mathematicians.

[60] Van Helden (1980) 153, 162   [61] OC **2**, 156

8 Force

'Recently I have invented a new construction. A clock that runs with such regularity that there is a good chance that if taken to sea it can measure the longitude.' This is how Christiaan concluded his routine letter to Frans van Schooten on 12 January 1657, as if it were a mere trifle.[1]

We shall follow the path that led him from this invention to the summit of his intellectual achievement, and witness the unique concentration of all his powers that served to propel him towards his greatest creative impulse in the autumn of 1659. It is also the path in which his personal life seems to recede, leaving little trace in his further correspondence. This is an opportune moment to quote from Schopenhauer's reflections upon genius:[2]

> Such a privileged person (as a genius) leads, alongside his personal life, another life that is intellectual. It is this life that gradually becomes his only goal, and for which he comes to consider the other as merely a means towards achieving it. This intellectual life especially comes to preoccupy him; it acquires, through the continual growth of insight and knowledge, a lasting cohesion and intensity; it moves constantly forward towards a more and more self-contained perfection and fulfilment, like a work of art in genesis.

Christiaan's invention consisted of attaching a pendulum to a clock commonly in use at that time, which had a so-called Foliot wheel; this was replaced by the pendulum, but still driven by a balance spring.[3] This was done in the final days of 1656. We do not know

---

[1] OC 2, 5   [2] Schopenhauer 299 (§52)   [3] OC 17, 3–13

whether he went on to design the cycloidal 'cheeks', but it was completed by him, and he has been accredited with it. The fork, as well
as the 'cheeks', can be seen in the pendulum clock that he had made
soon after, and for which his clockmaker, Salomon Coster, obtained
a patent from the States-General on 16 June 1657.[4]

He tended to play down his invention, because all the actual
components of his clock already existed. In August 1659 he wrote
to Chapelain that it was just a clever trick, and that 'it looks more
and more as though I have been blessed by good fortune with the
reputation this construction has given me'.[5] Ingenious wheel clocks
driven by weights had been around for centuries. And for decades,
astronomers had used pendulums, which they drove by hand. They
had acted promptly to Galileo's suggestion, in 1630, that the amount
of time taken by a weight at the end of a thread to move to and fro
would be constant, and they counted the number of swings that an
eclipse lasted.

Not only did all the components already exist, but also the idea
of coupling. This too was Galileo's. Christiaan was aware of it from
the time he assessed Johannes Placentius' application for a patent in
1655,[6] and had also seen Galileo's proposition to the States-General
in 1636 to determine a position at sea.[7] He merely put the idea into
practice. His clever trick was to enable the power of the clock to
drive the pendulum, and the regularity of the pendulum to drive the
regularity of the clock. Still, we must not underestimate his design.
The forked construction on top of the balance spring that drives the
pendulum, while at the same time allowing it enough freedom of
movement, is ingenious. Pierre Petit, a Parisian engineer, who wished
to learn about this invention in detail, wrote to him on 18 October
1658, 'It is not everyone who can invent something, then make it as
well!'[8]

---

[4] Leopold 225; OC 2, 236–237   [5] OC 2, 455   [6] OC 1, 318
[7] OC 3, 493   [8] OC 2, 253

We recall Henrick Bruno's rhetorical question in 1643, when he called Christiaan's passion for small instruments and machines inappropriate, 'Surely the Republic, who vests such high hopes in his birth, may expect him to follow his father into business?' Well then, with this instrument, the epitome of regularity, he could serve the Republic directly.

Didn't he write straightaway to Van Schooten, '*longitudines ejus ope definire posse*' ('with which one can determine longitude')? And wasn't the ability to determine a position at sea of vital importance to shipping traffic, which was the livelihood of the Republic?[9] Jemme Reinersz (Gemma Frisius) had already demonstrated in 1530 that an accurate timekeeper could make this possible.[10] But it would take until 1730 before instrument-maker John Harrison would design the first successful sea chronometer.[11] Christiaan's creation was an enormous step forward, yet fell short of achieving this goal. The unreliable materials available in his time were to let him down. Yet when he first made his discovery, he considered that he might well succeed ('*non parva spes*').

Perhaps it is superfluous to explain how a timekeeper can be converted to determine geographic longitude. Longitude, together with latitude, which follows from measurements of the height of the pole star (or of other stars in their passage through the meridian), convey locality across the earth.

Imagine that a ship on the ocean beyond Ireland passes the longitude of 15 degrees to the west of the current prime meridian. Because the longitude of Amsterdam is 5 degrees to the east of the prime meridian, the ship will be 20 degrees to the west of Amsterdam. As the circumference of the Earth is divided into 360 longitudinal degrees and the Earth rotates once in 24 hours, 1 degree equals a time-span of (24 : 360) hours, which is the same as 4 minutes. Therefore the time difference for the ship on the other side of Ireland with Amsterdam is

[9] Mahoney 234–270   [10] OC 17, 7 (n6)   [11] OC 17, 180; Sobel 61–74

$20 \times 4 = 80$ minutes, and because the ship is to the west, it is earlier than in Amsterdam.

Imagine now that the captain of that ship has a clock that maintains the time in Amsterdam, which is to say: it will indicate midday when the Sun in Amsterdam passes through the meridian. Then it has reached its highest point above the horizon in Amsterdam. Around that moment, however, the captain sees that the Sun is still climbing, and will reach its highest point only when the clock points to twenty past one. By calculating the difference with twelve o'clock, he then knows that the ship is 80 minutes or 20 degrees west of Amsterdam. If he mistakes the time by a few minutes when observing the sun, or if the clock is wrong by a few minutes, then he will make a mistake of about 1 degree.

Finally, think of a captain sailing a ship somewhere on a southern ocean. He has the Sun on his port side and observes that, according to his clock, it passes the highest point at twenty past five in the morning. By calculating a time difference of 12 hours, he discovers that his ship lies at 400 minutes or 100 degrees east of Amsterdam. From the Joan Blaeu map he knows that he must sail due north, with the wind and the current, to come out exactly at the entrance of the East Indian archipelago, which is where the East India Company does her trading.

It is worth mentioning here that the astronomical observation (the passage of the Sun through the meridian) necessary to convert timekeeping into longitude must be corrected according to variations in the length of a day. Throughout the year, this varies by about one minute, as a result of the variations in the Earth's speed in her path round the sun and the inclination of the equatorial plane with the ecliptic. This was already known in ancient times, but only after the invention of the pendulum clock could it actually be measured.

One year after Coster obtained his patent, Christiaan published *Horologium*, a brochure on his pendulum clock.[12] This booklet, printed in large letters by Adriaan Vlacq, came out somewhere

[12] OC 17, 41–73

between 8 June and the beginning of September 1658. At the same time he personally presented Johan de Witt with editions for the States of Holland. It is just as well that he did not wait any longer; clockmakers were beginning to take over his invention. 'Strange,' he wrote to Carcavy on 26 February 1660, 'no one ever mentioned these clocks before me, and now, all of a sudden, inventors are springing up everywhere.'[13] But of course this was not really strange at all; it was a money-spinner, and he could not keep on being ironical about it.

His application for a patent in France, the subject of his letter to Ismael Boulliau on 13 June 1658, no longer stood a chance, because of his late publication. 'For a long time I did not think to make much ado about [a patent] for my invention, and had it not been for my friends advising to take advantage, like Thales and his monopoly on olives, I would never have pursued it.'[14] Boulliau answered, 'I regret that my efforts with the chancellor [Pierre Seguier] to obtain a patent for you came to nothing. He refused three times, each time with the argument that all the clockmakers in Paris would be up in arms.' 'I no longer wish to think about it,' Christiaan wrote back on 25 July, and promptly transferred all his attention to his work on the cycloid.

Coster's patent, in which he had a share, did not benefit him much either. It was circumvented by the Rotterdam clockmaker, Simon Douw, who insisted that he had discovered a pendulum clock himself. He had even succeeded in obtaining a patent for it from the States-General, on 8 August 1658, more than a year after Coster's had been granted. The latter wrote furiously, 'By use of deceit and underhand means he has the impudence to present the same invention that was to be seen at Mr Huygens', and at other locations, as if this were quite in keeping'.[15]

On 9 October the matter even came before the court of Holland, Zeeland and Friesland, with Adriaan Pot as presiding judge. Christiaan defended Coster at this 'furious comparison'[16] and had Van Schooten

[13] OC 3, 26   [14] OC 2, 183, 185, 200   [15] OC 2, 244   [16] OC 2, 251

called as a witness, 'We meet after noon, but I should like you to be present before noon and take lunch with us, so that I may acquaint you with the whole matter and what has been said.' Pot sat on the fence and decreed on 9 December that Douw should pay Coster and Christiaan one-third each of the proceeds of his clocks.

It was better than nothing. But it was hardly satisfying, because Douw had wrongfully obtained his patent in the first place. Moreover, Douw was an unprofessional craftsman and had been dismissed the year before, for default, as city clockmaker of Delft. Christiaan was angered by the affair: 'I am becoming thoroughly distracted by this invention of the clock, and by the shabby tricks these swindlers play,' he wrote to Hevelius on 16 September.[17] Ten days earlier he had expressed himself even more scathingly to Wallis: 'This invention has meant much hard work for me, or rather, the unpleasantness caused by crooks [scelestorum hominum improbitas].' Christiaan's irritation was also evident in the letter he wrote to Lodewijk on 22 November:[18]

> Only yesterday did I receive your letter of the 14th with the appendix, for which I shall thank Cosijn Pieck [cousin Willem Pieck, member of the States-General for Gelderland who handled patents and was responsible for the tower clocks of Gelderland]; I shall repay him his double ducat, although I do not know from whence I shall obtain this within the coming eight days, out of the yield from the work in Gelderland. But in this I am duty-bound towards him. I have requested by letter the little flutes from Josijn [Josina had a shop for instruments in Amsterdam], but have had no answer, nor to the 2 or 3 other letters that I wrote to her and to her husband concerning metal-rimmed mirrors for my telescope. I do not know why the bitch is so stubborn, for she knows that I owe her some fifty guilders . . .

We do not know what actual financial benefit Christiaan earned from the patent, nor the number of pendulum clocks involved. In

---

[17] OC 2, 218   [18] OC 2, 277

1658 Coster may have built ten chamber clocks with spring drives, and perhaps as many as twenty from the time he obtained his patent until he died. Apart from this, he supplied existing tower clocks with a heavy (40 pounds?), long (12-foot) pendulum: the one in Scheveningen in January, and another in Utrecht cathedral, in October.[19] He died on 18 September 1659, after completing his adjustments to the tower clock in Breda. Christiaan and Salomon Coster had got along well together and after his death Christiaan showed no further interest in the patent. It would also take three years before he had the first sea clock made, in collaboration with Alexander Bruce (the Scottish Lord Kincardine).

In *Horologium* we find the description of a different pendulum clock to the one built in the spring of 1657. The main difference was the absence of 'cheeks' and the limitation in the pendulum swing. The 'cheeks' at the point of suspension of the thread were to make the time of oscillation independent of the swing. His idea was as follows:

If a pendulum had a wide swing, it needed more time to move back and forth than if it didn't swing as far. In addition, it appeared to need less time for the movement if the wire were shortened. If the thread was shortened, in a certain manner, during its swing, a pendulum could be made that would always take the same amount of time (tautochrone) to swing backwards and forwards, independent of its swing. In order to shorten the thread, he began to experiment with the 'cheeks'.

There were copper platelets on either side of the thread, at the point at which they were hung, which pointed straight downwards and deviated gradually further and further from the vertical line. When the pendulum struck out, the thread connected up with one of the plates and got wound up, so that its free length, which determined the pendulum swing time, became smaller. The weight at the bottom of the thread no longer followed an arch, but an unwinding (evolute) of the form of the platelet.

[19] OC 17, 31–34

Christiaan didn't know what shape 'cheeks' would be best for the tautochrone property. For his first clock he had sketched a flowing line, which went through three previously determined points, and he had probably bent the platelets to match this form. During the course of 1657, he carried out experiments to try and establish the exact form. In the same year he was also working on bending copper platelets for the micrometer of his telescope. But he was unable to discover any clear indication as to which form they should take. How could this have been done, for that matter, without an even more accurate clock? Then he decided to leave out the 'cheeks' and considerably restrict the swing of the pendulum, which brought it very close to being tautochronic. In 1658 he got Coster to make a second clock, which he describes in *Horologium*. But it ended in defeat.

In Christiaan's letter to Boulliau on 25 July 1658 – are we hallucinating when we see that Christiaan's rather emotional letter to Bouilliau, about the pendulum clock, ends abruptly in words about the cycloid?[20] A cycloid is the curved line followed by a point on the circumference of a circle as the circle rolls over a flat surface. This is the movement followed by the valve of a bicycle wheel as it moves along the road. When it is underneath, it stands still, then it moves forwards with increasing speed until it is on top, at which point it moves at double speed, steadily decreasing in speed until it stands still, once again, underneath.

We know too little about the subconscious workings of the mind and the role of emotions to rule out hallucination. Perhaps the defeat he had suffered in his struggle with the 'cheeks', his disappointment about his patent and – who knows? – his rising anger against swindlers all conspired to make him take refuge in his familiar reflex: withdrawal into himself. Seeking clarity in mathematics, he ended up with a cycloid.

As it happens, the 'cheeks' are cycloid-shaped. And the evolute of a cycloid, the path followed by the weight of a tautochronic

[20] OC 2, 200

pendulum, is also a cycloid. He had probably worked this out a whole year before he could prove it.

It was no coincidence that he was so preoccupied with the cycloid in the summer of 1658. Emerging from his retreat in Port-Royal, but under the pseudonym 'Dettonville', Blaise Pascal had challenged mathematicians to find the circumference, surface, centre of gravity and other characteristics of this curve. Did he himself answer? Christiaan got word of this from Boulliau on 21 June, in the same letter about the outraged clockmakers of Paris.[21] He thought it over, but didn't take part. The competition ended in hostility, when Pascal, in his October report, *A. Dettonvillius historiae trochoidis sive cycloidis continuatio* (Sequel to Report by A. Dettonville on the Progression of the Wheel or the Cycloid), refused to award the prize to anyone.[22] He did praise Christopher Wren though, who claimed that the circumference was four times as long as the diameter of the rolling circle, but without supplying proof. If this were true, then this would be the first time a curved line was rectified!

Christiaan received Pascal's report at the beginning of January, and was impressed by Wren's proposition on the circumference, 'It is the best that is to be found about that curve. For I distinguish between that which is difficult and that which is elegant. I found the proof superior &c.'[23] (This was his own summary of a lost letter, which he wrote to Wallis on 31 January 1659.) He proved Wren's proposition by applying a theorem of Archimedes on the sines of a circle and by his brilliant manipulation of the infinite sum of infinitely short cords. He must have applied all his mental capacities to do so. His continuous correspondence with René François de Sluse, too, on his 'pearls', shows how preoccupied he was with mathematics.

Before touching briefly on his other work on the cycloid, we shall first make a long detour. In the last, decisive round he would be challenged not by another, but by himself, and we shall need to show just how this challenge came about.

[21] OC 2, 185   [22] Yoder 77; (OC 2, 187)   [23] OC 2, 329

When Pascal received his copy of *Horologium*, which Christiaan had had delivered to him in October or November by a certain Du Gast, he wrote him from Paris on 6 January 1659:[24]

I have received your gift, which does me great honour; it was presented to me by a French nobleman who informed me that you had received him in the most obliging and refined manner possible. He even told me that you were well acquainted with him, but you confused him with myself. [Christiaan thought that Du Gast was the author of *Lettres provinciales* (Letters from the Provinces) that Pascal had published in 1656.] I assure you, Sir, that I am most pleasantly surprised, for I was unaware that my name was familiar to you, and I have restrained my ambition to be remembered into posterity. But I have been told that you even hold me in high esteem. I dare not believe it, and do not deserve it to be so, but I hope to share in your kind regard, for if I may deserve to do so by my deference towards you, then certainly I merit it no more than another. I am full of such regard towards you, and your latest work *Horologium* has added not a little to what has already gone before. It does you great credit, and better than the other. I have been amongst your first admirers. And I have believed to see even greater things to come. I should dearly like to come and visit you, but find myself incapacitated to do so. All that I am able to do is to send to you just as many copies of the report on the cycloid as you may wish. I hereby send you just a few summaries, for otherwise the parcel will become too large to travel by post. I shall inform the bookmakers that they are to be sent to you in a fitting manner. Do not imagine Sir, that in doing thusly, I in any manner wish to dispense with my duties towards you. I swear to you that it cannot be otherwise, and indeed it is with my whole heart that I acknowledge the good service that you have rendered to me in the person of this nobleman. He deserves

[24] OC **2**, 309

it more than I, even if you did not know him. Rest assured and all
these matters shall be attended to. I remain your ever humble and
obedient servant, Pascal.

The best comment on this letter is Pascal's Thought 427: 'If one
observes the natural style, one is full of surprise and amazement, for
one expects a writer and discovers a human being.'[25]
'The unknown nobleman [Du Gast] has not been able to convey to you the smallest portion of my esteem for you.' This is how
Christiaan begins his reply on 5 February, and he continues:[26]

> You could not know how much pleasure it gives me to receive the
> letter that does me such great honour. I am not able to do justice,
> in words, to my feelings in this matter. Allow me to say, however,
> that I believe myself to be happier than ever before, now that I
> have received this token of your friendship, and that such a
> possession means more to me than all that I have achieved thus
> far . . .

It is a highly polished reply, but we read three times in succession the word 'not': Du Gast has not, you cannot, I am not able . . .
Pascal hardly ever makes use of this style, and when he does, it sounds
more natural. It seems rather forced to begin like this.

We note that on his thirtieth birthday, on 14 April 1659,
Christiaan may not have been at home. 'Little happens here that is
of great interest', he had written to Lodewijk on an earlier occasion,
when discussing affairs on the home front.[27] He had still been at home
on the 9th, when he received a letter from Johan de Witt about some
correction work he was doing for him.[28] He also sent information to
Dirck van Nierop on the calculation of angles. We hear from him once
again on an unknown date later in April, when he informed his parcel
courier to Boulliau, Abbot Brunetti, that he had been in Antwerp.[29] He

[25] Pascal 330 (No. 290)   [26] OC 2, 340   [27] OC 2, 278
[28] OC 2, 388; Rowen 60–61   [29] OC 2, 389

wrote that he had seen a book at the Jesuit brothers about air pumps, and gave his comments on it. Evidently he was away for some time, and the only letter he wrote in April was to Brunetti. There were five letters in January, six in February, five in March, three in May, three in June and five in July. Not exactly hard evidence, but we may assume that he did not consider that this milestone in his personal life, his thirtieth birthday, was of any great importance. What was important was the air pump. His latest project.

The comment, in Pascal's Thought 165: 'The only universal rules are laws governing ordinary matters, in other matters lies diversity. How is this so? Because of their force.'[30]

The waves broke in foaming rollers above the dark groundswell. We follow one or two of them, in an attempt to find a pattern, but search in vain for anything of significance.

On 27 March Christiaan wrote to Jean Chapelain that the text for his book on Saturn was finished, and that, for some time now, the engravers had been working on the illustrations. 'I could never have imagined that it would cost me so much effort.'[31]

On 14 May he wrote to Boulliau that he had observed the eclipse of the moon on the 6th, and that it had run its full course at 38 minutes past nine o'clock, 'It differs little from your report, if we take into account the difference in the meridian (according to your chart, 9 or 10 minutes).'[32]

On 6 June he wrote to Van Schooten that nowhere in Leiden had he been able to buy the new edition of *Geometria*, 'Therefore I ask of you that you summon Elsevier [in Amsterdam] to send to me a stitched copy. Tell them to enter it on my account, and if it is not to be had there either, then at least Van Heuraet's page.'[33]

On 5 July he completed his assignment of *Systema Saturnium* for Leopold de Medici of Florence, whom Boulliau had recommended to him: '*Princeps Serenissime* [Most august Prince], in this small piece

---

[30] Pascal 470 (No. 299)   [31] OC 2, 379   [32] OC 2, 405   [33] OC 2, 412

of work I examine the farthest phenomena in the heavens, which can be observed only by ingenious means . . .'[34]

Also in July, but on a date unknown, he wrote to Sluse on rectification (curve length) of a parabola, which figured in the recently published *Geometria*, and which 'our young countryman Van Heuraet most cleverly found'.[35] Soon afterwards, thanks to the intervention of Petit and Charles Bellair, he received a calculating machine based on Pascal's design of 1642.

In August he wrote letters to Boulliau, Chapelain, Van Schooten and Bellair; in September, when his book on Saturn was ready to be sent out, he indulged in a renewed burst of letter-writing, this time to Carcavy, Sluse, Chapelain, Bellair, Daniel Seghers, Gregorius van Sint-Vincent and Boulliau. Finally, on 17 October, he wrote to Hevelius, who was entitled to an elaborate explanation to accompany the book.[36] Then, all at once, the surf subsided.

In order to discover what was stirring in the dark groundswell, we must delve into his notes.[37] To begin with: the pendulum. On 21 October 1659, Titan carried out an experiment based on one that Mersenne had once written to him about, and which was published in his *Reflexiones physico-mathematicae* (Physical and Mathematical Considerations) of 1647.[38]

Marin Mersenne had estimated the length of a second-pendulum. This consisted of a weight attached to a thread, the other end of which was attached to a hook in the wall. After the weight had been pulled sideways, as far as the level of the hook, it was released to swing along the wall to the other side. The swing was to last 1 second. By trying this out with varying lengths of thread, he found 3 feet to be about the right length. With this pendulum, he attempted to measure the height from which a ball falls in the space of 1 second.

For this purpose, he did not draw the weight upward along the wall, but perpendicular to it, so that the weight hit the wall half a

---

[34] OC 2, 432   [35] OC 2, 417   [36] OC 2, 498
[37] Yoder 16–43   [38] OC 16, 255–301

second after it was released. At the same time that he released the weight, he dropped a ball from a specific height above the floor. By trial and error, he found out the height from which impact on the floor and impact on the wall occurred simultaneously. According to him, this just happened to be 3 feet. Thanks to Galileo's discovery that the path travelled in a fall increases quadratically, he knew that the height for a fall in two half-seconds must be four times as great. That meant: 12 feet for a fall lasting 1 second.

What he did was to measure g, which is the abbreviation for gravity, or, to express it in full and as a present-day concept: the acceleration of gravity. The value of g is 9.8 metres per second per second. (This value applies only to the Earth; it is different for other celestial bodies.) The path that is travelled by an acceleration g in a time t is $gtt:2$. If $t = 1$, in the first second, any object will fall $9.8:2 = 4.9$ metres. Mersennes's rough estimation of 12 feet comes to 3.8 metres.

Giovanni Battista Riccioli, together with the Jesuit brothers in Bologna, had gone about it differently. He used a much shorter pendulum, which swung about six times to and fro in 1 second, and he counted the number of swings made with balls falling from different heights. He arrived at 15 feet (or 4.7 metres) for a fall of 1 second – slightly higher than Mersenne's. He published this in 1651 in his *Almagestum novum* (New Summaries).

Titan knew the work and attempted to find out why it differed from the value given by Mersenne. He repeated the experiment with the smack on the floor and the smack on the wall. On 21 October, the house on Het Plein shook: 'Little happens here that is of great interest.' Even in his notes he used Latin: '*Semisecundo minuto cadit plumbum ex altitudine 3 pedum et dimidij vel 7 pollicum circiter. Ergo unius secundi spatio ex 14 pedum altitudine*' ('The lead falls in half a second from a height of three feet and about one half, or seven inches. Therefore in a time interval of one second it comes from a height of fourteen feet'). As he used a pendulum the same length as Mersenne's, the answer was unsatisfactory.

The very same day, he went a step further. Pondering the effects in the fall and in his pendulum, he wrote a list of propositions, which he would work out straight away and set down three weeks later in *De vi centrifuga* (On Centrifugal Force). He was still trying to find a definition of g, because he was referring to the work of Riccioli and to the rule of the uneven numbers for the path travelled.

(The explanation of this rule, which follows from Galileo's discovery that the path travelled in the fall h, is given by $h = ctt$, if t is time and c is constant, according to our notation equal to $g:2$. After $t = 1$ is $h = c$, after $t = 2$ is $h = 4c$, after $t = 3$ is $h = 9c$, and so on. In the time from 0 to 1 the travelled path is therefore c, in the time from 1 to 2 it is $4c - c = 3c$, in the time from 2 to 3 is $9c - 4c = 5c$, and so on. In equal time intervals, the travelled paths increase with the uneven numbers 1, 3, 5 and so on. See also the explanation in the chapter 'Student'.)

Titan was not only trying to find g; he had also added the following theoretical statement to his list: 'Weight is the *conatus* (drive) downwards.' Although this did not appear to amount to much in itself, it refers to an obscure text that was written that same day, or very shortly afterwards:[39]

> The weight of a body is the same as the drive of rapidly-agitated dust particles, just as great and directed from the centre. Whoever holds the object, prevents these particles from escaping; whoever releases it, allowing it to fall, provides the opportunity for those same particles to escape from the radius of the centre [*secundum radium*, but he knew that it followed the tangent, didn't he?]. But because, at first, it moves away from the centre, following uneven numbers starting with 1, the strong thread [*corpus grave*, as it is freely translated] compels it with the same accelerated movement [*similiter accelerato motu*] towards the centre, so that in the beginning these movements – the movement of the particles

---

[39] OC 17, 276

away from the centre, and the approach of the falling object towards the centre – must be the same. Therefore if we know how far a body falls in a given time – for example, if it falls 3/5 line in 1/60 of a second – we can also work out the movement of the particles from the centre, which will also be 3/5 line in 1/60 of a second. At the same time, the speed of the particles is given by the radius of the Earth [*Hinc jam celeritas materiae data terrae semidiametro*; we shall return to this concept later on]. Hence we have centrifugal force [*vis centrifuga*; here the phrase word is used for the first time, at its moment of discovery] within a smaller circle. Now, however, we need to examine what exactly this consists of and what determines the extent of this drive. For a fixed time duration, the extent of the deflection will certainly depend on the speed of the revolution and the circumference of the circle.

Though the various components of the text may be obscure, when we look at the text as a whole, all becomes clear. We can see that gravitation and centrifugal force are related, and are even interchangeable in their accelerating effect. Circular movement is said to accelerate towards the centre, while this also applies to the fall (towards the centre of the Earth) with its uneven numbers. It is a brilliant piece of insight, which we now call the equivalence of gravitational and inertial mass and it is worth our while to examine just how Titan put them to work.

The fall of 3/5 line in 1/60 of a second only *seems* abstruse; it concurs with Riccioli's 15 feet in 1 second (there are 12 lines in an inch and 12 inches in a foot), and it is simply the acceleration by gravity on Earth. This does little to clarify the following obscure sentence, on the speed of particles. We ask ourselves whether he is referring to visible particles (stone, sand, water) that move around with the Earth, or invisible particles in a whirl. In the first instance it concerns the centrifugal force, and in the second, gravity, as explained by Descartes. In the second instance, Titan's allegiance to the latter would be proven.

The answer can be found in the next sentence, which is not about gravity at all. He has omitted to write what kind of force! Is the omission before or after the obscure sentence? If after (the second instance), then the transition to the following sentence is highly illogical, since this sentence is about the extent in which an identical centrifugal force changes if one goes from a large to a small circle. That is why we shall assume that the type of force was omitted *before* the obscure sentence (the first instance). The linguistic pairing with the previous '*Hinc jam*' should therefore not be taken too seriously. There was never any question of a whirl – or of allegiance.

For the rest, it is quite possible that Titan continually interchanged the two forces because, as he saw it, their effects *could* be interchanged. The vague and variegated concepts in this text (obtained from his earlier reading of Galileo, Descartes, Mersenne and Riccioli), show perfectly how suggestive confusion can be transformed into fresh coherence. We find this coherence in the introduction to *De vi centrifuga*, in which he announces his calculation of the force in its dependence on the speed of the circuit and the measurement of the path of the circle, the same discussion that Titan began on 21 October:[40]

If there are two bodies of equal weight, both attached to a thread, and if they, as a consequence of an acceleration, have the same drive causing them to travel along the prolongation of the thread in the same amount of time, then we assert that we feel the same pull [*attractionem*] in these threads, whether they be pulled upwards or downwards in whichever direction . . . And the pull must be measured in the initial movement, taken in an arbitrary small interval of time [that is, the immediate change in speed that we call acceleration] . . . Let us now consider the extent of the drive, from the centre, in bodies that are attached to a thread or to a turning wheel.

[40] OC 16, 259

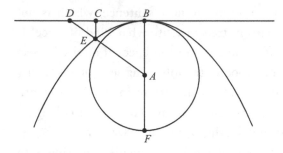

FIGURE 2

These are the decisive passages from the introduction. Titan wrote them down several days later. The fact is, they consist of the exact elements that he used for calculating the centrifugal force in the final days of October. He did so with the aid of a geometrical result, in combination with a physical argumentation.[41]

We shall first consider the geometrical aspect of the problem, which is the conventional part. He constructed a section of a parabola, drew the inscribed circle and added a few straight lines. It was the parabola of the accelerated movement (with uneven numbers) and the circle of the wheel, while the straight lines helped him to envision the change in speed in a small interval of time (see Figure 2). Parabola and circle touch at B. From B to F, a line runs through centre of the circle A across to the other side, a line that is also the axis of the parabola. Then a tangent line runs from B to D, and a line from D to A that cuts the parabola (not the circle) at E. Since the intersection E approaches B, it also comes to lie on the circle. Finally, one more line runs from E to C, perpendicular to the tangent line.

If E lies on the parabola, then according to the equation of the curved line we get

$$BC \times BC = CE \times BF$$

If E approaches B, then the triangles ECB and DEB obtain the same form, so that the long sides BE and BD are in proportion to the short

41 Gabbey (1980) 170–173

sides *CE* and *DE*. In addition, this approach means that *BE* shall be equal to *BC*. So it follows that for these short line sections we get

$$BE : BD = CE : DE$$
$$BC : BD = CE : DE$$

This means that *BD* can replace *BC*, if, at the same time *DE* replaces *CE*. Finally, *BD* will be equal to *BE* in its approach. So from the first equation it follows that

$$BD \times BD = DE \times BF$$
$$BE \times BE = DE \times BF$$

The last line is the geometrical result that Titan needed.

Now let us consider the physical aspect of the problem, which is the unconventional part. He had, of course, already concealed the physics in the diagram. He had in mind a lead ball, attached to the edge of a wheel turning at a constant speed on a vertical axis, so that gravitation constantly plays the same role, and can therefore be forgotten. He saw, however, the force exerted on the ball with the turning of the wheel as being the same as that exerted by gravity, because that too would cause acceleration.

That is why he drew, at some arbitrary point on the wheel, the path brought about by acceleration in the horizontal surface, and the form of that path is a parabola. He could draw such a parabola on any point on the wheel. The circle that touched the one parabola in the centre *B* could touch all those imaginary parabolas in *B*. He saw that this inscribed circle coincided with the wheel.

If the ball moved evenly along the circle and all effects were considered simultaneously, then the arch *BE* must be a measure for the speed *s*. In the above result *BE* × *BE* must be proportional to the square of the speed, *ss*. *BF* is the diameter of the wheel and therefore a measure for the distance *d* to the centre. It then follows from the previous result that *DE* must be proportional to (*ss* : *d*).

What does *DE* represent? Christiaan was trying to work out what would happen if the ball became detached from the wheel at *B*. Following the findings of Galileo and Descartes, he claimed that the ball with the speed *s*, which he had, would fly along the tangent line to *D* in the given time. (Not quite as far as *D*, actually, but to *D'*; *D'* is found from the intersection of the tangent line by the circle with *B* as centre point and *BE* as radius; however, *D'* coincides with *D* when very short times are considered.) So, without actually calling it by name, he posited conservation of momentum and, *nota bene*, in a rotating system.

If the ball was not released, it would have landed in *E*, and therefore the distance between *D* and *E* must give an indication of the difference between the fixed and the free state. Titan saw the distance *DE* as measure for the driving force of the ball to move freely: that is, the centrifugal force *f*. It was the force that he felt as a pull on his hand when, in a similar case, he swung a ball on the end of a thread.

This is how he arrived at the proportionality of *DE* and *f*, which together with the previously mentioned proportionality leads to

$$f :: (ss : d)$$

And he expresses this in *De vi centrifuga* as follows:[42]

> Proposition I. If two equal bodies travel along two different circles in the same time span, the proportion of the centrifugal force in the large circle to the small circle is inversely proportional to the circumferences, or the diameters. [He wrote 'equal' where we have written 'inversely proportional', which is less clear.]
>
> Proposition II. If two equal bodies travel along the same circle with different but constant speeds, the proportion of the centrifugal force on the fast body to the slow body is equal to the proportion of the squares of the speeds.

[42] OC **16**, 267 & 269

We can write the proportion $f :: (ss : d)$ as an equation with the aid of a constant m, the mass of the ball, so that $f = m (ss : d)$. The quantity $(ss : d)$ is expressed in metres per square of a second, and is therefore an acceleration.

In Titan's expression $f :: (ss : d)$, force, which until then had been a vague concept, became defined; it assumed accurate meaning and could be calculated. A speed s and a distance d could, after all, be measured with a clock and a ruler. However simple this expression may seem to us now, it was indeed one of his greatest findings. Perhaps even the greatest, if we judge it by the influence it has had on physics; it is at the heart of the dynamic law that every force causes acceleration, and at the heart of the general law of gravity.[43]

Thanks to J. G. Yoder's research into Titan's manuscripts, we know that this historic finding was made immediately after 23 October 1659.[44] On that day, as it happened, he repeated the experiment of the 21st. Apparently it frustrated him that he was unable to measure 15 feet, the figure of Riccioli, which he considered more correct than Mersenne's 12 feet, and which therefore played a role in his 'obscure' text.

To increase the accuracy, he lengthened the pendulum from 3 feet to 6 feet and 11 inches, which is 2.3 times as long. This means that he increased the pendulum swing by 1.5 times (he now knew Galileo's result that the squares of the times are proportional to the lengths; and $1.5 \times 1.5 = 2.25$). If Mersenne's 3-foot-long second-pendulum was correct, then it had a swing of 1.5 seconds, which would hit the wall at 0.75 seconds. In this way he could better hear the simultaneity of the smack against the wall and the smack on the floor.

But the height of fall in 1 second that he found was no closer to that of Riccioli. It was even less than the 14 feet that he had found previously: 13 feet and 8 inches. And this is the best part. With this new figure he managed to improve his calculations as seen in the

---

[43] Barbour 1, 451–497   [44] Yoder 23

handwriting of *De vi centrifuga*, which shows just how far he had come on 23 October. Right up to Proposition I and II!

It is highly surprising that in the midst of all this work, or just as he had begun, he still managed to find the time to write letters to Gregorius van Sint-Vincent and Gottfried Kinner von Loewenthurn about the outline of a parabola, and about Hendrik van Heuraet's quadrature.[45] He wrote to them about the surface of a parabola that was rotated on its axis, on dioptrics, and about the nonsense written by Gassend and Descartes on parhelions and haloes. The man from Prague received a thoroughly chatty letter that ended with a cordial *'Vale Kinnere praestantissime, et me ama.'* Let us translate this as 'May all go well for you, my good Kinner, and remember me.' But not a word about *De vi centrifuga*.

Or perhaps there was? After all, he did discuss the surface of a rotating parabola, which describes a concave dish. It is clear to us, but not to his correspondents, that he was already thinking of the parabolic pendulum. He would make this in the beginning of November, after having realised that g, or the height of fall in 1 second, could be measured with accuracy. But he started out with something simpler.[46]

If, as he thought, $(ss:d)$ is an acceleration such as g, and if both are equivalent to forces, according to the method of Stevin, one must be able to add them together in a parallelogram. If $(ss:d)$ were as great as g, then that parallelogram would become a rhombus. And if $(ss:d)$, in addition, were to act perpendicularly on g, then the rhombus would become a square. For a horizontal operating centrifugal force, such as in the above consideration of the wheel, the combined force would work under an angle of 45 degrees from the horizon.

This led him to change his pendulum. Instead of swinging backwards and forwards along a straight line he would have it describe a cone. The ball would go round in a horizontal circle. Then the thread that absorbed the pull of the centrifugal force and the force of gravity, together (this was a natural manifestation of Stevin's combination of

[45] OC **2**, 500, 503   [46] OC **16**, 302–311

forces) had to make a constant angle of 45 degrees with the horizon (or with the vertical axis of the cone).

Because the circle has a circumference of $2\pi d$ at a distance d from the vertical axis, the time for a circuit t follows from a division of the circumference by the speed s. And because in this pendulum the speed is determined simply by the gravitation g, namely by $(ss : d) = g$, it can be expressed simply in g and d. In other words, t can be expressed in g and d. If we carry out this calculation, we find:

$$t = 2\pi d : s$$
$$ss : d = g$$
$$ss = dg$$
$$dd : ss = dd : dg = d : g$$
$$d : s = \sqrt{(d : g)}$$
$$t = 2\pi \sqrt{(d : g)}$$

It is his formula for the pendulum time t, which is so famous that Umberto Eco refers to it on the very first page of his book *Foucault's Pendulum* and then, in true poetic fashion, proceeds to render it incorrectly.[47] By the way, this was not how Titan wrote it; he represented it only in proportions.

Once he had found it, one of the first things he used it for was to work out g. He took his pendulum of 23 October. The length was 6 feet and 11 inches (so $d = 2.17$ metres), and the assumed time of a quarter stroke, 0.75 seconds, came to $t = 3$ seconds for one to-and-fro movement. After rewriting the pendulum formula we get

$$g = 4\pi\pi(d : tt)$$
$$= 39.48(2.17 : 9)$$
$$= 9.53$$

Consequently, in his repetition of Marin Mersenne's experiment, he should have measured half this value, or 4.77 metres (15 feet and

---

[47] Eco 1

3 inches), for a fall height in 1 second. But on 21 October he found 14 feet, and on 23 October, even less, 13 feet and 8 inches. He had a problem.[48]

Something in the experiment was wrong; or else the assumed time was wrong. It is unlikely that he doubted the accuracy of his formula, even for a moment. On 15 November he repeated his experiment for the third time, with the same pendulum and taking the same time-span. This time, though, he added a couple of clever tricks. He joined the two balls with a string and then cut it through, so that they moved simultaneously. Then he used his eyes instead of his ears. He watched the moments in which the (cardboard?) ridges in a box moved on the floor and the parchment tore or dented against the wall.

This led him to measure a height of 8 feet and 7 inches for a fall in 0.75 second, which comes to 15 feet and 4 inches (4.8 metres) for a fall in 1 second. It is fairly suspect, for it was the answer according to his formula! We may have our doubts as to whether he was entirely impartial in his observations. He did write that the flaw in his observations was 4 inches, which, when converted to the 1 second, comes to slightly more than 7 inches (0.18 metres). The extent of this flaw is also suspect, as we shall see. The result of the third experiment, if rounded off, would then have been $g = (9.6 +/- 0.2)$ metres per second per second.

At the time he did this experiment, he had, in fact, already finished *De vi centrifuga*. The decisive moment for him came when he deduced from the movement of the conic pendulum ('*ex motu conico penduli*') that the height for a fall in 0.75 second is 8 feet and 9.5 inches. He wrote '*proxime*' (approximately) next to it, but when converted it turned out to be the same as $g = 9.8$ metres per second per second, exactly the right answer. It took him until 15 November to estimate the fault.

In spite of all Yoder's investigations, nothing is known on how he made the cone pendulum meet at an angle of 45 degrees, except

[48] Yoder 29–33

that this had kept him busy since 5 November. He must have seen that the thread of his pendulum would need to be made slightly longer to enable a circuit to be made in exactly 3 seconds. And we can be sure that his dish pendulums provided him with this insight.

In the first two weeks of November he was, in fact, considering other, more unusually shaped pendulums, in which not a thread, but a hollow plate served to keep the ball (which we shall now call a marble) running along the path of the circle. The plate was shaped like a parabola rotated around a vertical axis. The marble was to run freely inside, in a horizontal path, and the height of the path above the lowest point of the dish (the top of the parabola) depended on its speed. If it moved fast, then it chose a high path. If it moved slowly, it chose a lower one.

These cordless pendulums probably came about as a result of the practical problems he must have encountered with a conic pendulum meeting at an angle of 45 degrees. His first idea was to replace that pendulum with a marble that ran round in a conic funnel with a right-angled apex, so that the wall always made an angle of 45 degrees with the vertical axis. For this, he calculated that a marble that ran round in 1 second, without going up or down, would have to remain at a distance of 8.3 inches from the top (of the axis), later to be 9.5. He was not quite sure.

The difficulty with the second pendulum in the conic funnel is that the requirement that $(ss:d) = g$ is only met for one value of the distance d to the axis. Then the speed s is already fixed and equal to the circumference divided by the 1 second, or $s = 2\pi a$. The requirement thus leads to $4\pi\pi = g$, which means that the distance has to be $d = 0.248$ metres (9.1 inches) for the correct value of g. We see that Titan first arrived at a smaller figure, and subsequently at a larger one.

We also have no report of experiments with a second pendulum, which is understandable. There was no way in which they could be carried out. Just try to roll a marble at a set speed along a set path and, even more important, to keep it rolling. The path is unstable. The conic funnel was no more than a concept, the concept of the linear form. Perhaps he used his intuition, but certainly because he was not yet sufficiently familiar with g, he turned to a quadratic form: the parabolic dish.

In the parabolic dish it is not necessary for $(ss : d)$ to be equal to g. It can be either smaller or greater. In the bottom of the dish, where it slopes only slightly upwards and there is hardly any force to propel the marble inwards, $(ss : d)$ will be smaller than g; but in the upper part of the dish, where it slopes steeply upwards and there is much greater force to propel the marble inwards, it is greater. We shall now work out the balance of the acceleration on the parabolic dish to see what surprises were in store for Titan. To avoid the use of geometrical diagrams, we shall use present-day mathematics.

As long as the marble moves along a horizontal path, the accelerations are perpendicular to one another, so that they are added along the diagonal of a rectangle (not necessarily a square). This diagonal is always perpendicular to the tangent plane of the parabolic dish at the location of the marble. This means that the angle between the tangent plane and the horizon plane is as great as the angle between this diagonal and the vertical.

Now the tangent of the first angle is given by the derivative $(\delta h / \delta d)$ of the equation for the parabola $h = cdd$, where h is the height, d the distance to the axis, and c a constant. The tangent of the second angle is given by the ratio of $(ss : d)$ and g. If we equate them we discover a surprising relation between d and s, in which the time t of a circuit is no longer dependent on the distance d:

$$h = cdd$$
$$(\delta h / \delta d) = 2cd$$
$$(\delta h / \delta d) = (ss : d) : g$$
$$2cd = (ss : d) : g$$
$$g2cd = (ss : d)$$
$$g2c = ss : dd$$
$$\sqrt{(g2c)} = s : d$$

$$d : s = 1 : \sqrt{(2cg)}$$
$$t = 2\pi d : s$$
$$t = 2\pi : \sqrt{(2cg)}$$

The parabolic dish pendulum is thus tautochronic. This is what we call serendipity. Titan was not looking for this especially, but in the beginning of November he had discovered a tautochronic pendulum! From whichever height he rolled a marble inside it, the circulation time would be the same. He tried it out and saw that it was true.

His experiments with the dish pendulum were sheer perfection, utterly original and highly informative. This must have led him to discover, for the first time, the correct value of g. He already knew the value of the constant c, which was the measure of the curvature of his parabola. He could also accurately determine the time for the circuit t, since there were many revolutions to count during the period (one minute, for example, or a quarter of an hour) in which the marble kept rolling. Due to friction, the marble lost a certain amount of speed and fell downwards while rolling, but this did not alter the circuit time. What kind of dish he actually made, and how and what he did to measure and count, we shall never know. We are left merely with concepts, calculations and sketches, but no report.

His general remarks on the practice of the parabolic pendulum do give us some insight, though: 'By inducing a minute movement of the dish – that is to say, if the top of it is caused to describe a negligible circle – the marble can continue to move. If its rotations would be counted, an exact measurement of the time would be possible, with a more accurate pendulum.'[49] 'If' and 'would', he writes. Didn't he count then?

What he actually felt in the face of all this was, 'I stepped out alone into the emptiness, and was the first to do so' ('*Libera per vacuum posui vestigia princeps*').[50] It was a quotation from Horace, which he wrote on the cover of *De vi centrifuga*.

[49] OC 16, 308   [50] OC 16, 302

Five days after completing this work, on 20 November, he wrote to Ismael Boulliau: 'I observe no change in the star of the Swan. As far as the gentlemen in Leiden are concerned, I do not know who is keeping watch. It is scandalous that no one there observes eclipses, or anything else . . . I get up at three o'clock to see if the eclipse of the moon is taking place according to Lorenz Eichstadius' predictions . . .'[51] It was a breaker on the shore. The groundswell was by no means spent.

He had had to cheat – we cannot call it anything else – to ensure that Mersenne's experiment, which he had tried out once again on 20 November, would come out 'right'. He knew this. Honesty, known as integrity in the world of science, drove him towards new discoveries, equally astonishing as those that went before. At the time he wrote: 'The time to pass through the quadrant of the circle is being sought, but I doubt whether it can be found.'[52]

By looking at what Christiaan had already discovered, we can see the shortcomings of Mersenne's pendulum. After all, the acceleration of the ball along the quarter circle (the quadrant) could not be constant. When it was released, it was only g that operated, because the ball still needed to gather speed and $(ss : d)$ was negligible; but with the strike against the wall (ss) contributed considerably. If the acceleration changed, the speed along the circle would also be forced to change.

On 1 December, Titan wrote on a small corner of a sheet of paper covered with notes, 'What is the ratio of the time of a minute pendulum swing to the time for a fall from the height of the pendulum?'[53] It was another of his clever brainwaves, to put it mildly. By asking what was the ratio of a very small swing (whereby the acceleration hardly changes) to a wide swing (whereby the acceleration does change), he understood how the answer was to be obtained: step by step, with rules to calculate small changes and then, with the help of other rules, to add these together.

---

[51] OC 2, 509   [52] OC 16, 303   [53] OC 16, 304

This is how he arrived at mathematical methods which we now call differentiation and integration, but which he carried out using pure geometry. We shall not attempt to reproduce them. Through trial and error and curious detours, which indicates to us something of his wandering spirit – as if, after *De vi centrifuga*, he simply stopped looking – he stumbled upon the answer that he doubted could ever be found. It was complex, but these very detours led to more findings, and to the most important discovery of all. Serendipity. This answer to an unasked question was quite simple.

On returning from his ramblings through this labyrinth of mathematical enigma, an Odyssey lasting the whole first week of December, he wrote to Van Schooten:[54]

I have been busy with a new discovery, which I have made these last few days, to make my clock more accurate than it already was. [This is not what actually happened, it was pure coincidence.] I think you know that I had first provided these clockworks with two curved plates (AB, AC), between which the hung pendulum moved. This enabled each pendulum swing, which would otherwise not be isochronous, [which is the same as tautochronous] to be executed in the same time, just as I have explained in my book [*Horologium*]. Now, however, I have found out what I never expected to discover: the actual form of (AB, AC) that makes the pendulum swing with such accuracy. I have determined this form with geometric argumentation, and I shall instruct the clockmakers as to how they may draw the curve for themselves, without the slightest difficulty. I am confident that this discovery shall afford the greatest pleasure to the highly astute Van Heuraet, for to me it is surely the most fortuitous of discoveries I have yet chanced upon.

This letter is dated 6 December. He did not actually say so, but he had chanced upon the cycloid. After making his calculations in

[54] OC 2, 521

January, he was fully familiar with its characteristics. Now that, in December, he had considered changes in the quadratic curves (parabolas) which had seemed useful to calculate average speeds, the term for a tangent line sprung to mind. Without a trace of excitement he writes, *'Hoc autem cycloidi convenire inveni ex cognita tangentis ducendae ratione'* ('But I have found that this [curve] touches a cycloid, according to the known method of drawing its tangent line').[55]

So he recognised the cycloid from its derivative, as we now say. This was admirable. But he owed the discovery as much to his insight into the fairly constant acceleration of the pendulum (over a small arch), and the fact that his rough calculation that it would have the same speed (over that arch) was too crude. That is why he looked at quadratic curves, much to the delight of Van Heuraet: *'Sine quibus motus aequabilis in cava cycloide inveniri non poterat'* ('Without which the regular movement in the hollow of the cycloid could not have been found').[56]

The cycloid apparently described the path in which the acceleration is proportional to the distance to the centre. From his previous work he knew immediately that this concavity was an unwinding of a reciprocal concavity (convexity), or, to put it in precise terms: that the path of a tautochronous pendulum was the evolute of a cycloid. That portion of the proof was complete. Hence the spontaneous triumph in his letter to Van Schooten. On 15 December 1659, having abandoned detours and scrapping the approximations, his proof was finally perfect.

What Titan felt was 'Great things which no single genius has ever explored' (*'Magna nec ingenijs investigata priorum'*).[57] This time he was quoting Ovid.

---

[55] OC **16**, 397   [56] OC **16**, 398   [57] OC **16**, 406

9 Temperament

For days on end it had been pouring with rain, but on the great day, 20 April 1660, the sun shone down on Het Plein. Party-goers lent a helping hand to the open carriage carrying the bewigged groom. In the church of Saint Jacob awaited a richly powdered bride.[1]

Two days later, father Huygens described the wedding of his daughter Suzanna to Philips Doublet, her first cousin.[2] After the sermon, people climbed up onto the pews to see the ceremony; on leaving the church the bride threw two hundred pennies to the eager crowd, and upon her arrival home, the same sum in candies. While the masses scrabbled outside, inside it resounded with the joyful smacking of kisses amongst the guests. Forty-two sets of cutlery awaited them, complete with knife, fork and spoon, as well as glassware, plate and napkin, all set out upon the white damask of the L-shaped table. Jacques, the master chef, had been brought in specially, together with three assistant chefs and five helping hands, to prepare the swine's head and the 127 partridges, capons, turkeys, pheasants and hares, all stuffed with lamb's meat and lardoon, and astonishing quantities of pastries made with sugar and marzipan.

At four o'clock sharp, the music started up. The bride was the first to be led to the table, where she and the groom sat down with their forty-two guests to indulge in the vast spread of meats and pastries, until nine o'clock. After a while they moved on to 'drinking as you fancy'. This entailed two guests alternately kissing and sipping from each other's glass of wine, all coyly concealed behind a napkin and accompanied by vigorous banging of plates with the cutlery by their table companions. No bawdy drinking songs. In the meantime,

---

[1] Rietbergen 181–189   [2] OC 3, 67

600 candles – strongly perfumed to outdo the aromas of the meat – were brought in to light up the great hall for the ball, a ball that would keep the band playing until well into the small hours, well after the bride had been escorted, amidst further revelry, to the decorated bedchamber, where ribbons, wreathes and shoelaces were removed.

'When the girls . . . refused to leave, the bridegroom put on a night cap and pulled on a dressing gown over his clothes. In this manner he chased away the girls, who thought he wore nothing underneath it. There was much romping and rollicking upon the stairway . . .'[3] This according to father Huygens' account to Béatrice de Cusance, the Duchess of Lotharingen. He had paid out 3544 florins.

Titan was there. Did he join in the game of 'drinking as you fancy'? We don't know. He certainly enjoyed himself. Two days after the wedding, he wrote to Boulliau that the guest of honour, the French ambassador, 'for part of the night joined in the merrymaking of us younger ones', then went straight on to add, 'I regret somewhat the two days I have missed [le temps que j'ay perdu] due to these ceremonial follies.'[4]

In 1668 Constantijn would marry Suzanna Rijckaert, fourteen years younger than he, but not before he had fathered a daughter by Isabella Dedel the year before. Similarly, in 1674 Lodewijk married Jacoba Teding van Berkhout, also fourteen years younger. But Christiaan . . .

Suzanna and Philips Doublet produced many children. Geertruyd (1661), her first, was weak, as were many of the others; only Constantia (1665), Philippina (1672) and Philips (1674) survived to adulthood. Constantijn produced (the illegitimate) Justina (1667), and Constantijn (1674, 'Tientje'). Lodewijk produced one more Constantijn (1675), Lodewijk (1676), Paulus (1677?), Maurits (1678?) and . . . Christiaan (1680).

Le temps perdu.
Lost time.

[3] Ibid.   [4] OC 3, 65

Apart from science, Christiaan only really had time for music.[5] A year later, he and Constantijn adapted a piece by La Barre, to be practised on the harpsichord, 'The Gigue is surprisingly fine, but rather difficult, so that we are unable to agree on how certain passages should be played, although we have been studying it these last three weeks . . .'[6] Three weeks!

La Barre was the Parisian organist he had met in the summer of 1655. From his days in Paris, he also knew Thomas Gobert and Jacques de Chambonnières. In The Hague he knew the organist Quirinus van Blankenburg; in Amsterdam, François Hemony. But he never mentioned the truly well-known composers of that time, Johann Jacob Froberger and Jean-Baptiste Lully. He himself wasn't much of a composer. We do have sixteen bars by him of a courante in E minor, but these are bland.[7] No one would bother about them, if it were not for the fact that it was Christiaan Huygens who had written them.

On one of the sheets of music that must date from this time we read: 'Straying into the realm of unfamiliar notes is inappropriate, for in that way one quite forgets how one began. This weakens the force and beauty of the melody. [*Hoc enervat vim ac decorem cantus.*] Listeners dislike continual modulation.'[8] As a listener, he himself wasn't fond of drama – in the passion of Lully's opera *Alceste*, for example, which he might well have heard.

He didn't like contrapuntal art either:[9]

I wish that our composers [*melopoei*] would not pursue always the most artificial and complicated, but what appeals most to the ear. For what is it to me that people have regard for close imitations, which they call fugues, or singing canons, if this makes the melody less free and pleasing? Artists claim to be delighted by them, not by the sweetness or the harmony, but by the artistry of the composition. They are clearly mistaken in their judgement of music. The aim of music is to caress the ear with sound, not to subject it to art. Indeed, these are different things entirely.

---

[5] Cohen (1980) 271–301; Rasch (1996) 53–63 [6] OC 3, 413 [7] Rasch (1986) 43 [8] OC 20, 127 [9] OC 20, 125

Could he only experience art in its expressive forms, or not there either? Mathematics, we think, mathematics was his art.

Le temps perdu.
We have our doubts . . .

There is an unfinished piece of music from 1662, or earlier, which bears a relationship to his mathematics. If we fix this at the time of Suzanna's wedding, then it belongs with his work on the hyperbola: the integral of a hyperbola is a logarithm, and that logarithm is the leaven of the article. If we do so, it is also in accordance with the tasteful orchestral music that undoubtedly contributed to the general merriment of this occasion.

Titan elaborated on suggestions that he found in Mersenne's *Harmonie universelle* (Universal Harmony) of 1636.[10] These were suggestions on how natural overtones should fit into a harmonic sequence:

That lost moment when he became engrossed in the problem of tuning.

Tones that sound in the ear are the result of pulses on the eardrum made by changes in the air pressure known as air vibrations. They can be induced by causing a string to vibrate. This is never one simple vibration; besides the fundamental frequency it also has frequencies that are twice, three, four times as great. The higher frequencies belong to the natural overtones.

When placing these tones within an octave (that is, between the fundamental tone and the first overtone, of which the frequency is twice as great as the fundamental tone) they become divided by a whole number, so that the ratios become less than 2. It is assumed that, in the ear, this division takes place naturally. In this way we obtain ratios for the five harmonies that are pleasant to the ear:

---

[10] Cohen (1980) 295

$3:2 = 1.500 =$ fifth

$4:3 = 1.333 =$ fourth

$5:4 = 1.250 =$ major third

$6:5 = 1.200 =$ minor third

$5:3 = 1.667 =$ sixth

They do not form the regular sequence referred to in the harmony designation of mediant (third), subdominant (fourth), dominant (fifth) and sixth. There are gaps between 1 and 1.200 to fit in a second and between 1.667 and 2 to fit in a seventh. But which ratios do they require?

For a second, $7:6 = 1.167$ could be used, although it is questionable whether the seventh overtone sounds strongly enough to be heard. This would apply to a greater extent for a minor sixth with $8:5 = 1.600$, for which the eighth overtone should be audible. But it is highly doubtful whether the ninth overtone is audible at $9:5 = 1.800$, a ratio that could be used for the seventh. Another question is whether the ratio of 1.167 for the second and the 1.800 for the seventh fit into the regular sequence. There is, therefore, room for discretion.

Such discretion constituted a problem in the seventeenth century. Violinists could position their fingers on the strings to find a sequence with a second and a seventh that was pleasing to the ear. They could build up this sequence independently from the fundamental tone of their melody, so that they could vary the tonic, and in baroque music, which did not shrink from high drama, this was desirable. Harpsichordists, on the other hand, could not. A fixed tuning of strings in these keyboard instruments restricted them, which meant that they could not vary the tonic without falling into a sequence that was out of tune.

This can be demonstrated by means of a numerical example. If they played a fifth on the tonic doh (i.e. sol against doh) they heard the pure ratio of $1.500:1 = 1.500$. If they tried to play that fifth with a modulation of the tonic doh to the tonic sol (i.e. re against sol) they heard the ratio 2.333 : 1.500 = 1.555, which was flat. Here we assume, as in the above, that 1.167 is the correct ratio for the second in the sequence on the tonic doh. If we double this, we get 2.333, which is the ratio for the re that sounded in the flat fifth. If the ratio were smaller, for example, 1.125, then the flat fifth can become sharp, because twice 1.125 is 2.250 and 2.250 : 1.500 = 1.500.

Polyphonic modulating music for keyboard instruments requires not discretion, but a tuning system that is fixed and of equal temperament, which corresponds reasonably well with the five nat-ural harmonies. By 'equal temperament' we mean that the tones, which follow one another in an octave, remain in constant accor-dance with the system of equal temperament, or that the relationship between their frequencies is always the same. Then whichever tonic is chosen, the scale will always sound accordingly.

In their search for such a relationship, it became clear to Gioseffo Zarlino and Francesco Salinas in 1570 that an octave of twelve notes was better than one of eight.[11] So they added tones which the ear – this is still the assumption – could not naturally hear. According to our understanding, the relationship must then be the same as the root to the power of 12 of 2, which is 1.0595. (The frequency of the first tone is therefore 1.0595 times that of the tonic, for the second it is 1.0595 × 1.0595 = 1.1225 times, for the third it is 1.0595 × 1.0595 × 10595 = 1.189 times, and so on.) This is not how they calculated; they made estimations, but they came fairly close to the twelve notes that are still in use:

1.0595	doh♯		
1.1225	re	(1.167?)	second
1.189	re♯	(1.200)	minor third
1.260	mi	(1.250)	major third
1.335	fa	(1.333)	fourth
1.414	fa♯		

[11] *Ibid.* 284–285

1.498	sol	(1.500)	fifth
1.588	sol♯	(1.600?)	minor sixth
1.682	la	(1.667)	major sixth
1.782	la♯	(1.800?)	seventh
1.888	si		
2.000	doh	(2.000)	octave

If we compare the scale with natural tuning, we see that the fourth and the fifth, which can be easily heard, can also be easily calculated. Both thirds and both sixths come off slightly less well, while the seventh works out rather low.

Now we come to Marin Mersenne's suggestions. As long as there were only a few keyboard instruments being built for these twelve tones, so that the practice of making music was not yet fixed, it was easy enough for people to adjust this tuning. Mersenne proposed that Zarlino's and Salinas' ratios should be calculated and not estimated. He also suggested that the seventh was too low, although he did not think that the ratio 9 : 5 could be heard. And finally, he considered that a scale with thirty-one notes would be more compatible with natural tuning than a scale with twelve tones.

Around 1555, Nicola Vicentino had built an 'archicembalo', with thirty-one keys per octave. It is a mystery how he came by the number thirty-one; perhaps from the (highest) number of days in a month. But although the modest step from eight to twelve notes may have led to pleasing results, the step from eight to thirty-one was evidently too great. Francesco Salinas thought the instrument sounded terrible. Apparently Vicentino ordered the keys into groups of six, and he said in his will that the tuning was of equal temperament, 'con le quinte & quarte alquanto spontante, secondo che fanno li buoni Maestri' (with the fifth and fourth being more or less spontaneous, just as with the great masters).[12] We can easily find the tones. The relationship of the frequencies of successive notes should be given

[12] *Ibid.* 299 (n43)

by the 31st root of 2, which is 1.0226. In the same way as in the above, we get the following scale:

1.0226		
1.0475		
1.0693	doh♯	(1.0595)
1.0935		
1.1182	re	(1.1225)
1.1435		
1.1694		
1.1959	re♯	(1.189)
1.2229		
1.2506	mi	(1.260)
1.2788		
1.3080		
1.3373	fa	(1.335)
1.3675		
1.3984	fa♯	(1.414)
1.4300		
1.4623		
1.4955	sol	(1.498)
1.5293		
1.5639		
1.5993	sol♯	(1.588)
1.6354		
1.6724	la	(1.682)
1.7102		
1.7488		
1.7884	la♯	(1.782)
1.8289		
1.8702		
	si	(1.888)
1.9125		
1.9557		
2.0000	doh	(2.000)

When we compare the scale with the natural harmonies, we see that not only the fourth (1.333) and fifth (1.500) are well matched, but also both thirds (1.200 and 1.250) and both sixths (1.600 and 1.667), while the seventh (1.800), although still a bit low, has come closer to the 'right' number. We have placed the notes from the scale of twelve, which more or less correspond, in the second column; only in the case of si did this not work.

Titan did the same, also using logarithms. Later he would go to great trouble to explain what logarithms were, and how they worked. This was fully justified, for it was quite new in his time.

His series of numbers proved the accuracy of Mersenne's propositions: the equal temperament could be calculated, the seventh was too low, and generally speaking, the thirty-one-tone scale was better suited to the natural harmonies than the twelve-tone scale. He added proudly that he could even hear the natural seventh![13]

But how far can Christiaan actually be regarded as the (re)inventor of the thirty-one-tone scale? He claims emphatically to have discovered the number thirty-one on his own and that Mersenne's suggestion (who, in turn, had it from Salinas, who in turn got it from Vicentino) played no part. Is this to be believed? Did he find out which scales corresponded to natural tuning and then calculate that this would only work with thirty-one? There is no evidence of this in his notes, but it could well be so.

Most likely, he divided the logarithms of 1.200 and 1.250 (for both thirds), and 1.333 and 1.500 (for the fourth and fifth), and 1.600 and 1.667 (for both sixths) by the logarithm of 2.000 (for the octave), subsequently multiplying these quotients by numbers over twelve. The first figure for which all six answers came anywhere near whole numbers was thirty-one:

$$0.0792 : 0.3010 = 0.2630 \ (\times 31 = 8.15 \text{ by } 8)$$
$$0.0969 : 0.3010 = 0.3219 \ (\times 31 = 9.98 \text{ by } 10)^*$$

---

[13] *Ibid.* 291

$$0.1249 : 0.3010 = 0.4150 \ (\times 31 = 12.87 \text{ by } 13)$$
$$0.1761 : 0.3010 = 0.5850 \ (\times 31 = 18.13 \text{ by } 18)$$

$$0.2041 : 0.3010 = 0.6781 \ (\times 31 = 21.02 \text{ by } 21)^*$$
$$0.2218 : 0.3010 = 0.7360 \ (\times 31 = 22.85 \text{ by } 23)$$

The values marked with an asterisk, which correspond to the major third and minor sixth, work out perfectly, as we have already seen above.

How does it work for other (relatively) smaller numbers? It is not difficult to calculate that they can bring some, but not all six harmonies, in the region of a whole number, at the same time. Only with twelve is it fairly successful. For this, the answers are, respectively: 3.16 by 3; 3.84 by 4; 4.98 by 5; 7.02 by 7; 8.13 by 8; and 8.85 by 9. We can thank Titan's fine sense of hearing for the fact that he rejected the possibility of twelve; he was able to distinguish the minutest differences in tone pitch. He spoke of anguish to the ear.[14] A scale of twelve notes was too excruciating for him, but one of thirty-one was perhaps bearable. It was his own temperament that rebelled against equal temperament.

Was he planning to build an 'archcembalo' that was not terrible, but on the contrary, pleasing to the ear? How should all the keys be put together and how should they be played? He did not abandon the idea. But this lost moment spent on logarithms passed by, and apparently feeling rather lost, he abandoned this unfinished work.

Perdu.

Lost.

Christiaan was deeply afraid of losing any of his work. It was at this time (12 April 1661) that he asked Constantijn to send him some lenses:[15]

---

[14] OC 20, 159   [15] OC 3, 13

I request that you go up into my room [we would love to know which of the four this was], to which Cousin holds the key. You will find the key to the larger of my two cupboards in the box on my table, which also contains Pascal's machine. When this is opened you will need to release the bolt at the top, where I have placed the lenses of the large telescope.

Most of his books he concealed under a curtain; his notes, in closed drawers. Often he worked in the quiet seclusion of the night, scribbling his hermetic Latin with a glass stylus, in the patch of light afforded by the candle that burned between a mirror and a projection lens.

He was plagued by headaches and infections. He took medicines, but these gave him little relief. Doctor Van Liebergen paid him frequent visits for purging and bloodletting. We read about all this in the countless letters exchanged amongst the family since autumn 1661, although not in his professional correspondence, and before this time it would not have been otherwise.

A few examples:[16] 'For the past three days I have suffered from a sore throat,' he wrote on 11 January 1662, 'but last night it was so bad I was unable to sleep.' And a week later, 'I can neither read, write nor think, for if I do so my head immediately begins to hurt.' Exactly one year later, 'First I had toothache, then a headache, but worse than ever, and now an upset stomach.' Or on another occasion, 'I must abstain from every exercise of the brain' ('Il faut que je m'abstienne de tout exercise de cervelle'). And, 'Despite all this, I am even more laconic than usual.'

Certainly he was not laconic all the time. Andreas Tacquet, his friend from Antwerp, wanted to convert him to the Catholic faith, and after the wedding sent him a book about the Eucharist. As if the Eucharist were one more headache that kept him from exercising his brain, he answered:[17]

[16] OC 4, 10, 11, 288   [17] OC 3, 104

If you wish to influence me with arguments, then it is very much
to be doubted whether you shall find them, for they must be
greatly convincing in order to persuade me in a matter of such
consequence. But you come up only with books, as if these on
their own are authoritative arguments, books of which the text
can be falsified and written by people who may be mistaken. How
far this stands from the power of persuasion afforded by
mathematical proofs!

He wrote this on 3 August 1660.

He was still brimming with mathematical proofs. Even after 15
December 1659, he didn't spare himself. The breakers were surging,
in long waves. He continued his calculations on hyperbolas; he was
working on a general theory on evolutes, while also writing out the
texts of all his discoveries in a new edition of *Horologium*.

In fact, he wrote three-quarters of this new edition in the beginning of 1660. It was to be called *Horologium oscillatorium* (The
Pendulum Clock). Due to various other distractions it was four more
years before he managed to complete the missing quarter. And then
it took a further nine years of procrastination and revision before he
finally had it published, in 1673.

In size, depth, and rigour, *Horologium oscillatorium* is his greatest work.[18] One is tempted to call it his life-work, but in a strict sense
this is not true. Such a designation would not only detract from his
earlier and later works, but it also does not yield up everything he
had to say. In fact, it stops when he is about to disclose his findings
of 15 November. The full description of this discovery, *De vi centrifuga*, would share the fate of that other masterpiece, *De motu* – to
be published only after his death.

Horologium oscillatorium consists of five sections. The first section is
a revised version of his pamphlet *Horologium*, in which the mechanics

[18] OC **18**, 69–368

of the clock are described in detail, down to the revolving chain to provide the power. The set of plates along the pendulum also features in the work. The weight is discus-shaped, so that when it swings, it encounters only minimal air resistance, and there is a movable weight along the thread with which the pendulum time can be adjusted, to the second.

The second section consists of a number of propositions on the fall: first the free fall, then the fall along a slope, and finally the fall along curved paths. It leads up to a proposition that a body falling along a cycloidal path always reaches its lowest point in the same amount of time, regardless from which point on the path it begins its fall. In other words, this is the proposition that the cycloid is isochronous.

The third section describes the evolute theory. This theory gives the mathematical relationship between curved lines with which, among other things, the span of these curves can be determined. With the help of these results, it can be established that the shape of the set of curved plates, which is described in part one, must also be cycloidal.

The fourth and longest section discusses the true physical pendulum, and no longer its mathematical approximation. He added this section in 1664. Here he takes into account that the mass is not concentrated in one point at the bottom of the thread, but is distributed across the discus-shaped weight around that point and also, to a small extent, along the thread. In other words, it discusses the determination of the midpoint of the pendulum.

The fifth section discusses the idea of the isochronous pendulum that revolves around a vertical axis in a parabolic dish. This is followed by a list of thirteen theorems on centrifugal force, without any explanation or proof. Nothing points to any backlash after all the effort he put into this achievement. He kept on working at his calculations. He never stopped observing the planets. He went on writing letters. And he also made new plans to visit Paris. 'My writings on the clock were finished long ago,' he informed Chapelain in September, 'but that is no reason to have them printed before I leave.'[19]

[19] OC 3, 118

That summer he was occupied with criticism of his *Systema Saturnium*. Boulliau had reservations, Riccioli's were of a more serious kind, but these were only expressed in letters. The Florentine astronomer Eustachio Divini, however, actually went as far as to publish his criticisms. Christiaan heard news of this from the Jesuit in The Hague, Nicolaas Heinsius, who in turn had got it from the Florentine Jesuit Carlo Dati, secretary to Prince Leopold, to whom the book was dedicated. On 25 May 1660 Dati wrote to Heinsius, *'Che Eustachio Divini scriva contro, o per meglio dire sopra l'opera del Signor Hugenio deve esser vero, ma perche egli e uomo idiota la far con l'aiuto del Padre Onorato Fabry Giesuita Franzese che la metta in latino.'* ('That Eustachio Divini wrote against, or rather about the writing of Mr Huygens may be true, but because the man is feeble-minded he did it with the help of Father Honoré Fabry, a French Jesuit, who translated it for him into Latin').[20]

'Be wary of its fierce style; this is characteristic of the society.' These were the words of another member of the society, the Roman Jesuit Guisony.[21] He also cautioned him on another difference in culture, 'As is customary in this country, Eustachio demands that all believe him blindly, without reservation, or else they are to be reviled.' When we read Divini's *Brevis annotatio* (Brief Annotation) after this rather portentous announcement, it is not too bad. It even contains some praise. But, Divini begins:[22]

> In *Systema Saturnium* I am accused of not having observed
> something, but rather of having made it up. Therefore, I shall first
> refute the accusation of deception. Then I shall proceed to point
> out a couple of errors in the book, with moderation, as befits a
> Christian gentleman. [It is an allusion to the name Christianus
> Hugenius, in which he constantly spells Hugenius as Eugenius,
> which means 'Well born'.] I shall conclude by describing the
> system of Saturn, but without that figment of imagination of a
> ring, which Eugenius has not observed, but rather has made up.

[20] OC 3, 83   [21] OC 3, 101   [22] OC 15, 404

Divini is slightly less detached in his comments on errors in *Systema Saturnium*. For example:

[Eugenius] has made mistakes in his proportions. Another might be tempted to enquire whether the errors might be attributed to the telescope or to the rich imaginations of the writer. As for myself, I would not cast doubt upon the honesty of such a learned gentleman. Therefore I consider that all the errors to which he admits should be attributed to a flaw in his telescope. [No errors are admitted to in *Systema Saturnium*, but the possibility for error is acknowledged.]

He would have done better to buy one of my telescopes, Divini brags, for they are available throughout the whole of Europe. Let impartial parties look at both our telescopes and decide which is better. He goes on to poke fun at the notion of inhabitants on Saturn, 'This can better be refuted with laughter than with argument. Moreover, it is against Catholic doctrine, which acknowledges that man originates only from Adam.'[23] Against Catholic doctrine. This is an echo of 1633, when Galileo's scientific findings were condemned by Rome.

We can be detached in our reactions, but Titan, of course, could not. He must have read the *Brevis annotatio* in August 1660. How angry was he? Here, before the eyes of Prince Leopold and the rest of the world, was an underhanded attack on his work and person. Divini may not have enjoyed any great fame, but Honoré Fabry did. For the rest, he could deny having made errors of observation, but not that the idea of a ring was a fantastic theory and that the possible inhabitation of the planet was against Catholic doctrine.

He replied straight away with an indignant *Brevis assertio* (Short Assertion), which he had printed in September by Vlacq, together with Divini's attack.[24] 'I am deeply disappointed, for I see only that they fail to contest my observations with proper arguments . . . while they

[23] OC 15, 419   [24] OC 15, 439–467

openly accuse me of making these up.' This is one of the first sentences in his letter to Prince Leopold. He was playing off the translator against the writer, and he said of the latter, Divini, that he was a nincompoop. 'Finally, I request of Your Highness that you will not hold against me my freedom of language in this dispute. It is due to a provocation that I have thus demeaned myself in this quarrel. But nevertheless, nowhere have I violated the rules for a just defence.'[25]

In the eyes of Jean Chapelain, he had won hands down. But as usually happens in such disputes, his *Brevis assertio* did not mean that this was the end of the matter. For years to come, his defence would figure in his letters, until Eustachio Divini was abruptly brought to silence. Not by words, but upon being shown the ring through his brother Giuseppe's newest telescope. That was in January 1665.[26]

In January 1665 Isaac Newton, a twenty-two-year-old student at Cambridge, was working out the problem of a circling body. He reasoned that the body constantly collided against a circle circumscribed by parts of tangential lines, 'And soe if body were reflected by the sides of an equilaterall circumscribed polygon of an infinite number of sides (i.e. by ye circle itself) ye force of all ye reflections are to ye force of ye bodys motion as all those sides (i.e. ye perimeter) to ye radius.'[27]

The ratio of the circumference of a circle and its radius is 2π. By force of body movement he meant the speed s (in fact, the quantity that we now call impulse). What he calls the force of the rebounding was therefore $2\pi s$, it being the total force with which the body was rebounded in a circle. We obtain the force that operates at any moment by dividing by the time of a rotation. That time is given by the circumference of the circle $2\pi a$, in which a is the radius, to be divided by the speed s.

This is how the unknown student discovered that the rebounding force must be equal to $2\pi s : (2\pi a : s) = ss : a$. He did not know, nor

[25] OC 3, 132   [26] OC 5, 176; OC 15, 400–401   [27] Westfall 150

could he know, that Titan had reached the same result just five years before, though in a different manner, for the force in a circular movement, and that he had called it centrifugal force. The question was, should Titan make haste?

On 12 October 1660 he set out on his travels:[28]

> Left on Tuesday at 2 o'clock with Wolfsen; departed at five-thirty from Delft, where I first visited Van der Wals' library, and met with N. Becker and the husband. At 9 o'clock in Delftshaven with the ship; on board, among others, was Mr Hagens, agent of the Prince de Ligne, who had not made my acquaintance and spoke of clocks and of Huygens. On De Zwaan there was only bread to be had, as provisions on board.

It was the voyage he had spoken of earlier, to Chapelain. He would arrive in Paris on 28 October, this time for the entire winter season, until 19 March 1661. Then he would proceed to London, where he would stay until 14 May.

. His *Reys-Verhael* is far too long to relate in full and comment on. We offer two quotations, a short one and a long one, which we will summarise.

The day after his arrival, he switched language, 'Visited Chapelain, Brunetti was *à la campagne* [out of town]. Sent a letter to M. le Premier. A woman with a page's pantalon under her skirt came and performed all kinds of tightrope jumps at our inn. Visited Bosse, who told me *quel homme le curé de S. Barthelemy était. Escrit à P.* [what kind of a man the priest at Saint-Barthelemy was. Wrote to father.]'

The following is a translation of his journal for a week in December:[29]

---

[28] Brugmans 119–177   [29] *Ibid.* 135–137

5: Thévenot sent me the observations from Florence on fumes that descend in a vacuum. I went with Marlot and De la Chaise to dine at the Duke of Roannez. Pascal came round.

6: Hired a harpsichord. Saw Gobert and Auzout. Did not find B. F. Du Mont.

7: Saw Thévenot, where Frenicle was residing. There I was shown Fontana's observations. Together with Vlaerd, dined at the Monmor circle. Skeleton of a . . . where you could see all the nerves, veins, arteries, the heart, the eyes. It was sustained with copper wire wound round with silk. Rohault read aloud his experiments with water that rises in narrow tubes. Earthenware. Pecquet's theory that food is dispersed throughout the body through the nerves. Laurent informed me that Miss Ouwerkerk passed away in England.

8: Studied Vaulesart's quadrant, extremely clever. Message from the Duke of Luynes and letter from Brunetti. Conrart came to look me up, spoke of Miss Perriquet. Went with Conrart to visit Van Beuningen. Went to Boreel.

9: The Duke of Luynes sent his carriage to fetch me. There I came upon the Duke of Saint-Simon. He spoke of the inventions to measure distance, and for the use of clocks for longitudes. Dined with the Prime Minister, Gentillot was also present. He promised to introduce me to the King. De Bautru. Did not find Chanut, Roberval, Sorbière. Saw Miss la Barre.

10: Du Mont dropped by, took with him Laurent's psalm copies. Wrote to father and brother Zeelhem.

11: Saw Laurent and his rather messy inventions, and one or two contrivances for conic sections. Dined with the ambassadors. Van Beuningen showed me the book of Saumaise against Milton, in which he abuses the Batavian Heinsius. Met Miss Van Gent. The young lady Ottersum. Hauterive dropped by. La Roque. Marlot.

12: Saw Chapelain, Conrart, Roberval. Did not find Carcavy. Took a walk in the Luxembourg Gardens. The Duke of Roannez came to look me up, together with Pascal. Discussed the force of rarefied steam in their guns, and flying. I showed them my telescopes.

A whirlwind of activity![30] And so it goes on, day in, day out. He reaped the harvest of five years of correspondence and five years of research; that much is clear. It was not unusual for him to speak to ten people in one day, and in the end he probably knew more about architectural heritage, art collections, theatres, plays, composers, concerts, ballets, painters, engravers, shops, workshops, clockmakers and lens polishers in the city than any Parisian ever did.

He also got to know a great deal about relations within the various Paris circles. Quarrels had disrupted the academies of Mylon and Montmor.[31] Shortly before his arrival, Claude Mylon had died, which also spelt the end of his academy (the old one, of Mersenne). Paul Scarron had just died, on 20 October. Young talent, in as far as it was around, did not present itself. Roberval had quarrelled with Montmor, and no longer showed his face. Boulliau no longer showed his face either, but this was because he was with Johannes Hevelius in Poland.

Although the academies were no longer what they had been in 1655, Christiaan still managed to dig out the gentlemen mathematicians – outside their circles, too – and the lady: on 29 December he dined in the company of Marie Perriquet, at the home of Valentin Conrart. When he had found them, they began inviting him: Rohault on 13 and 17 November, and again on 20 and 21 December; Thévenot on 7 December and several times after that; Auzout on 8 January; Petit on 24 January and on plenty of other occasions (which resulted in his falling in love with his daughter); Amproux, Clerselier, the free-thinker Alexandre d'Elbène, the Bishop of Laon César d'Estrées . . .

He no longer needed introductions from the Dutch embassy; just the day before he left, he paid a visit to the intendant Tassin.[32]

[30] *Ibid.* 45–53   [31] Mesnard 37–39   [32] OC **22**, 562

Insofar as this was necessary, he obtained introductions from Jean Chapelain, who was clearly keen to do his best to keep him in Paris. This gentleman was more of a poet than a mathematician, and at sixty, in 1656, he had attempted to make his name famous with *La Pucelle d'Orléans* (The Maid of Orleans). The epic was slated, but his influence at court hardly suffered. He was a good friend of Jean-Baptiste Colbert, who expected to be appointed as head intendant of the royal household.

Titan, as we've read, went to see him straightaway: 'Visited Chapelain.' They undoubtedly discussed how he could best introduce himself. On 22 December he had an audience with Louis XIV, to whom he most likely offered a clock: 'Waited long in the chambre of the Lord of Villequier, Captain of the Guard. Splendid bed.'[33] But beyond this, there is no mention of it in the *Reys-Verhael*.

From that moment on his name played a role in the politics of court. Colbert would request grants for poets and mathematicians, which led to the royal gratuities of 1663, most of them a sum of 1200 pounds.[34] Christiaan was awarded one for his clock. Amongst the seven other foreigners on the list were Nicolaas Heinsius and Gerardus Vossius.[35]

Naturally, Chapelain enticed him along to 'his' academy, which met almost every Tuesday. Through Chapelain, or the academy, he met the learned women Madame De Bonneveau and Marie de Guerderville. It was Pierre de Carcavy who introduced him to the closed world of Port-Royal, in which Pascal moved.[36]

This also enabled him to meet the Duke of Roannez and the Duke of Luynes, the knight de Méré, Mitton, Monconys and the two pious Jansenists, François de La Chaise and du Bois, who would publish Pascal's *Pensées*. It was de La Chaise who showed him round the ostentatious residence belonging to Marshall Charles de La Meilleraye, not far from the store for munitions and weaponry, and the Palais-Royal, the home of the widow of Charles I; the new palace on the

---

[33] OC **22**, 542   [34] Roger 41–43   [35] OC **4**, 390   [36] Brugmans 48–49

river, the Louvre, had just been completed. Meeting up with so many, he could hardly miss coming across Ménage, the man against whom Chapelain bore such a grudge.[37]

Gilles Ménage had set up his own literary circle, and was sufficiently important to be the object of satire in Molière's *Les Femmes savantes* (The Learned Women). Christiaan first came there on 19 January. On the 29th he was taken to the circle of Madeleine de Scudéry, where he met Segrais and La Menardière. On 5 March he was there again. It was here that he met the Marquise of Rambouillet, the Abbot of Villiloin, Marolles, de Launoy, La Mothe le Vayer, and Boisrobert, who promised to introduce him to Ninon de Lenclos . . .

He was so much in his element that he sang the praises of Paris to Heinsius, 'It is certain that I could live nowhere more happily than in this city, whose exquisite inhabitants and their singular friendliness bind me to her more and more' ('*Scio nunquam jucundius quam in hoc urbe me victurum ubi lectissimorum hominum consortium, ac singularis humanitas magis ac magis is dies devinciunt*').[38]

On 7 and 8 March, he drew Marianne Petit. She was the daughter of the verbose engineer Pierre Petit, whom he had visited on several occasions, when he had witnessed her lively management of her father's household and heard her witty tongue. He found the portrait difficult, and it took him until the 10th to finish it, only to need more changes made on the 13th. The young beauty was unapproachable, and made him out to be a heretic. Still, she fascinated him.

Did she fit into the plans he had in mind for Paris? Did he think he could conquer her aversion and win her favour? Somewhat clumsily and long-windedly, as if unable to find the right words, he wrote on 21 July, when he was back in The Hague:[39]

Dear Lady, I do not know if you remember the man who had the honour of visiting you on several occasions last winter, who had a

[37] *Ibid.* 50   [38] OC 3, 838   [39] OC 3, 304

modest hand with a crayon and wished to draw your portrait, but only partially succeeded. When that man departed, he crossed over the sea to England from whence, due to the intervention of I know not how many ill winds, he was unable to send you his news. At that time he thought to return to Paris, to beg your forgiveness for his absence, but his sorry fate that may, however, change for the better any day has, to his regret, prevented him from doing so. It is his brother [Lodewijk] who shall bear this letter to you, but he will not make excuses, unless you suspect him, unjustly, of being as incapable of keeping his word and his promise, since he is of the same blood. For the rest he shall be not a little envious of him, since he shall have the pleasure of seeing you, for he cherishes in his memory all the beauty in your face and the charm of your words. If, after all, you should have a soft spot for him, then he may hope to gain your favour [*Si vous aviez de la bonté de mesme, il espereroit de pouvoir obtenir sa grace*], but knowing something of the matter, he does not expect too much, and does not presume to imagine anything more if you, Dear Lady, do not know that he, whatever may come of it, is your most humble servant, Chr. H. van Zuylichem.

This letter lacks entirely the playfulness of that earlier one, in which he excuses himself to a young girl whose portrait he had drawn (August 1657). It is written in earnest. Suddenly he was in love, and hopelessly so. She would not have him.

Marianne was thought to be an only child. She had devoted her life to her father, taking the place of a mother, who had died young (?). Up until 1673 father and daughter were referred to in one breath. Her outbursts of anger were ill-concealed and the father – whatever part he might have played in this – did not dare to oppose her. 'La Signor Marianne becomes greatly ferocious.' Titan could well imagine such a scene when he wrote to his brother on 31 August 1662. But she triggered something off within him that he was unable to understand,

and with all her anger he was deeply attracted to her: '*rien qui me charme si puisamment . . .*'[40]

Her answer to his suit – we cannot read his letter of 21 July in any other way – must have been very cool, because Christiaan responded to Lodewijk with the remark that he would let it be. But he also wrote, 'I am too much in love to remain so far away. What would become of me if, instead of politeness, she addressed to me a few friendly words? I am well acquainted with my weakness.'

Even if he did not stand a chance himself, he could not bear her to be courted by others. Lodewijk saw her frequently in 1662, and kept him informed. Hardly relieved, more in annoyance, he wrote, on 25 May, 'Unbelievable that she refuses such a favourable proposal. What kind of a fortune is she waiting for?'[41] Horrified, on 1 December he writes, 'Confound it, what is she thinking of, is it pious despair or madness that leads her to take such a step?'

When Christiaan was once more in Paris, he would make one last attempt to change her mind. On 28 December 1663, he reached the conclusion, 'Her thoughts are directed towards prayer and the cloister, which makes her stubborn and over-conscientious.'[42] The question is, why did he so readily and unaccountably fall in love with a woman who was unattainable, a woman who seemed to be emotionally scarred? From the scarce information on this unrequited love affair, it appears that he did return to Paris. This was on 3 April 1663. We shall briefly pass over his equally important time in England and its aftermath in The Hague, and finish this episode in Paris. Why did he return?

His father wished so. He had been sent to Paris in October 1661 to carry out an assignment for the States-General. The widow of Willem II and mother of Willem III – the 'child of the State' about whom her mother-in-law had more say than she herself – this Henrietta Maria

[40] OC 4, 213   [41] OC 4, 136; 271   [42] OC 4, 479

had finally got her own back. She was also sister to Charles II, who had become king of England at the end of Cromwell's Puritan republic in May 1660. She considered herself now in a strong enough position to get back her son, and she made Louis XIV co-regent. But this did not get her much further. The French king saw in her actions an argument to attack the Orange Principality, and to demolish the stronghold there. And then, at the beginning of 1661, during a visit to Charles II, the Grim Reaper came for her. It was up to father Huygens to try and retrieve the land of Orange for the Oranges.[43]

He was forced to beg. Formally it was an issue of violated rights, and in this he was in a strong position; but in reality, it was all about French power, and in this he was weak. His only argument was that it would be to the French advantage to be on good terms with the wealthy Republic. His first audience with Louis XIV was on 4 November 1661, but it would take until 21 April 1665 for him to be successful. Only then he could announce to the parliament of Orange that the princely family was restored in its rights, that the army of occupation would leave, but that, in order to placate the king, the governor would be Catholic.

When father Huygens left The Hague, family life there fell apart. Suzanna had left earlier to live with Philips Doublet, and now, not only was the father leaving, but Lodewijk as well, because it was his duty to keep him company. The two of them found lodgings in the vicinity of Saint-Germain, in the home of an old friend of the father's, Sébastian Chièze, who was actually part of the household of . . . Marianne Petit. When Lodewijk had had enough of his role as go-between, the learned son was forced to acknowledge his defeat. Constantijn was let off the hook; he had to take his father's place in the auditor's office of the Oranges and manage the family property. This resulted in a veritable avalanche of family correspondence.

[43] Smit 263–269; Strengholt 101–102

Let us describe both the serious and light-hearted themes that
Titan broached in his letters to Lodewijk. The first: 'I am deeply sad-
dened at the death of Mr. Pascal [on 19 August], who was peerless,
although as far as mathematics is concerned, he had died much earlier. I had always cherished the hope that he would one day overcome his infirmity and recommence the studies in which he so
excelled.'[44]

The other theme figures in a number of letters. On 20 July:[45]

> There is a great deal of fruit at Hofwijck, apples, cherries and
> melons, the first of which we have just eaten. Yesterday I went
> with Signor Chieze [visiting from Paris] and Tet [Constantia le
> Leu de Wilhem] to Aunt Geertruyd at her country house, where
> there is also an amazing abundance of cherries, and even more at
> the house of Mr. Van Leeuwen. There we read poetry by your
> poet [Gaston Babtiste], a hundred for 15 pennies, with their bawdy
> tales.

Lodewijk wrote back on how his father spent his spare time,
for outside of the season there was little opportunity for negotiation.
The old Huygens had found his way back to Ninon de Lenclos, and
tried to fool Lodewijk into believing they played the lute together –
something he took with a grain of salt. On 27 July he received this
response:[46]

> I know quite well who Mademoiselle de Lenclos is. Monsieur
> Boisrobert has promised me three times over to take me to her,
> but each time there is some matter that intervenes. He showed
> me a picture in which she is quite naked. Then she was reasonably
> attractive, but not any more. It is said that she now associates with
> respectable women, now that, due to her age, she has given up her
> former profession. It is certain that my father visits her only out of
> love of music, and nothing will induce me to think otherwise.

[44] OC 4, 213   [45] OC 4, 179   [46] OC 4, 183

In fact Ninon continued her life of well-paid *fille de joie*, which she began as a young girl in 1636, right up until 1690. Father Huygens' visits to her were hardly innocuous. Witness the following verse, in which he sings her praises:[47]

Elle a cinq instruments dont je suis amoureux,
Les deux premiers ses mains, les deux autres ses yeux;
Pour le dernier de tous, et cinquieme qui reste,
Il faut etre galant et leste.

Five instruments of love are hers,
The first two are her hands, the other two her eyes,
But for the fifth she keeps till last,
One need be gallant and foolhardy.

Is this how he (sixty-five) got the costly doxy (forty-two) – to continue in the same vein – down on her back? The quatrain became notorious. Bernard de Fontenelle published it in his 'Recollections of Miss de Lenclos' and attributed it, correctly, to the Huygens who 'was ambassador of the States-General in France between 1661 and 1665'. But in the following century, Voltaire confused the ambassador with the scientist. In his youth he had visited the elderly Ninon and seen the verses. He republished them to prove that eminence in the sciences and eminence in the art of poetry do not go together. By now they were known by almost everyone. One century later, in 1855, François Arago accused him of bad taste, in his short Huygens biography. He should never have parted with his quatrain![48]

Had Titan known in March 1663 that he was to be companion to his father for as long as fourteen months, he would have refused to take over from Lodewijk. But having arrived in Paris, and hearing that his father was unable to make any headway with his negotiations, he could not in any decency turn back. Promptly he fell ill: 'I have undergone bloodletting so as to be faster rid of an influenza . . .'[49]

[47] Worp 9, 8 (+n)   [48] Arago 321   [49] OC 4, 68

When is a diplomat not making headway? Only when he achieves less in his consultation than he did in the one before – not when he has to wait for a long time for the next one. Father Huygens certainly did have to wait a very long time before the king, or Jean-Baptiste Colbert, or some lesser minister, would consent to speak with him, but the Orange cause was a relatively paltry matter for Paris. Yet the ease with which he could get something done, and then the same ease with which it was overruled, gave him food for thought.[50]

Meanwhile, he killed time. Now he had the company of his exceptional son, but this did not mean he accorded him any exceptional rights beyond delivering messages or accompanying him on social visits or to the theatre. He did allow him to attend the academy of Montmor each Tuesday, and for the rest, he pacified him by letting him off to visit London, outside the season, from 7 June until 1 October.

Perhaps the father was trying to spare his son's feelings towards Marianne Petit when he moved their lodgings to Little Moses House on the Rue du Petit-Bourbon, also in Saint-Germain. Titan did not even keep a diary; this was not the life he chose. We pick an example, at random, from one of his letters to Lodewijk: 'Yesterday [21 February 1664] we watched the ballet at the Royal Palace, beautiful and quite splendid, especially the entry of the ladies, 22 altogether and most of them superb. It was so warm that afterwards I had to change my shirt.'[51] Arousing spectacle and sweat: the physical sublimation of intellectual exertion. That winter he fell ill once again, for two weeks.

On 23 November 1663 he reported, 'Cousin writes to Signor Padre that there is a new fashion, that young persons have all their hair cut off and well-nigh cover their entire person with a wig, and then on top of that a small hat. '*Je voy par la que nous sommes tout de mesme*' ('From this I observe that we are all the same').[52] So was he.

[50] Strengholt 102–103   [51] OC 5, 30   [52] OC 4, 435

J'AIME LES CORPS NUS (I enjoy naked bodies)
ON DEVIENT LAS D'AMOUR (Love brings lassitude)

These are rebuses that Constantijn sent his brother to solve that autumn.[53] If brother Zeelhem, who at the time enjoyed the favours of three women all at once, hoped to arouse his brother with suggestion, he was in for disappointment. Christiaan sent back a rebus on stargazing.

No wonder he called this a futile (oiseuse) year. It was lost time, as far as science was concerned. After a long time spent in uncertainty about his father's plans, he could write to Lodewijk on 23 May 1664: 'Whatever transpires, I am making my preparations to leave next week for Holland, for I have permission to do so.'[54] At last he was free to go . . .

> Expecting nought, no hope to end in sorrow;
> Greater the joy when thanks be not expected:
> Only arrows unseen can wound you.

. . . with the verses dedicated to him by a patient father.[55] As a poet, the man had not yet exhausted his resources, but as a diplomat he had. At the end of June he set forth to London; in order to retrieve the cause of Orange, concessions had to be made.

It is very likely that father Huygens attempted to exert his influence on the position destined, in Paris, for his son. In the next chapter we shall argue that this influence worked against his interests, rather than to his advantage. Chapelain and Colbert were looking seriously at ways to build up the academy of Montmor substantially to become a royal academy for the sciences, a process in which the son could possibly play a key role.[56]

Immediately upon his arrival on 3 April 1663, he was approached about the plans for a new academy, though indirectly: Montmor and Sorbière asked him to think about regulations to prevent the rows

[53] Keesing (1983) 70   [54] OC 5, 68   [55] Worp 7, 61   [56] Roger 42

that had occurred in the past about religion and politics, and which had defined scientific aims. He had barely returned from his trip to London, when Colbert handed over to him 400 pounds of the previously nominated gratuity of 1200 pounds, with the message that he could, in future, count on a similar gratuity each year. It would take years for the plans to be carried out.

And now that London has been mentioned once again, it is time to report on his visits to that city:[57]

2 April 1661 (*Reys-Verhael*): Too much wind [in Gravesend] to take a flatboat, we were advised to hire a larger ship with a crew of six, costing 20 shillings. Vice Admiral Lawson accompanied us, showed us a yacht under construction at Woolwich for the Duke of York, and a small model of it at the house of the master carpenter. We saw a great number of boats upon the river. Arrived in London at 4 o'clock. Took a carriage, I was to look up Prince Maurits [who was a neighbour in The Hague], and found lodgings in the area.

11 April: Went with Bruce (who wished to try out sea clocks) to Gresham College, where the scientist Godart showed us the experiments with liquids. He mixed two transparent ones, which gave a black ink, and by pouring in a third, the liquid became transparent again, without any residue. Two others gave a milky substance . . . , others foamed violently and grew warm, while still others turned to all kinds of colours. It concerned mainly the essence of tartar and of vitriol. Dined in a tavern with Robert Moray [president of Gresham College, which Charles II raised to the status of Royal Society the following year), Paul Neal, Lord Brouncker, Boreel, Vermuyden, Godart and . . . They refused to let me pay. After dining, we observed in the laboratory the experiment with lead, which was weighed with the beaker, laid in the oven, pulled out of the fire and then turned out to weigh around 1/100 more. Boyle joined us.

[57] OC **22**, 567

12 April: Boyle called on me and we discussed at length [his experiments with the vacuum]. In the afternoon, Oldenburg [secretary of the college], took his leave of Bruce.

3 May: was the crowning of the King in the chancel of the church of Westminster. First the Bishop of . . . held a lengthy sermon, while the King sat before the altar. After this the Bishop of Canterbury anointed His Majesty, who was disrobed down to his red satin shirt . . . , then the King placed himself upon the throne opposite the altar where all the nobles came one by one to kiss his hand, accompanied by music consisting of voices and instruments. This all by way of hearsay, for I was with Reeves in order to observe the passage of Mercury over the sun, which I saw. But it must be over thirty years since Gassend could also have seen it.

Back at home he made his own comparison (9 June to Lodewijk): 'I had little pleasure of my visit to London . . . The stink of the smoke is unbearable and most unhealthy, the city poorly built, with narrow streets having no proper paving and nothing but hovels . . . , little going on and nothing compared with what you see in Paris. The people are gloomy, the well-to-do are polite but unforthcoming, the women have little to say and are not anywhere near as they are in France.'[58] This first visit, the result of his desire to meet Wallis and Wren, kept him once again from finishing his *Horologium oscillatorium*.

Alexander Bruce came up with the idea of compensating for the rocking of a ship by placing a steel ball inside a copper tube, with the clock inside the ball. Its pendulum might swing irregularly, but it would keep going.[59]

The idea lay fallow for some time, but in 1662 Christiaan had two such clocks made by Severyn Oosterwijck in The Hague. (He called his new clockmaker by his Christian name only, as if he were a servant.) In November he hung them from the beam of his room, and

[58] OC 3, 275   [59] OC 22, 587–592

however much he shook and shoved them around, they kept going. He managed to send them off to Bruce in London before leaving for Paris.

Already in 1663 Captain Robert Holmes took them with him to Lisbon, then to Guinea, and after that across the Atlantic Ocean. All went well. Bruce was on that voyage, and found that their longitudes corresponded well with the known readings. It was encouraging.[60]

On the second journey, in 1664, a disaster was averted. After leaving Saint-Thomas (São Tomé) for the African Coast, Holmes had sailed 2400 sea-miles along the equator in a westerly direction, before heading back north east. When the water supplies threatened to dry up he was advised to sail west again, towards Barbados, because according to the log it was still another 300 sea-miles to the Cape Verde Islands, and a good three days' sailing. According to the clocks it was 90 miles. But Holmes went ahead and hit land the following afternoon.[61]

Robert Boyle gave Christiaan another idea.[62] Shortly before 1661, this son of the Duke of Cork and his assistant Robert Hooke had built a pair of air pumps, after hearing about the work of Otto von Guericke and Caspar Schott. When Christiaan saw the pumps, he immediately became enthusiastic.[63]

No sooner had he arrived back in The Hague than he had a machine for the vacuum made, in accordance with his own design. The pump consisted of a wooden cylinder, 14 inches high and 2.5 inches across, in which a piston under a layer of water could be pushed upwards. With a sealable tube, he attached the part beneath the piston, where the air had been thinned, to the cavity in a glass apothecary's bottle that stood upside down on a flat slab. He filled in the cracks with a mixture of wax and turpentine.

The machine was ready in November 1661.[64] Already in his first experiment 'a [fastened] bladder remained swollen' for a whole night,

---

[60] Mahoney 252–253   [61] OC 17, 230–233   [62] OC 17, 258–260; Sparnaay 24–31
[63] OC 3, 359   [64] OC 17, 305–330

which means that he had effectively stopped the leaks. He carried out the same experiments as Boyle and Hooke, but when he proceeded a bit further, he quickly saw something new.[65]

He poured water into a long glass tube that had been well cleaned beforehand, turned it upside down in a vessel of water, and pumped the air away above the vessel. For this purpose, a tall bottle had had to be made. The water in the tube should have sunk to the water level in the vessel, because neither in the tube nor the bottle could air exert pressure, but he saw that the water in the tube 'refused'. This was the result of forces acting between molecules, but he could not know this. He was thoroughly puzzled.

In the New Year, he looked at other effects. He wrote to Lodewijk:[66]

> It is due to a leak in my machine that I have not carried out experiments with lettuce and suchlike. It has not been in use for some time, and because it is quite a feat to repair I have not yet bothered. But I should like to have it fit for use as early as possible, were it only to provide you with food for conversation with those highbrow gentlemen. I am unable to recall the experiment with the shrimp, or else it is I who has the memory of a shrimp . . .

Constantijn left the room during the experiments with mice and small birds.[67] A year later Christiaan was in London once again. We quote from his diary:[68]

> 22 June 1663: My experiment with purified water in a vacuum succeeded two or three times: in a seven feet high pipe, the water remained at the same level, without sinking. Mylord Brouncker, Mr Boyle, and many others were present. Their machine stands the other way round to mine, and is completely immersed. The tap has a long piece of iron, sticking out of the water, to turn it

[65] OC 17, 321–325; Snelders 107–108   [66] OC 4, 144   [67] Keesing (1983) 61
[68] OC 22, 599

open and shut, and the air hole is in the suction tube, in which is a
small stick that goes up and down with the sucker. Met there Mr
Sorbière and the Abbot de Beaufort. Received afterwards a gilded
metal clog of 600 gl. from the King, which was carried in by the
President on a purple damask cushion. Also the Society
coat-of-arms. They had me sign the Society book.

The city offered him other delights. On 10 September:[69]

> Went to M Midd. [Miss Middleton, a London beauty], began on her
> profile, her brother had been acquainted with Crommon in
> Antwerp. In the afternoon M. Evelyn came, and then P. [papa] with
> Boreel. I stayed until evening. Wrote to her on Thursday 13
> September. Continued to draw on Friday 14th, ate at Bret.
> Saturday 15th drew some more, dined there, bade farewell. Wil
> Mr. H. kome again and you wil be very welcom, you doe
> understand that doe not you.

[69] OC 22, 600

10 Weight

He had moved into rooms. A servant stoked the fire under a mantelpiece of wood, which has been made to look like marble. In this warm room, with its colourful wall hangings, he was able to carry out his studies; in a yellow-painted recess, he could perform his experiments. But now he was standing in the bedroom, with a red wool bedspread and red-and-green brocade on the walls.[1] He got dressed in his best clothes. Mr Huygens wouldn't come again. He looked out of the window at the Palais Mazarin that largely blotted out the ruddy evening sky, and to the left he saw the gardens of the Palais-Royal. You do understand that, Paris, don't you?

Now, three centuries later, if anyone were to look for these rooms, they would be unable to find them. The area, though, is easily recognisable. The Palais-Royal still stands, and the Palais Mazarin has been replaced by the National Library which fills all the space between Rue Vivienne, Rue des Petits-Champs, Rue de Richelieu and Rue Colbert. To the east of Rue Vivienne, there is now a rather untidy conglomeration of houses built around a number of inner squares and narrow streets, but previously there stood the distinguished mansions belonging to Colbert. On the corner of the Rue des Petits-Champs, which runs parallel to the Palais-Royal, stood the residence of this head steward of the royal household, and beside it, the slightly smaller building that housed the state library, containing 62 000 books and 12 000 manuscripts.

Christiaan lived above the library, right next door to the family of Pierre de Carcavy, who had become a librarian. Whenever he joined them there for dinner, which was often, he'd spend the rest

---

[1] OC **10**, 727

of the evening there playing backgammon.[2] Now he knocked on his neighbour's door, to fetch him. They made their way down the stairs, to the room where the Académie Royale des Sciences was to hold its first meeting. Carcavy, the host, would chair the meeting, while he, the undisputed doyen of European science, would put forward the proposals. It was 22 December 1666, nearly six years to the day since his audience with the French king.[3]

There is no report of the sitting, as neither he nor Carcavy had thought of appointing a secretary; only later would Jean du Hamel be nominated to this post. Nor is there any royal decree for its institution, or notary's act on property or housing; a word from Colbert would have been sufficient. He greeted the guests seated around the table: Adrien Auzout, Bernard Frénicle, Edmé Mariotte, Claude Perrault, Jean Picard and Gilles Personne de Roberval, the only person in Paris who, figuratively speaking, was similar in stature to Titan. Most likely there were more. Although we have no report, we can be reasonably sure that immediately after the chairman's opening declaration, he addressed those present: 'Gentlemen, I have made a list of subjects that we shall discuss. I am aware of your interest in celestial phenomena. I therefore propose that we begin with an accurate measurement of the polar shift . . .'[4] He must have said something along these lines, because a week later he would work out how accurately the height of the pole star could be measured with the aid of a sextant, and within two weeks, the academy decided to have a more accurate instrument made for this purpose. The gentlemen, to whom the cause of this shift was still a mystery, nodded their assent, their curly wigs glistening in the candlelight.

Prior to Christiaan's appointment as scientific director – this best describes his position – quite a lot of political wheeling and dealing had taken place.[5] His supporters had been aware that he was not averse to

[2] Vollgraff 628   [3] *Ibid.* 625–627   [4] OC 21, 32–33
[5] Brugmans 54–66; Roger 41–48

the idea, probably in 1660, when he visited Chapelain, and certainly in 1663, when he discussed regulations with Montmor. But whatever he may have felt about it, his own wishes counted for little. What really counted were the wishes of Jean-Baptiste Colbert.[6] But this powerful man needed time to win the trust of the headstrong king. And even when he did, it required all his resources to persuade Louis XIV, to whom French honour was paramount, to consider a foreigner as director for 'his' academy.

Initially Colbert had had little say in such matters. When the all-powerful Cardinal Minister Jules Mazarin died in March 1661, the king finally felt free to pursue his own wishes. Up until then he had been obliged to do as he was told, and now, having reached the age of twenty-two, he wished to rule alone. Mazarin had proposed that his best administrator, Colbert, should succeed him as principal advisor. The king waived aside his proposal. Colbert might serve him, but then only as a menial who would attend to money matters. In 1661 he became head steward of the royal household, which in practice meant France.

Louis XIV was a lavish spender. For his ostentatious palace in Versailles alone he needed 300 million francs (15 million pounds). Court expenses were equally high, because costly entertainment was obligatory; and of course, the expense of war, to which he was honour-bound. It was up to Colbert to discover the means of scraping together the necessary money from this less-than-affluent country.

Which he did. As a merchant's son from Reims, he made an excellent bookkeeper. He had no trouble at all sitting behind a desk for fifteen hours a day. From there he managed the entire state finances. If he considered the interest on a loan to be too high, he simply lowered the value of the loan. It is most likely that, thanks to this kind of devaluation, he managed to reduce the state debt to less than half its size. Tax collectors were informed that all their returns were carefully checked, and that all of these were insufficient. He threatened

[6] Boissonnade 1–36

them with legal action if no improvements were forthcoming. Also, he introduced new taxes.

When the farmers of the Bourbon province joined forces and the army had to be called in to put them down, he ordered that four hundred of the strongest rebels be sent to the galleys, and the rest to the wheel, the gallows, or the stake. Anyone who protested against his policies, he had thrown into prison. If the king had ever hesitated to put down opposition so harshly, Colbert saw it through. In this manner, provincial parliaments that enjoyed the right to turn down taxes ended up approving every one of them that he chose to impose. After ten years the province of Burgundy, which, when he came into power, had protested against producing 1 million pounds in tax per year, paid more than 3 million, without so much as a squeak.

But the merchant's son recognised that there was a limit to how far he could exploit the peasants. The wealth of the Republic fascinated him, and he thought he understood where it came from. That was why he strove for a trade in products based on industry, yet securely established in the hands of the state; nothing was to be outside of his control. He set up companies in foreign trade, had canals dug and roads and bridges built. In addition, he recruited skilled workers from all over Europe to develop the mining, manufacturing and weaving industries. In other words, he managed to convince the king of the benefits of investment. But his supervision of the entire process was oppressive, and its development was frustrated by his desire to have a finger in every pie.

According to this policy it was also expedient to keep scientific development under state control, and to attract the most outstanding scholars from abroad.[7] He insisted that the Royal Academy of Sciences occupy itself with technology. It went without saying that this academy, as well as that of the expressive arts (set up anew in 1663), and architecture (founded in 1671) all served to extend the glory of the king. The members were expected to dedicate all their work to him.

[7] *Ibid.* 28–29

If subservience was paramount in the director of the Academy –
and Colbert's policy cannot be interpreted otherwise – then the presence of father Huygens in Paris must have stood in the way of his
son's nomination. Christiaan's erudition and practical genius might
have worked in his favour, but his father's tough negotiating must
surely have worked against him, for it was plain to all that he served
as his father's errand boy.

Moreover, if the ambassador was using him as an enticement to
regain land taken from the House of Orange – which is highly conceivable – then Colbert would have understood that Louis XIV would
never accept any such connection between these two issues. It would
be perceived by the king as relinquishing land in exchange for the
nomination of a foreigner, an unacceptable slur upon the honour of
France. However it may be, negotiations for his appointment only got
under way once the matter of Orange was settled and the father had
left Paris in the summer of 1665.[8]

In June of that year, Colbert had ordered Pierre de Carcavy to
approach Christiaan on the matter of the new Academy.[9] Thévenot
had written to him a good six months earlier, in November 1664,
that such an invitation was, in fact, a foregone conclusion.[10] The
gentlemen–members endured the same suspense as the man who was
tipped to be their director. His excitement was not in any way diminished by the long wait, 'This is better and far more satisfying, that
I be seated upon a horse and paid by a king, than that I remain idle
in this country for the rest of my days.' ('*Rectius hoc et splendidius
multo est, equus ut me portet, alat Rex, que de viellir icy dans le pais
sans rien faire*').[11] This is how he wrote of it to Constantijn. Which
horse he would have, and how he would be paid, was still unclear that
summer.

It turned out to be self-contained lodgings above the Royal
Library and, as far as payment was concerned, a salary of 6000 pounds

[8] Smit 267–269   [9] OC 5, 375   [10] Hahn 59–62; OC 5, 152
[11] Vollgraff 621; OC 5, 375

a year.[12] He had requested this sum, in consultation with his father, in September. For a director of the Royal Academy, 6000 pounds was not unreasonable: the ordinary members of the Academy got 1500 pounds, ordinary professors at the Royal College 600, and top officials such as state secretaries 20 000.[13] He tended to see his 6000 not as a salary, but as a liberty that was allotted to him by Colbert. He had every reason to see it that way, since the minister treated him as a protégé on the many occasions he was asked to come down and give advice on various practical matters. Yet he found it a scandal that Cassini, who was appointed later as a member of the Academy, got 9000 pounds.

It was not until 18 February 1666 that Carcavy was able to write to Christiaan that Colbert had agreed to his proposal of September (which has not been preserved).[14] There had been a confusing flurry of letter-writing between Montmor, Chapelain and Carcavy, who all contested for the honour of having been the one to have made him the invitation. And naturally, his father had been unable to resist having his own say in the matter. H. L. Brugmans finds it hard to understand why Colbert hesitated for so long,[15] but if the king did not wish there to be any perceived connection between his appointment and the cause of Orange, as we suspect, then this space of a year was necessary. The mistrustful negotiator for Orange requested from Carcavy, on 25 February, that he be granted a word from the king. It arrived in the form of a letter from Colbert to the father, a letter that has been lost; but the letter in answer to it has been preserved:[16]

> Sir. I have received your testimony, with all the regard that is befitting to one that comes from your hand, and from the King, [and I have read] all that His Majesty intends for the person of my son. At my advanced age [sixty-nine], Sir, I may rightly assume that my children will surely witness my death. But the

---

[12] OC 7, 88 (n); OC 5, 389   [13] Roger 46   [14] OC 6, 18 (n5)
[15] Brugmans 65   [16] OC 6, 22

conversation of this child especially is both highly pleasing and precious to me, for he possesses great talent in those sciences that interest me, and does not boast of it. [I therefore do not wish to lose him.] However, if I weigh the honour of such benevolence bestowed upon my child by such a great King, then I perceive that the scales do not weigh in my favour. Therefore I will humbly submit to the orders of His Majesty, who honours me with such indulgence that I could not be ungrateful, and I will consider that he extends that same graciousness towards the son as the father deems himself fortunate enough to have received. Here especially, Sir, I feel fully reassured, for you have deigned to write to me that you shall take him under your protection. I take the liberty to say that he shall amply repay this generosity with his services, and that I, for my part, will do all in my power [to please you].

The talented child did not leave for Paris until 21 April 1666, and then it was for good.[17]

So as not to while away his time at Het Plein, doing nothing – this was what he had said – as if, up until that moment, his life had been empty, and only now, in service of the great king, had it taken on any substance. It is hard to say whether this is how he really felt, but in fact, during those years of 1664 and 1665, while he was waiting in suspense for his nomination, his work contains little that is new. It was extensive though, and mainly practical.

The work consisted of making improvements to his glass-polishing techniques, and he also devised useful lens combinations, which brought him into contact with Baruch Spinoza. This philosopher lived in Voorburg at the time, close to Hofwijck, where he wrote his *Tractatus theologico-politicus* (Religious–Political Treatise), meanwhile earning his keep by polishing lenses.[18] Three long letters from Spinoza from the beginning of 1666, which appeared to be written to him, were in fact to Johannes Hudde.[19] The reply that

---

[17] Vollgraff 625   [18] Israel (2001) 218–229, 275–285   [19] OC 6, 2, 24, 36

Christiaan is supposed to have written to him on 20 February has never been found. Five letters, written to Constantijn from 1667 and 1668, show clearly that his brother found Spinoza, as polisher, too much of a competitor to be told everything. He seldom referred to him by name, usually just calling him 'the Jew'.[20]

In a small town like Voorburg they must have seen each other often, but Christiaan kept Baruch at a distance. He was polite enough to lend his books to the strange Jew, and proud enough to show him his latest calculations of spherical aberration but, as scion of a rich aristocratic family, he felt himself elevated above this poor craftsman, this outcast of the Sephardic community of Amsterdam.[21] It may not even be accidental that the letter of 20 February 1666 has never been found. Christiaan probably never deigned to write to him. On his part, the secretive philosopher hid his doubts about Christiaan's calculations, doubts that he expressed freely to the secretary of the Royal Society.[22]

Spinoza was certainly not to be let in on the secrets of the eyepiece with which colourful but interfering rings in the image of the lens could be diminished. This achromatic 'Huygens eyepiece' consisted of two convex lenses, with focuses at distances $f(1)$ and $f(2)$, one behind the other in a mounting, at a distance d. He discovered that the image became practically colourless when the distances came to:[23]

$$f(1) + f(2) = 2d$$

and he proposed that $f(1)$ be made twice or three times as large as $f(2)$. In spite of this success, Christiaan still did not understand what caused the colourful rings. Robert Hooke had published his *Micrographia* (Descriptions of Small Matters) in 1665, a book that described how such rings could be conjured up by pressing convex lenses together. In November of that year, he copied Hooke's experiments, which he had so admired.[24] He tried to discover the relationship between the

[20] OC 6, 155, 158, 164   [21] Keesing (1984/85) 113   [22] Israel (2001) 246–252
[23] Acloque 184; OC 13, 254   [24] Vollgraff 617–618

radius of the rings and the distance between the lens surfaces. First he used two lenses, but later on he saw the advantage of having one single lens with very little curvature on a flat glass plate. We shall return later to his thoughts on the nature of light, and on colour as an aspect of light.

The dramatic success of his sea clocks close to the coast of the Cape Verde Islands motivated him to make certain of the rights, after almost forfeiting them on his house and tower clocks. In the summer of 1664, he had already devised an improvement in the accuracy of his clock, which was desirable, especially to determine position at sea.[25] To avoid losing the drive provided by the large weight, which had to be lifted now and then, he hung an extra small weight in the chain work, a *remontoir* that was raised every 30 seconds by the large weight. This small weight ensured a more even pressure on the wheels.

In the autumn, having made this improvement, he put in a request for a patent. At the same time he arranged for the publication of Holmes's report.[26] The French version, which was important for his position in Paris, came out on 23 February 1665 in the *Journal des Scavans*. He soon obtained his patents: the Dutch one on 5 December 1664, the French one on 5 February 1665, and the English one on 3 March of the same year. Licences were granted to Severyn Oosterwijck in The Hague, Isaac Thuret in Paris and Abraham Hill in London. Oosterwijck completed his first sea clock in August, and his second in November. They cost 300 guilders apiece.[27]

Then came the question of who was going to buy them. There was little enthusiasm. In January, in order to promote their benefits, he wrote the brochure *Kort onderwijs aangaende het gebruyck der horologien tot het vinden der lenghten van Oost en West* (Short Tract on the Use of Clocks for Finding the Longitudes of East and West).[28] He translated it into French, while the Royal Society in London would

---

[25] *Ibid.* 610; OC 17, 183   [26] OC 5, 204–206
[27] OC 22, 87; OC 17, 157–177   [28] OC 17, 199–235

take care of the English translation. Johan de Witt used his influence to convince the Dutch fleet to buy Oosterwijck's clocks, but on the condition that Christiaan went along on a trial voyage.[29] That was in 1667, when he had commitments in Paris, so could not meet the condition.

Colbert may have set up French companies, but they did not yet see any great benefits in the clocks for French navigation. He therefore delayed his translation of *Kort onderwijs*, so as to reinforce it with information on sea voyages that would interest the French. Such a voyage was that of the Duke of Beaufort to Crete, where his squadron was to fire at the Turks. It took until March 1668 before the Duke left Toulon, and Colbert only heard of the results in November 1669.[30] According to the reports made at sea, the clocks made by Thuret and taken on board on the insistence of the Academy represented, fairly accurately, the longitudes on the Mediterranean Sea. But this wasn't really saying much, as they were sailing on calm and familiar waters. That is why he pinned his hopes on the results of Jean Richer's forthcoming voyage to Cayenne.

In London results were, if anything, even less encouraging. On 27 February he sent *Kort onderwijs* to Robert Moray for translation.[31] Unfortunately, one week later, the day after he had received his patent, a (second) sea war broke out between England and the Republic.[32] This time, fighting took place at sea close by the town of Lowestoft, and didn't cease until 1667, after Michiel de Ruyter and his men set fire to the English fleet that lay anchored at Chatham. The conflict didn't do his cause any good. The translation of *Kort onderwijs* was not printed in the *Philosophical Transactions* until 1669, and Abraham Hill received no orders at all.[33] The English had become sceptical. Holmes's experiences had indicated that the metal of his clocks might be susceptible to air (or as we would say today: to temperature). By 1670, they insisted on waiting for results from Richer on Thuret's clocks.

[29] OC 6, 129, 167; OC 17, 177 (n6), 197 (n)   [30] OC 6, 501
[31] OC 5, 246   [32] Israel (1995) 768–774   [33] Mahoney 268

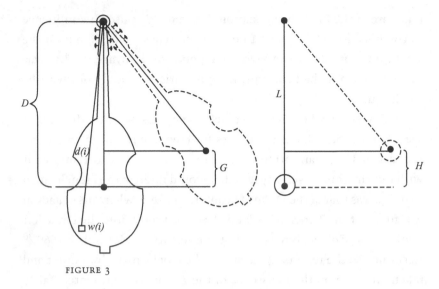

FIGURE 3

The situation was a discouraging one. Shortcomings in his theory also became apparent. He knew only too well that his calculations in the great *Horologium oscillatorium* described not a true pendulum, but an ideal one. So in June 1664, he set himself the task of making not only improvements in the drive mechanism, but also calculations for a true pendulum. This resulted in his theory for the centre of oscillation of the pendulum.[34] Marin Mersenne had asked him for this before, but at a moment when he was not yet ready for it. Now that practice demanded it, he was able to write it up without any difficulty, also because, in the meantime, he had been busy working out methods to do so.

He drew an arbitrarily shaped body that could swing round an axis at a distance D above the centre of gravity (Figure 3). He divided the body into small numbered parts with a weight $w(i)$ and a distance $d(i)$ to the axis, where i represents the number ($i = 1, 2, 3 \ldots$). Together, the parts added up to the total weight W, or

$$\Sigma w(i) = W$$

---

[34] Gabbey (1980) 195 (n39); OC **16**, 374–376, 470 (n6)

where Σ is the symbol for the sum over all numbers i. He compared this true, physical pendulum with the ideal, mathematical pendulum that had a length L, fell from a height H, reached a speed S at the bottom and then rose again to height H. So he had to compare height G, which the centre of gravity of the physical pendulum crossed as it fell and rose, with H. The ratio of G and H should then be equal to the ratio of the distance D and the length L, or

$$G : H = D : L$$

This was his first result.

Now he looked at the movement of the true pendulum, according to his argumentation of 1652 and 1659. He proposed that the parts would be detached when the pendulum moved through the lowest point. Their speeds $s(i)$ must then be related to S in the same way as their distances $d(i)$ are related to the length L, or

$$s(i) : S = d(i) : L$$

The parts would rise to heights $h(i)$, which Galileo had discovered to be proportional to the square of their speeds $s(i)$. If $h(i) :: s(i)s(i)$ were applied to the ideal pendulum, then it follows that

$$h(i) : H = s(i)s(i) : SS$$
$$h(i) : H = d(i)d(i) : LL$$
$$h(i) = d(i)d(i)H : LL$$

The second rule follows from the first by filling in the result that has just been found for $s(i) : S$. When the pieces reached their highest point, the height G to which their centre of gravity rose comes to

$$WG = \Sigma w(i)h(i)$$
$$G = \Sigma w(i)h(i) : W$$
$$G = \Sigma w(i)d(i)d(i)H : (LLW)$$
$$G = IH : (LLW)$$

This was his second result. The third line follows from the second by filling in the result that had just been found for $h(i)$, and in the fourth line I stands for an abbreviation of $\Sigma w(i)d(i)d(i)$. He did not give I a name, but I is now known as the moment of inertia (when weight equals mass).

The equation contains of two characteristics of the ideal pendulum, namely the height H and the length L. Because L by definition determines the centre of oscillation of the pendulum and H has an arbitrary value, he eliminated H. For this he used his first result, which can be written as $G = (D : L)H$, and found

$$IH : (LLW) = G$$
$$IH : (LLW) = (D : L)H$$
$$I : (LLW) = D : L$$
$$I : (LW) = D$$
$$I = DLW$$

In plain language: he could find out DLW by calculating I, which is to say that for all the parts, he had to add up the products (of their weights with the squares of their distance to the axis). If he also knew the weight (W) and the distance of the centre of gravity to the axis (D), then he knew the distance of the centre of oscillation of the pendulum (L).

Naturally, after he had discovered this general relationship, he calculated the moment of inertia (I) of special body shapes with regard to special axes. On 10 October 1664, while fully occupied with these calculations, he wrote to Moray that he had results for triangles, half circles, and cylindrical wedges (ungulae).[35] As the number of special cases is unlimited, he thought up ways of acquiring some kind of overview. By means of geometrical transformations he could see I as a square, as a cube, or as being dependent on the centre of gravity of curved surfaces of bodies. In addition, he discovered that the centre of

[35] OC 5, 120

oscillation of the pendulum and the axis are interchangeable, without any change in the time-span of the pendulum swing, if the product $D(L - D)$ is constant.

In Paris he would never again be able to do work of such depth and precision. In that city, where he arrived in May 1666, peace and quiet would be hard to find, and the continuity of his scientific work would be disrupted. Had he been aware of this beforehand? But even if he had requested the freedom to work as he had done at home, he was uncertain of his obligations. He was already well acquainted with that city, with her civilised chaos of music and song, her warm playhouses, sensual and exotic ballets, and galleries filled with wall hangings, red carpets, silver vases and risqué paintings. And when he wished to get down to work, although the lime trees in the Tuileries were already in bud – the facade with its chalk-stoned pillars was reflected in its full glory in the river – and beautiful women made eyes at him as they wandered amongst the trimmed flowerbeds and glistening fountains, he must have thought restlessly of those single ladies from The Hague who resided here and clamoured to receive him. But he could have known that his protector, the head steward, would expect him to attend all the parties at court, where he would be asked for advice on patents, and on the waterworks of Versailles, or on the opulent observatory that was under construction on the ramparts of Saint-Jacques. And besides all this, as 'supreme head' at the Academy – it was Ismael Boulliau who had spoken of it as: '*omnium caput*'[36] – he must constantly prove himself. He must engage with other great brains, especially the sly and snappish Gilles de Roberval, but also with Heinrich Oldenburg, secretary of the rival society in London, who never stopped trying to lure him out into the open. He would have to write him endless letters, but even more to family in The Hague, who wouldn't let him go. And if he did have any time to himself, he had to think of his friends, of his neighbour Pierre de Carcavy,

[36] OC 18, 4

of Adrien Auzout who was too good for this world, and of Claude and Charles Perrault, those erudite brothers whom he would often go and visit at their country home at Viry.

Let Christiaan tell, in his own words, how Paris seduced him from science, how he was finally forced to come out of his shell. The source of his words is abbreviated. P(6)6 indicates: from a letter to P (in month 6, meaning June) and year six of the decade, meaning 1666; C stands for Constantijn, L for Lodewijk, P for Philips Doublet and S for Suzanna; L(5)7 therefore means: from a letter to Lodewijk of May 1667; A cites from, or refers to, minutes of discussions at the Academy.

> P(6)6: I still had a bit of a headache, as is customary the day after, but Mr Colbert had obtained an invitation for me to come to Saint-Germain, before the King was to leave. . . . As the Council departed, he introduced me to His Majesty, who spoke many obliging words to me, though I was unable to answer him with anything of worth, for not only was I agitated about my audience, I also had a fever.[37]

> C(6)6: For three days now we have known that the fleets have engaged in battle and that the cannon shots were to be heard along the whole coast. We haven't had much news yet, except that a badly battered English frigate has arrived in Oostende, and that the rest of the squadron was beaten by the Dutch.[38]

> P(7)6: Make sure to tell our dearest Ida and Constantia that I have finally found Lipcot, and then in a most abominable state, without clothes or shirt. He was bathing in the river, where their humble servant was also to be found and, stark naked as we were, we hardly recognised one another.[39]

He observed the eclipse of the Sun, on 2 July 1666, in Colbert's gardens, together with Auzout, Buot, Carcavy, Frénicle and Roberval, and decided that the Moon had no atmosphere.[40] During that sweltering

[37] OC 6, 40   [38] OC 6, 47   [39] OC 6, 68   [40] OC 6, 58–66

summer, he probably also drew up recommendations on work topics for the Academy. Colbert looked them over and in the margin wrote 'bon' beside each proposal. The text begins as follows:[41]

> In my opinion the most important and most useful occupation of this meeting is to work at physical science according to the plan of Verulam [Francis Bacon]. There is no better topic for research, and nothing is more useful to know, than the origins of weight, warmth, cold, magnetism, light, colours, the compounds of air, water, fire and all established matter, the breathing of animals, the development of metals, stones and plants, all matters that man knows little or nothing of.

He endorsed the fundamental precept of the Academy: 'During all meetings, members shall discuss neither the mysteries of religion, nor matters of state.'

> L(8)6: I have discovered here, at the painter Jacob van Loo's, all the Caron cousins and the two daughters of Gangel, who are also well-endowed. The latter, and the eldest Caron [Suzette, then seventeen years old] have had themselves painted together, in the same portrait.[42]

> P(9)6: The latest letters have brought bad news: The damage at the Vlie [the English had set fire to Ter Schelling], the nullification of the verdict on Cornelis Tromp, the betrayal of the State [Henri Buat, page to the court of The Hague, had been caught in a conspiracy with the English against Johan de Witt], and the threat of the plague . . . How does our friend [Elizabeth Buat-Musch] fare under this misfortune that has befallen her beloved spouse?[43]

> P(11)6: I would not have waited so long with my answer if I could have written without getting a headache . . . I have asked my sister for a copy of your portraits by Netscher, though I do not know what made the good Lord inspire him.[44]

---

[41] OC 6, 95; OC 19, 255–271   [42] OC 6, 76–77   [43] OC 6, 81   [44] OC 6, 83

L(12)6: You would not reproach me if you knew how fast time flies here in Paris, especially for people who, like me, are diligent and have much to occupy them. Indeed, after my illness I worked as little as possible, so as to recover fully . . . Here is the letter for Mr De Witt that you requested. You could deliver it, whether or not the matter [the assessment of mortality] still has his attention, for I have adapted the recommendation.[45]

At a meeting of the Academy in January 1667, it was reported that 'Mr Huygens filled two cannons with water and then sealed them well. The water froze and caused them to split open, with much creaking.'[46]

L(1)7: Tell me something of the marriage of Mr van Ouwerkerck [Hendrik van Nassau] and details of his sweethearts. My kisses and congratulations for Miss Post. O digno conjuncta Viro [if only she had married a worthy man].[47]

Next month at the Academy they cut open a calf's eye and surmised that accommodation occurs as a result of a deformation of the lens, but Mariotte dissuaded them. He spoke of the history of astronomy.[48]

P(2)7: The Duke of Roannez has requested me urgently for information on those little mills with which people in our country pump water out of their gardens . . . He wishes particularly to know whether they turn to face the wind without people having to intervene, and also whether they use buckets on a chain . . . You were very late in sending me the hair sample, but the wig I have had made is exactly the right colour . . . One month ago I heard that Mrs Buat could arrive at any moment, but I see that there is little haste.[49]

Henri Buat had been condemned to death for high treason.[50] When the sentence was carried out on 11 October 1666, father Huygens composed the following verse:

---

[45] OC 6, 91   [46] OC 19, 336   [47] OC 6, 102   [48] Vollgraff 634; OC 13, 787–790
[49] OC 6, 102 (113)   [50] Rowen 128–135

PLATE 1. Constantijn Huygens and his clerk, by Thomas de Keijzer, 1627. © The National Gallery, London.

PLATE 2. Constantijn Huygens and his wife Suzanna van Baerle, probably by Jacob van Campen, ca. 1635. © Royal Picture Gallery, Mauritshuis, The Hague.

PLATE 3. Constantijn Huygens and his children, by Adriaen Hanneman, 1639. © Mauritshuis, The Hague.

PLATE 4. Manor house, Hofwijck, drawn by Christiaan Huygens, ca. 1658. © University Library, Leiden (Ms. Hug. 14, f. 5r).

PLATE 5. Portrait of Christiaan Huygens, by Caspar Netscher, 1671. © Haags Historisch Museum, The Hague.

PLATE 6. Medallion portrait of Christiaan Huygens, by J.-J. Clérion, 1679. © Museum Boerhaave, Leiden.

PLATE 7. Portrait of Christiaan Huygens, by Pierre Bourguignon, ca. 1688. © Photo M. Dooijes, Royal Netherlands Academy of Arts and Sciences collection, Amsterdam.

PLATE 8. Engraving taken from the portrait by Netscher, by G. Edelinck, made between 1684 and 1687. © Museum Boerhaave, Leiden.

Here lies a guilty man, deprived of head and neck.
When alas he sinned, he kept his neck but lost his head.

L(3)7: Concerning the matter of widow B.: I have recently met
friends of hers, but also enemies who think that she co-operated in
the scandal of the deceased. But these folk know nothing of the
matter; it is already bad enough. [The twenty-eight-year-old
widow B. arrived on 8 April 1667.][51]

P(4)7: You appear to be good in the building of a carriage, for you
were able to manage with that messy sketch I sent you at the
beginning, to help you make a success of your calash. The manner
of hanging that I indicated in my last drawing should not be
neglected, because that is important for the suspension, especially
if the road is uneven and the wheels become stuck.[52]

L(5)7: I have written to no one by ordinary post, because I haven't
received any letters either. Signor Padre says that it is the fault of
brother Zeelhem. [That month the father forced Constantijn to
abandon the idea of marrying Isabella Dedel, who was pregnant by
him.][53]

The summer of 1667 he spent discussing mathematical problems.
First he turned his attention to the calculation of logarithms, of
the greatest and smallest value, and of tangent lines. In July he
produced new work on parabolic and hyperbolic bodies, and in
September he discussed small permutations of cords stretched by
weights.[54]

L(8)7: Why are you so keen to know how our widow is faring?
Why do other Dutch people wish to know? Do they speak ill of
her? I assure you that all goes well up until now and, even if I
should wish to do so, I could not speak ill of her. But she certainly
does not suffer under the prospect of being courted by so many
gentlemen who wish to marry her. All in all, she looks well and

[51] OC 6, 117   [52] OC 6, 126   [53] OC 6, 130   [54] Vollgraff 635–636

people believe her to be immensely rich. I am somewhat acquainted with the matter, but in her present situation I think that she is enjoying herself too much to commit herself to one suitor. Do tell me, please, what people say of her, and also about the cousin here. [The husband of Constantia Caron-Boudaen, the cousin in question, had left for a long sea voyage from which he would not return.][55]

C(9)7: I admire your energy to start again with telescope-viewing. If it were not for you, I would be without a companion. You help me enormously.[56]

A(10)7: Gilles de Roberval has brought up the rules of collision for discussion.[57]

L(11)7: There is little of importance to recount on lady B. That trip after the wedding in Rambouillet signifies nothing. She had never seen the young man before, but he did visit her afterwards; just once, she claims. I never came across him there. She swears by all that is holy that she has no thought to remarry, and certainly not a Frenchman. But you know what such women's talk is worth.[58]

C(12)7: Now I am working on experiments with circular movements, for which I have had a round table made that turns on an axle and has a hole drilled in the centre. I believe, with these experiments, that I have discovered the origin of weight, and that is quite something.[59]

On this table he placed a vase of water in which he floated shavings of sealing-wax. He turned the table, with the vase on top, until the water with the shavings revolved just as quickly, then stopped turning the table so that only the water in the vase kept turning. The shavings that had been driven to the edge of the vase now moved in a spiral towards the centre.

---

[55] OC 6, 136   [56] OC 6, 151   [57] OC 19, 95, 181 (n2)
[58] OC 6, 160   [59] OC 6, 164

P(1)8: Mr Colbert sends most inventors to me and I am to judge whether they are any good, which is seldom.[60]

For three sittings of the Academy in January 1668 he spoke about the rules of collision. He had to, now that Gilles de Roberval had brought them up for discussion. This was finally the moment to publish *De motu*. Perhaps he planned to, because there is a copy in the academy archives that is not in his handwriting. Presumably, then, he no longer kept it concealed. But he was too much absorbed at the time in the theory of weight to really chase it up properly. Moreover, he must also have been engaged in the improved version of his vacuum pump from The Hague.[61]

> S(1)8: I already knew of the adventure of Miss Ida, and I laughed heartily about it with the aforementioned lady Buat. That poor little suitor is more to be pitied than the young lady, and still she finds that he does not deserve to fall in the water with her. . . . You promised me to write to Suzette. I have already informed her that you will, so be sure to think of it and do so.[62]

> A(3)8: An experiment is also being carried out with the pneumatic machine of Mr Huygens; after having worked with the machine for some time and having pulled out the sucker, the carp's bladder under the glass is swollen and a large fly has expired.[63]

> L(4)8: I have heard nothing about our widow [Buat] intending to marry, but I have not seen her for a full two weeks. If she intends to have a Duke – there are those here who ascribe this wish to her – then there are all kinds to be had. She moved house three or four days ago, and I am told that she intends to remain here for the summer. . . . I have just shown Mr Colbert and his wife and daughter my experiments with the vacuum. He was most satisfied with them. I was busy all morning making my preparations for them.[64]

---

[60] OC 6, 173   [61] Vollgraff 637   [62] OC 6, 176
[63] OC 19, 200   [64] OC 6, 210 (212)

In April, a book by James Gregory inspired him to explain that the quadrature of the circle and the hyperbole was impossible. And so he kept his work on weight, *De gravitate*, which he had written that month, to himself.[65]

> L(5)8: The marriage of cousin Caron – the eldest, for the other one is not yet ripe for a husband [*non matura viro*] – would have already been arranged and performed had she and her mother been able to decide. At any rate, they inform me that there are three more candidates at present who ardently press their suit. I have advised her to take Mr van Vredenburgh, for he is by far the best match. . . . As regards the marriage of our widow B with some duke or other, that is merely gossip. I do not believe that she will ever marry here. I have just seen her and sought information from her in this matter. She tells me she intends to return in a month.[66]

In May, he was summoned by Colbert, together with Auzout, Carcavy, Picard and Roberval, to come and assess the plan of Reussner from Neustadt. It consisted of an extremely simple way in which to measure longitude at sea. After a week of discussion with Reussner, they rejected the plan.[67]

Then, in July, he gave his comments on Buot's assertions about wheels on an uneven road: 'I should not have waited so long with instructions for the suspension of my calash. I was satisfied with my first experiment on the road, but my second one, carried out along half a mile of the uneven road to Saint-Germain, showed my carriage to be far less comfortable.'[68]

> P(7)8: I had wished to recount to you the festivities in Versailles, but I had no time to do so, and the brochure that I had been promised by Monday has not yet arrived. [More than 3000 guests were invited to these festivities, held on 18 July 1668, to celebrate the annexation of the Free Duchy of Burgundy, Franche-Comté,

---

[65] Vollgraff 640   [66] OC 6, 217   [67] OC 22, 218–226   [68] OC 6, 237

into the kingdom of France.] I found the fireworks the most impressive of all; never have I seen so many rockets fired into the air at the same moment. Molière's comedy, in which a peasant who had married a lady was deceived, was a sloppy and trivial piece of work. But the theatre was magnificent, as were the other two octagonal-shaped halls made of wood, which were decorated with leaves, wreaths of flowers, paintings and fountains, the one for the celebration, the other for the ball. They were about 60 or 70 feet across and raised up all along the length of the paths of the garden. There were so many people that even the King had difficulty in finding places for his womenfolk in the theatre; this meant a good number of men were obliged to stand up. In order to attend I left home at 5 o'clock in the morning and did not return until 7 o'clock the next morning, having withstood extreme heat and then, during the night, extreme cold, without sleep and without much to eat. It was a comfort to discover that everyone else found it equally taxing.[69]

L(8)8: Mrs Buat informed me yesterday that acquaintances in The Hague had written that cousin Caron is already divorced, and that Mr Vredenburgh has left her at the request of his parents. I can hardly believe it, when I see that he is, more than ever, devoted to the beauty. [A year later she would marry François de la Ferté.] Mrs Buat has fallen quite ill from bathing in the river, which has also befallen several others.[70]

L(9)8: Your report [on the marriage of Constantijn to Suzanna Rijckaert] is pleasingly elaborate and entertaining. I wish that there were as many people here with whom I could feel at home. Here there is only Mrs Buat and Miss Boreel.[71]

In the autumn of 1668 he entered into an extensive correspondence with James Gregory on quadratures.[72] In the same period Adrien Auzout, who had made important contributions to the building

---

[69] OC 6, 245   [70] OC 6, 249   [71] OC 6, 257   [72] OC 6, 272–289

of the observatory, left for good to Italy. It was with him that Christiaan had spent a whole year carrying out measurements for Saturn's ring, together with Buot and Picard; he was sorry to see his friend leave.[73] Did he have no say in the matter? Colbert summoned Giovanni Cassini from Bologna for the observatory, at a salary of 9000 pounds per year.[74] An insult? He laid the matter before his father.

> L(11)8: Listen to the oracle who declares the latest fashion. At court they wear only black worsted with lace. Lately the King was dressed in just such a suit that almost covered his trousers, with lace neither on the sides nor on the belt, but not all copy him faithfully in this. They are also wearing tight waistcoats of black velvet with finely wrought buttons made of gold . . . They talk here of great changes that the Prince has been making to his house, and say that the Council has been dissolved. Tell me of what you know. [Amalia van Solms had handed over control of all properties belonging to the House of Orange to her grandson, Willem III, on his eighteenth birthday.][75]

On 20 December 1668 father Huygens wrote to Colbert with a request to raise his son's salary, considering that 'it bore no relationship to his capacities'. In vain.[76]

In March the following year, he published his rules of collision. A year earlier he had considered publication of *De motu*, but only took action late in 1668. Oldenburg, with whom he had been in touch about Gregory's work on quadratures, informed him that Wren and Wallis had submitted their rules of collision to the Royal Society. Suddenly he was in a hurry. On 5 January 1669 he sent the first part of *De motu* to Oldenburg, but shortly afterwards he decided to make an excerpt. These *Règles du mouvement dans la rencontre des corps* (Rules governing Movement in Collision of Bodies) appeared in March

---

[73] Vollgraff 627; OC 21, 9   [74] Van Helden (1996) 100   [75] OC 6, 290
[76] OC 6, 303

in the *Journal des Scavans* and the Latin translation appeared in April in *Philosophical Transactions*.[77]

> L(4)9: For the last three days it has not ceased to freeze and to snow.[78]

In April! A sign to mark his fortieth birthday.

In May that year, he made a report on observations whereby a block of oak was secured by a rope in the Seine. The vibrations, which he recorded using a clock, were a measure of the force the flow exerted on the block. He reckoned that this force was proportional to the square of the rate of flow.[79]

> L(5)9: I am forwarding to you this letter from Mr Berni, whom I have got to know at cousin Caron's . . . A thoroughly good fellow, and with money, so I am told. It looks as though one of the little Indians [daughters of Constantia Caron-Boudaen, who believe their father to be in The Indies] might make a catch of him, though he does not seem likely to be ensnared just yet. It surprises me that the tea still hasn't arrived.[80]

> L(7)9: Just what do you mean when you say that I show your letters to Mrs Buat? She is pulling your leg. I have shown her nothing. Indeed, it is you who neglects to inform me about your own wedding plans. It was Miss Boreel who told me on her return that your Climene was the little B. She was also pulling your leg with her chatter about my love for cousin C. [Constantia Caron-Boudaen]. Nothing is further from the truth. Nobody here would dream of saying such a thing. I call on her only once a month . . . This time I have nothing for father. Just tell him that Mr Charas has sent me the book [on lightness and weight] by Sinclair, an impudent writer. And tell him, too, that my harpsichord is a great success and that it would take something quite momentous to tempt me away from it.[81]

---

[77] Vollgraff 650; OC 6, 383–385, 431–433   [78] OC 6, 400
[79] Vollgraff 652   [80] OC 6, 441–442   [81] OC 6, 471 (473)

Table 1

Month	1666	1667	1668	1669
January		2	4	3
February		4	3	3
March		3	3	6
April		5	5	3
May		2	2	2
June	4		6	1
July	4	3	5	3
August	3	1	3	3
September	1	3	2	2
October		3	2	4
November	3	3	5	4
December	3	6	5	2
Average	2.6	2.9	3.8	3.0

This is sufficient to convey some idea of his concerns and activities. But it is not enough, if we truly wish to become familiar with his life in the Paris of the Academy. Although he says a great deal, he does not say everything, and some letters are missing, not only those to his father (which we know existed, because he refers to them in letters to his brothers). To substantiate these assertions, and to get some idea of what is missing, we provide the table above of the number of preserved letters that he wrote each month since his arrival in Paris (Table 1). On average, then, he wrote about three letters each month. But sometimes he wrote many more, sometimes less, or none at all. He wrote more in December 1667: three to Lodewijk, two to Constantijn and one to Philips, all on business matters, and to catch up on news. He also wrote more in March 1669: to Colbert (about a patent), to Gallois (about his publication in the *Journal des Scavans*), to secretary Jean du Hamel, Moray and Oldenburg (on the priority of Wren and Wallis and of Gregory), and to his brother Lodewijk (about a lens and about tea); there was simply a lot going on at the time. The six letters of June 1668 hardly attract our notice in that year's crop.

There must have been something unusual going on in the months in which he wrote fewer letters. The one September letter is written on the third of that month, so right in the beginning; we have already quoted from it (P(11)6). After that he suffered terribly from headaches, and was unable to write anything for the entire month of October. Was this due to an infection that he had caught in July while swimming in the river? This is hard to believe. Did this incapacity at the start of the season coincide with unpleasant discoveries? That not he, but Carcavy, would be chairman of the Academy? That, apart from this, few if any arrangements had been made? We do not know, as he remains silent on the matter. It is certainly odd that the members, who had initially met on 2 July, should wait until 22 December before meeting again.

The hiatus of June 1667 is easier to explain. In May he heard from his father that Constantijn was causing problems (L(5)7). He must have heard more in June, and he would certainly have corresponded on the matter, but these letters are missing. His father, the patriarch, had solved the problem for Constantijn: an illegitimate child was preferable to a pregnant bride, even if Isabella were the daughter of his friend Dedel a hundred times over. But an illegitimate child in the family was also no minor scandal, and that is why all the letters pertaining to it have been destroyed, including those from Paris.[82]

In August 1667 he thanked Lodewijk for informing him of the scandal – information he would otherwise only have received from Elizabeth Buat. It is the only letter he wrote that month, and the Academy did not meet. Perhaps that one letter provides an answer to the question of what is going on. After all, his reference to the merry widow is not unsympathetic. If she (with such a reputation) still felt rather alone, and if he (with such a father and such a brother) and at thirty-eight years old, allowed himself certain liberties, then they

---

[82] Keesing (1983) 71–74, 183–184

might well have crept into bed together now and again. But this does not explain his silence. Was he perhaps, in August, considering a future for the two of them? After all, it is clear, from the fact that Gilles de Roberval took over the reins (A(10)7), that he was preoccupied with matters other than the Academy. Even though she did not want to tie herself down, and probably no more than he did, they still met faithfully. He referred to her as someone with whom he could feel at home (L(9)8). And when she revealed to The Hague that he was in love with another, she might perhaps have felt resentful.

This brings us to the scarcity of letters in June 1669, when he wrote only to Heinrich Oldenburg (on the publication of *Kort onderwijs* in *Philosophical Transactions*). His denial of an affair with Constantia Caron-Boudaen is strong, but does not need to be taken any more seriously than the assertion made by Elizabeth Buat, who was usually well informed and refers here to a friend she had known since their youth together. His letter preceding this month of silence – we have already quoted from it (L(5)9) – gives food for thought. The fact is that it is not only about a suitor for a daughter, and tea, but also about Latin verses:[83]

> *Me viro primum ant' omnia Musa,*
> *Accipiant caelique vias et sidera monstrant –*

Verses? It is as if he has almost forgotten all about them. A dozen more followed, equally incomplete and slapdash. What was he trying to say with these untranslatable fragments, which seem to belong in the fourth book of Virgil's *Aeneid*? That Constantia would lose her seafarer François (*nota bene* to tea, because that is what his company was after) just as Dido lost her Aeneas? That her seafarer is already half-forgotten? *Unde tremor terris*: So the earth trembles, the sea gathers force, the shadow of a night –

Constantia was mother to Suzette and four other children. She regarded him as an uncle to her children, but in the shadow of a night in

---

[83] OC 6, 440 +

June 1669, we think, she also saw in Christiaan a husband for herself. Confusion? Yes, that month after the delirium of verse fragments, he fell silent. Love? Possibly, for he denied it too strongly. Restrained? Certainly, for he kept it to 'once a month' (L[7]9) and she waited two years before moving to a house in the same neighbourhood, in the Rue de Richelieu. This is what we are able to surmise on the basis of a few clues; but conclusive evidence is missing. Even if this were to be found in letters that he wrote to Constantia, these have been lost.

There is more that is significant about June 1669. The harpsichord with the thirty-one strings in the octave, which he must certainly have ordered to be built six months earlier in a Paris atelier, was now ready, and there was little that could tempt him to leave it (L[7]9).

It has not been preserved. From his later descriptions of it,[84] it appears to have been a harpsichord with an ordinary keyboard, 'for people would become confused with the large number [thirty-one] of white and black keys'. But this keyboard with twelve keys (the seven white keys, and five black half-keys in between), could be shifted slightly, no further than one key, over the thirty-one strings of his harpsichord. These strings were tuned with the aid of a logarithm table and one loose string, and they were plucked by batons no wider than a fifth of a key.

The following construction was, as he said, admired and imitated by the 'greatest masters'.[85] He had a slide made to move the (wide) keys along the (narrow) batons, and hooks that ensured that a key that had shifted to the next position would pluck the baton beside it. In the first position the keys plucked the strings with the number 0, (2), 5, (8), 10, 13, (15), 18, (20), 23, (26), 28 and 31 – the numbers for the black keys are in brackets. In the second position, strings are plucked with number 1, (3), 6, (9), 11, 14, (16), 19, (21), 24, (27), 29 and 32 = *1; in the third position strings with numbers 2, (4), 7, (10), 12,

[84] Vollgraff 653; OC **20**, 160–161   [85] Rasch (1986) 69–85

15, (17), 20, (22), 25, (28), 30 and 33 = *2, and so on, until the lowest key in fifth position plucked string 4.

Was this all? To be sure, he now had an instrument that was easy to play and sounded truer in the even temperament than a twelve-tone instrument, but he could not play a modulating melody on it. To give an example using the minor sixth: if string 0 gives the tonic, then string 20 will give a minor sixth. With a very simple modulation towards the dominant, 18 will become the tonic. The next minor sixth will be given by string 18 + 20 = 38 = *7, and that string could not be struck in the first position! There was, of course, no question of his shifting the keyboard to the third position while playing, where the *7 could be played.

Such an impoverishment cannot have escaped him. Is this why he so detested modulations? In comparison, the new harmonies that could be conjured up with his thirty-one notes seemed to offer untold richness. Whether we find them beautiful or ugly [to be honest: ugly], in his shadow of the night, he can only have heard beauty. And if he had given free rein to his fantasy, the harpsichord would certainly have had three octaves and about a hundred strings.

When had Colbert last admonished him to produce finally a proper volume on clocks, with a well-turned dedication for the King? In 1669 the Earth trembled in more than one respect.

Now we come to his work on the origin of weight. It was the only new work with which he, as director of the Academy – or whatever he actually was by then – had attempted to lead the way. But he was not present when Gilles de Roberval – who seemed to wish to make use of the event to celebrate his sixty-seventh birthday – brought up the subject, on 7 August.[86]

'Mr Roberval publishes nothing, but discovers much, and he is a persuasive speaker with convincing arguments,' according to the secretary of the English embassy.[87] His birthday speech was certainly clever. First he referred to 'writers who take refuge in some highly

[86] Costabel (1986) 21–22   [87] OC 7, 8

refined material or another that moves with great rapidity and nestles easily amongst the particles of greater bodies, and then in such manner as to cause them to move either upwards or downwards'.[88]

He objected, however, to this Cartesian whirl theory. He considered it possible that weight did not originate in movement: 'Writers who hold this opinion attribute a characteristic to the body which pushes it downwards. Others wish to believe that there is a reciprocal pull between the particles of the body, so that they unite, as far as they are able, to form a whole.'[89] Nevertheless, he found that experiments were necessary before he went to the trouble of 'understanding fully the fundamentals and the origins of weight'. On 14 August Frénicle backed him up by saying that an influence exerted from a distance was conceivable, for after all, a magnet was capable of attracting iron filings.[90] Then came Jacques Buot.

'Mr Buot looks like the great twelve-foot-wide stone ball that they have placed in Les Tuileries, not to roll, but as a massive ornament.'[91] This, once again, according to the secretary of the English embassy. On August 21 this same Buot attacked Roberval, 'Certain people [!] have established the causes of weight. Ethereal particles are most easily removed [from the centre] and must therefore push back large particles [to the centre] by exercising urge to overcome their original resistance.'[92] In order to better to understand this urge, he went upstairs, where 'certain people' were doing their homework for Colbert.

U. Frankfourt and A. Frenk write in their biography of Huygens that he was seriously ill in the month of August 1669,[93] but there is no proof of this. According to the minutes taken by Jean du Hamel, he had spoken on the usefulness of measuring the force exerted on mills. This was his last contribution to a series that began in 1667 with fifteen lectures, and a further sixteen in 1668, but in 1669 there had only been five.[94] If he had declared himself sick, then this was

---

[88] OC **19**, 628   [89] OC **19**, 629   [90] OC **19**, 630   [91] OC **7**, 8   [92] OC **19**, 631
[93] Frankfourt & Frenk 125   [94] Taton 60–61, 65 (n11)

a pretext. Did he need to finish *Horologium oscillatorium* first, and simply decide to leave them all to it, as far as the debate on weight was concerned? Alternatively, was it possible he could no longer stand the sight of Roberval, or was he perhaps fed up with the whole pack of them?

Whatever may have been the reason, Jacques Buot managed to persuade him to give a lecture. On 28 August, he read aloud to the Academy his *Discours de la cause de la pesanteur* (Discourse on the Cause of Weight).[95] He must already have completed the work, for you cannot complete such a text in only a few days. Also, it is certain that he discussed all eleven paragraphs, otherwise they would not have been able to criticise them all the following week. The first two, which are short, read as follows:

> In order to discover a comprehensible cause of weight, one has to see whether it is possible in nature to assume only bodies consisting of the same matter, without a quality or urge to move towards one another, but only differing in size, form and movement; whether it is possible, I say, that certain of these bodies still tend towards the centre and remain together there, which is the most ordinary and the most important phenomenon of that which we call weight.
>
> The simplicity of the basic principle, which I have just given, does not leave us with much choice in this research, for one can see immediately that nothing in the shape or smallness of the particles can have any consequence that resembles weight, which is a tendency to move that can probably only be caused by a movement. That is why all that remains is to try and discover how this operates, and which particles are involved.

It was to be a tour de force. Clearly, he did not wish to compromise Descartes' heritage, and he was obsessed with the idea that circular motion produces force. This idea led him to make a

---

[95] OC **19**, 631–640 (OC **21**, 427–499)

calculation – practically the only one to be found in the *Discours* – of the whirl of fine particles that must determine weight on Earth.

This went as follows. The speed s of the whirl around the Earth should follow from his formula for the acceleration g (which, in 1659, he had determined at 9.8 metres per second per second) in a circular course across the surface of the Earth at distance d from the midpoint (which he knew by calculation; the average value of d is 6 371 030 metres). In this way, he obtained for s a value of 7927 metres per second, because

$$ss:d = g$$
$$ss = gd$$
$$ss = 9.8 \times 6\,371\,030$$
$$ss = 62\,436\,094$$
$$s = 7927$$

To bring this value into perspective, he compared it with the rotation speed of the Earth. The speed with which the Earth rotates at the equator is derived from a division of the circumference, which is $2\pi d$, and the time-span of a natural day, which is $24 \times 60 \times 60 = 86\,400$ seconds. For this he came up with a figure of 463 metres per second. The whirl speed would therefore have to be $7927 : 463 = 17$ times as great. If bodies on the equator were to move as fast as the whirl, he reasoned, the collisions of the whirl particles would have no effect on them, and they would therefore weigh nothing.

This calculation established the limit of possible weight effects, but that was all. He would have to forgo quite a lot of what preceded this. First of all, the proposition (in the third paragraph) that motion in a circle is just as natural as that in a straight line. And secondly, the proposition that collisions of fine particles with crude inert bodies necessitate the latter to move towards the centre of the circle.

He lacked a good argument for the first proposition, because it was against the conservation of impulse. It need not concern us here whether he knew this term in the modern sense, but whether he and his contemporaries knew that movements, by nature, persisted in the same direction. This he knew for sure. Moreover, in his experiments with pendulums, he had experienced the force needed to cause a body to deviate from that direction. By proposing that circular motion was natural, he was proposing, at one and the same time, that centrifugal force was natural, too. That is why it is hard for us to understand why he was able to accept circular motion as natural, but not the force (or the weight).

For his second proposition he obtained his arguments from his experiment with sealing-wax shavings in rotating water, because he realised that there was no existing theory for this. In the case of circular motion, the effect of collisions (which was only barely predictable for easy motion along a straight line), must have been practically impossible to predict. One needed experiments! He suggested that shavings of sealing-wax could be compared with the crude bodies, and the (invisible) constituents of the water with the fine particles.

The question is whether this comparison actually applies. What he observed in the experiments, previously set out in letter C(12)7, is not only explained by the effect of the water on the shavings. The vase, too, plays a role. Sealing wax is heavier than water, so that the shavings will slide around on the bottom of the vase. By their friction against the bottom, their movement will be more impeded than that of the water, which rotates at the same distance from the axis. And because of their friction with the water, they will be drawn towards where the water also moves more slowly, and that is towards the axis. But wood shavings, which float freely, will tend away from the axis.

He did his utmost to understand what effect the water would have on the shavings. He looked for it in the greater turbulence of collisions of water particles further from the axis, since these moved faster, causing the outer edge of the shavings to be more strongly affected than their inner edge. The net effect of this, as he thought to be able to prove with words, was a thrust towards the axis, not a torque whereby the shavings would turn on their own axis. But wherever he was unable to make use of clear concepts and mathematical method, he got caught up in logical nonsense.

The *Discours* ends up being a rather careless circular argument. Perhaps he realised this when he wrote his final sentence, 'As long as there are no phenomena to be discovered that may dispute it, then nothing stands in the way of the truth of our assumption,' which really says nothing at all. Did he no longer wish to give the matter any thought?

II Crisis

Gilles de Roberval was hardly the man to handle him with kid gloves. This shrewd farmer, who had managed to become a member of the Academy of Sciences because he knew the difference between assertion and proof, had callous hands.[1]

'It seems to us that certain matters in the writings of Mr Huygens must be either proved or clarified.' This is how he opened his attack on the *Discours* on 4 September 1669, with Edmé Mariotte as aide-de-camp. We shall summarise this attack in several quotations and, for clarity's sake, add Christiaan's defence immediately following them. But that defence didn't actually come until 23 October. It had taken seven weeks of reflection.

Gilles:[2] 'To begin with, he rules out all attracting and repelling features of nature and introduces, without any proof, a world in which only size, form and movement count. Besides this, he insists that one movement gives rise to another, but then he must clarify the first, which is just as difficult, if not more so, than those features. Then he assumes a circular motion; however, many assert that the circular motion in itself is not natural at all, but the consequence [*le droict*] of occurrences.' His words were well chosen.

Christiaan:[3] 'My answer to the first objection is that I rule out attracting and repelling features, because I seek an understandable cause for weight. Moreover, I find that it says little or nothing to assert of falling bodies that such a body, or the Earth, possesses an attracting feature. And as far as the [fine] particles with a size, form and movement are concerned, I fail to see how people can say that I introduce them without proof, for the senses [*les sens*] inform us that

---

[1] Costabel (1986) 22; Taton 60   [2] OC 19, 640   [3] OC 19, 642

they occur in nature. My answer to the second objection is that we know for sure that moving particles are capable of setting others in motion, and that I assert only this, without seeking to discover what has been the means of creation of the first motion. To the third objection, my answer is that I did not say the circular motion was natural, but that it occurs, which cannot be denied.' His reply was constrained.

Gilles:[4] 'The conclusion he draws in paragraph 4 seems not to be inevitable, but is merely what people wish to find; one may argue about the experiments used to prove it, and he cannot, in any case, base his conclusion on them, as will be seen. On paragraph 5, much may be said. Amongst other things, that the continuously repelled body E [the shavings that serve as an example] are not propelled towards the axis according to any variance in the exposure of its surface, as shall presently be seen. The effect of his apparatus [the vase with water and shavings], of which he speaks in paragraph 6, can be explained by the weight of the water, aided by the movement of the vase that first accelerates it, then slows it down. Such alternation gives rise to a double perambulation of the water, the one around the axis, and the other between the wall of the vase and the axis. This means that as the vase turns, the water is higher at the wall than at the axis, causing it to flow towards the axis . . .' This last was incorrect.

Christiaan:[5] 'To the sixth objection I answer that . . . one has only to mix shavings with the same weight [as water] in the water to see whether they follow the perambulation we have asserted; but experience shows that this is not so.' He was unable to find a truly convincing answer to the rest: 'My answer to the fourth objection is that the experiment with the perambulating water shows that bodies are drawn towards the axis, and that I believe the conclusion to be quite indisputable. To the fifth objection, my answer is that the reason I offer, as to why the body E is propelled towards the axis, is perfectly clear, and that evidence is not to be found to the contrary.' It could not have been more blunt.

[4] OC 19, 641   [5] OC 19, 643

Gilles presented two more circular arguments and a contradiction, but each time with the courteous, 'It appears that . . .' Because Christiaan either denied them, or attempted to evade the issue by offering fresh assertions, his own proposition became less convincing. Had he no doubts whatsoever?

After he had delivered his defence at the Academy, Jean Picard was allotted the task of assessing it. It must have been a thorn in the flesh for the good-natured astronomer, and just what he made of it on 30 October is unknown, as minutes of the meeting are missing. Possibly Jean du Hamel, the secretary, who could hardly have had his own opinions on this matter, summarised the assessment once more on 6 November. He said at this point that weight might be attributed to 'the body', but that the elaboration that Huygens and Buot (!) had given, based on the Cartesian principle, was probably correct.[6] Mariotte repeated the following week 'that one could assume a natural tendency whereby the bodies move towards others'. Claude Perrault, who closed the debate on 21 November, gave preference to the whirl theory, according to which there is the 'whirl of ethereal particles that goes from west to east' and another 'that goes from south to north'.[7] So Christiaan did have his friends.

He was a guest at the home of this Perrault and his brother Charles in September and the beginning of October. Claude was the architect of the observatory, which was erected without wood or steel; Charles supervised the royal building works and was paymaster. Their modest castle lay about 20 kilometres upstream on the Seine, against the hills of Viry.[8]

> Sandy banks bordered the river; later, floating mists ascended, the sun broke through, the hill that accompanied the flow of the river on its right side sank lower and lower, but close by, from the other side, another rose up.[9]

[6] OC 19, 644–645   [7] OC 19, 645   [8] OC 6, 497   [9] Flaubert 1, 4

Did Claude, who was also a poet, invite him to stay immediately after the attack of 4 September? He must have written his defence in the tranquillity of Viry. He also found time to measure the speed of sound, and to devise a clock for the castle fountain. He sent black earth, which could be found at the new bridge close by, to brother Constantijn for his glazing work.

The canny peasant farmer did not leave it at that. After the debate, he continued to disturb his peace. When Christiaan was working on the final version of the fourth part of *Horologium oscillatorium*, Roberval got hold of the text. Did he obtain it from the printer, or from the author himself? We do not know exactly what happened, because Jean du Hamel did not write any detailed or dated minutes from 18 December 1669 onwards until 1675. We only know that in 1670 Roberval gave a spirited and brilliant (*'fuse & subtiliter'*) commentary on the theory of the centre of oscillation.[10] He would have referred to his own work from the 1650s and 1660s, but it is certain that he knew the text of *Horologium oscillatorium*, because he attacked it.

We shall consider why it took until June 1671 before Christiaan responded, in *Responsa ad objectiones Robervallii contra demonstrationes nostras de motu pendulorum* (Answers to the Objections of Roberval against our Proofs on the Motion of Pendulums).[11] A. Gabbey offers the plausible theory that, besides these formal *Responsa*, there must have also been formal *Objectiones* from Roberval, but that these have been lost. It seems almost certain that the Frenchman presented these in the beginning of 1670, and that this fresh attack proved too much for Christiaan.[12]

Titan broke down.

I arrived at 3 o'clock to find him in bed, his head propped high on the pillows, and only a servant was in the room. After I had greeted him and sat down at his bedside, he sent the servant away. When we were alone he regarded me, then proceeded with such

[10] Gabbey (1982) 73–74   [11] OC 18, 439–456   [12] Gabbey (1982) 75–76

words of politeness, in which he expressed greater esteem and affection than I deserved. I saw that his condition was far from favourable. His weakness and pallid countenance made absolutely clear to what degree the illness had eroded his health. And, as if that were not enough, I saw something worse, something that no eye can penetrate, and no sense can discern. It was a dissolution of the spirit, an unbelievable need for sleep, which he understood as little as those who attended him. Because he did not know what was to happen, he assumed the worst. He predicted that his time had come.[13]

This happened on Saturday 22 February 1670, one month after he fell ill during the heavy frost. It is written by Francis Vernon, secretary at the English embassy, who reported to Oldenburg on the well-being of the Academy members. He had already called a couple of times but was sent away. Thinking that he was 'neare to the very Point of Death', Christiaan requested that the secretary be summoned after all.

According to the report, the scientist from The Hague handed over to him *De motu* and other unpublished works, for he wished these to be sent to the society in London. Not to the Academy. He saw this going to rack and ruin, just like himself, for it was 'mixt with tinctures of Envy' and entirely dependent on the favours of one minister. The spectre of death drove Christiaan to surrealistic gesture. He 'caught hold of a stack of papers arranged upon the bed, held them aloft and asked whether he should not request a candle with which to seal them in my presence. But after a while he said, "No, I shall not seal them, I give them to you as they are." '[14]

Titan could't sleep. Perhaps in the beginning, he just seemed over-tired. 'The doctors prescribed him rest; he should speak to no-one.' Later they saw that rest was not the answer. It even appears from a letter dated 17 March that Vallot, the royal physician, was

---

[13] OC 7, 9–10   [14] OC 7, 12, 11

called. It did not help either that he drank only thinned milk (according to the wisdom of the day, full fat milk contributed to melancholy). They came to the same conclusion as the family doctor in The Hague, who said one month later that he suffered from 'melancholia hypochondrica vera et mera' ('real and unadulterated depressive melancholia').[15]

> Melancholia is characterised by severe despondency, by lack of all interest in the outside world and feelings of self-esteem, which manifests itself in self-reproach and sometimes in delusions of future punishment. This picture becomes clearer to us when we consider that mourning manifests the same characteristics, with this one difference: there is no occurrence of disturbance in feelings of self-esteem.[16]

He took to his bed. If he did get up, he had himself carried by his servant.[17] He would not write a single letter until October. What thoughts he had, or what he felt, is as good as unknown.

Lodewijk, who travelled from The Hague to find out what was wrong, informed the family. Alas, the family's concern went as far as destroying his letters, apart from a list that could be a will, 'My brother has informed me on 13 May that the changes in the maritime clocks consist of . . .'[18] Even so, we can guess something of his message. In fact, Constantijn, who was deeply affected by the fate of his brother, wrote back to Lodewijk on 22 May with the advice, 'to admonish him in your own words, and in a seemingly casual fashion, not to deviate from the straight and narrow, for apparently he is afraid of vicars'.[19] And on 29 May he wrote once again, this time on the authority of Doctor Van Liebergen, 'If he is still eating, as you say, then a desire to live will overcome that poor humour, and you need not be perturbed by this feverish display of emotion that accompanies this crisis. This is quite normal (estant chose fort ordinaire).'[20]

---

[15] OC 7, 22   [16] Freud 74   [17] OC 7, 35   [18] OC 7, 26
[19] OC 7, 28   [20] OC 7, 29

The symptoms of the illness were already well defined in his time: severe guilt feelings – the admonishment to keep to the straight and narrow indicate their nature, and the fear of vicars, their intensity – and sleeplessness, the unbelievable need for sleep. What had not been established was the cause. Even now we are no wiser; but perhaps we can shed some light on it.

Let us make a comparison. Just as he could only understand the cause and origin of weight by following Descartes, so can we only understand the causes and origin of melancholia if we follow Freud. In the first instance, weight is a form of motion, in the second, melancholia is a form of mourning. We make the comparison because it contains a hidden meaning: just as the idea of Descartes turned out to be too simple, so can the idea of Freud also turn out to be too simple. But is there anything that can be better understood?

Since the manifestations of melancholia and mourning share common characteristics, their origins must be related. It is a loss. But while in mourning it clearly involves the loss of a loved one, in melancholia the loss can only be guessed at. It could lie in the disturbance in feelings of self-esteem, which is where melancholia distinguishes itself so strikingly from mourning. The loss should therefore be sought in the person himself. The assumption is: that which is lost is a cherished part of the Self.

This assumption is tested as follows: in mourning, the longing to possess the loved one must be suppressed (in the language of Freud: the appropriations of the libido must be withdrawn). At first, this is so distressing that one attempts to shut oneself off from the reality of the loss, but after a while one usually heals. Also in melancholia, the longing to possess the loved one within the Self – which is now split in two, as it were – must be suppressed. This, too, must be distressing, but the pain cannot lessen with time, since all the parts of the Self remain present in the same person. 'The sleeplessness, which is so characteristic of melancholia, is evidence of the impossibility of realising the necessary, all-encompassing withdrawal of appropriations that must take place before falling asleep.'[21] And because the Self that remains behind is incomplete, it is compelled to regard itself as worthless and rejected: 'The melancholic reviles himself/herself and expects to be rejected and punished. He demeans himself to others, and pities his nearest and dearest for their association with such an unworthy person as himself.'[22]

Christiaan would indeed express himself in such terms, years later, when he was mortally ill. Fortunately, the evidence has been preserved. We can only guess at what he may have moaned or screamed or cursed on his sickbed in 1670, but the fact that Lodewijk's letters have disappeared does make us wonder.

Difficult questions arise: which part of his Self might he have lost? The intellectual part, for which he saw his personal life as being merely a means to serve?[23] And how did this happen? Was it the result of Roberval's intellectual criticism, which he was unable to cope with? Did this Self always have to be the cleverest? Or not?

Even in question form, the answers are still too dogmatic, and they are certainly superficial. But at least a bit more of the hidden Titan is revealed. He broke down when he put aside his greatest achievement, De motu, in order to conform to the compulsive simplification of a Cartesian world – when he tried to see the world as his father did. And there is something more to be discovered beneath the surface. He allowed himself to be carried like a child that was lifted onto its mother's bed: 'Come here, my dear little man, and let me kiss you.'

The depth of his melancholia was the depth of his mourning. But there is nothing to indicate that he ever allowed himself to consider this. That summer, as soon as he was able to write again, it was

---

[21] Freud 84   [22] Ibid. 76   [23] See pp. xxiv–xxv above

QUI SUIT LA RAISON JAMAIS NE SE TROMPE
He who follows reason is never led astray.

in a rebus for Constantijn.[24] No irony intended. The intellectual Self was on the move again. It had already begun to stir when, in the small hours of 27 May, Christiaan observed the phase of Saturn, together with Giovanni Cassini and Jean Picard; and for this he would have had to leave his room.[25] Perhaps even more telling is that, at the end of May, he spent a bit of time calculating a centre of oscillation – in pencil, which means that he was probably still in bed, but he carried on from exactly that point where the intellectual discussion had been interrupted.[26] Even so, that summer there was still no question of his being pronounced cured.

On 9 September 1670 he arrived with Lodewijk in The Hague, and would not return to Paris until 12 June 1671.[27] During his sick leave he arranged his business matters with a Perrault (Charles, we presume, for he was paymaster) – not with Carcavy.[28] He continued to receive his regular salary, so Colbert must have agreed, and what is perhaps even more significant: on his return it turned out that Colbert had given him an extra room.[29] Was this merely a favour granted by the minister, or did he do so to make amends?

In that year, Christiaan was only a shadow of the scientist who knew what he was after. He hardly managed any correspondence; he spoke with Johan de Witt about the calculation of annuities,[30] and in the spring, drifted along the Rhine and the IJssel with Jan Hudde, studying sandbanks at the request of the States-General.[31] We need to realise this when we look at his finest portrait.

It is the first one that portrays him as an adult (Plate 5). Since the family portrait of 1639, he had grown to be a giant in his field, but now that we finally have a second portrait of him, we still see a child, slightly haughty, vulnerable and – we cannot get away from

---

[24] Keesing (1983) 79   [25] OC 7, 31   [26] Vollgraff 657   [27] *Ibid.* 660
[28] OC 7, 36   [29] OC 7, 83   [30] Vollgraff 664   [31] *Ibid.* 663 (OC 7, 60–78)

it – infinitely sad-looking. Caspar Netscher was the painter, and the intensely private look he happened to capture is something 'to which I know not what the good Lord inspired him'. It must have been painted in the winter, or the early spring of 1671.

The letter that he wrote about maritime clocks on 4 February, probably to Jean du Hamel, after he had read of the fiasco during the stormy crossing to Cayenne, is also far from cheerful. The head of the expedition, Jean Richer, had adjusted the clocks carelessly and even dropped them, he complained, but if only that had been all: 'I see that, for want of just a drop of oil in the hinges, they have been ruined. People have gone to very little trouble to ensure the success of this experiment.'[32]

Whether or not it occupied his thoughts is hard to say, but there was another crisis in the making that might have deeply affected him. In an ostentatious show of power, Louis XIV had come in May 1670 to inspect Dunkirk, Lille and other recently conquered forts in the southern part of the Netherlands. It was not yet known that he had already negotiated a secret coalition with Charles II against the Republic – a stab in the back for De Witt's foreign policy – but from that moment on, his plan to conquer the Republic was openly discussed. Father Huygens, who had spent most of the time in England while Christiaan was in The Hague, clearly understood the seriousness of the threat when he wrote to his Parisian friend, De Behringen, 'If, during this visit of the King, you do not wish to receive your friends [by which was meant: my sons], think carefully, for if once insulted, we are irreconcilable.'[33]

Louis XIV, however, was still unable to embark on his great war. After the peace treaty with Spain – only recently signed in 1659 – Hugues de Lionne, the minister of foreign affairs, wanted stability. Colbert had finally got the national budget onto an even keel, and was hardly inclined to squander his hard-earned savings on a war. If the king wished to win the support of the members of his state council,

32 OC 7, 54   33 OC 7, 23

he would first have to convince them. He also wanted to extend his coalition. A. Sonnino has shown that, contrary to assertions made later by French historians, his designs were far from laudable, and that François Michel de Louvois – the favourite minister of war – and his cronies, were fishing in troubled waters.[34] Half Europe seemed to be up for sale. Not only the English Charles II, but also countless German princes and their diplomats were keen to be bribed, and went out of their way to plunder the French treasury. And the strange thing about this century was that some of them were able to produce perfectly cool, analytical justifications for doing so.

In 1670, a lawyer of the Elector of Mainz came up with an alternative plan: if the king wanted his war, he reasoned, wouldn't it be better if he conquered Egypt instead of Holland?[35] That would cost less and yield more. The Elector, who had seen how Louis XIV had occupied his neighbour Lotharingen, saw possibilities in such diversionary tactics, and approached the king to see whether he might be interested in such an alternative. That turned out to be the case, and the lawyer was sent to Paris. His real name was Lubieniczy, but his germanised name was Leibniz. When he arrived in Paris in 1672, Lionne had died, and that left only Colbert in opposition. The cards were shuffled.

Gottfried Wilhelm Leibniz had more to offer.[36] His father had been professor in Leipzig, so that he had been exposed to scholarly books from a very early age. In 1663, barely seventeen years old, but with miraculous powers of comprehension, he managed to produce a dissertation on the individual. But in Leipzig he was considered too young for a doctor's title. He left Leipzig and went to Altdorf, in the hills near Nuremberg. There he made such a good impression with his second dissertation, *De casibus perplexis in jure* (On Complex Court Cases), that he could immediately become professor on the law faculty. But his ambitions lay elsewhere. He was also interested in philosophy and

[34] Sonnino 49–69, 132–154   [35] Aiton 37–39   [36] *Ibid.* 9–34

in mathematics, and he wished to see something of the world. This is why he entered into the service of the Elector of Mainz with his group of erudite friends. In 1669, one of them showed him the issue of the *Philosophical Transactions* containing the laws of collision. He understood them, but also claimed to understand how they could be derived.

Oldenburg wrote, in his letter to Christiaan on 18 November 1670:[37]

> I do not know, Sir, whether you are acquainted with a certain Doctor Leibnitzius from Mainz. He is advisor to the Elector there, but also makes a study of philosophy, especially with speculations on the nature and characteristics of motion. He claims to have discovered the same principles of the rules of motion that others, so he says, have only proposed, without a priori proof. Because I strongly doubt whether you have seen any of his work on the subject, I confide in you what he has recently written to me about it.

Oldenburg then sent this obscure explanation to The Hague. Christiaan, who was still keeping concealed his a priori proof in *De motu*, did not respond. But Leibniz would manage to track him down in Paris. To learn mathematics.

Paris, 9 July 1671:[38]

> According to brother Moggershil's letter, you [Lodewijk] will now be back in The Hague. And perhaps il Signor Padre has also returned, although I am inclined to think that he is still on his way, due to the slow pace of the negotiations with England. That is why I have not written to him this time. I rather regret not being in The Hague to welcome him, but if I had waited until that moment, and much longer if he had wished me to keep him company, I would have been summoned back here.

[37] OC 7, 46   [38] OC 7, 79

I understand from the letter of the aforementioned brother that Mr Zeelhem's fever is starting to abate, and that he is able to enjoy the daily trips in the carriage. It is most fortunate that he is recovered before the autumn. Please tell me what you know of the matter of Romf [secretary at the Netherlands Embassy], whether he is applying to work for the Prince [Willem III], or whether he will work for the State. For the time being at least, he is not expected home.

My lodgings are more spacious now, thanks to the addition of the downstairs room, with which you are already familiar. As you can see in the plan that I sent to brother Moggershil, I shall have it made into a pleasing abode.

On my arrival here, I discovered that Carcavy and his son had played a most shabby trick on me. I do not know whether I told you that just before my departure here, the boy asked me whether I wished to sell him my carriage. I asked 100 *écus* for it. Because he did not wish to pay this sum, the transaction did not take place. I locked up the cushions and the curtains in my study and took the key with me. During my absence, these gentlemen used my carriage, drove it into another and, in order to obtain the cushions and the curtains, broke open the study door, which was bolted from within. What cheek! They probably thought that I would not return. Now they are at a loss as to what to do, considering that I have promised them that I shall complain to Colbert, and I shall certainly do so.

This afternoon the entire court here awaits the King, who has returned prematurely because the young Duke of Anjou is severely ill and said to be dying. Four days ago our good, eccentric Mr Monceaux went the same way. I have asked the brother-in-law for more details. Do you hear anything more of the matter of de IJssel? It is most peculiar, after being of such service to the gentlemen, and going to such trouble on their behalf. They do not even thank us, nor do they pay what is still owing. I think one of these days I shall write to Mr Hudde on the matter.

We are fortunate to have this letter, written on his return to Paris, in its entirety, but we have to make do with fragments to suggest what followed. They come from letters to Lodewijk on 21 August 1671 and 6 May 1672:

> I have obtained some sort of compensation for the use of the carriage, but we now exchange the frostiest of looks, and I visit them on matters of business only. For some time now they have also been engaged in a row with the Perrault brothers. While I was in Holland, brother Claude and Carcavy shouted abuse at one another several times during the meeting.[39]

> People must know of our heroes: Frénicle, Roberval, Buot, Borelli, Cassini and captain-at-the-helm Mr Carcavy. We conduct our meetings in the familiar atmosphere, and I go to no particular trouble to accommodate myself, for it cannot continue in this way. If I come across Mrs or Miss Mariotte in the garden, I greet them, but I never approach them. Mother [Carcavy] uses every means to reconcile the Perrault brothers with her Polyphemus, but she does not have any chance whatsoever. Never has so much war paraphernalia been seen: streets filled with packed carts, mules loaded with men and horses fully harnessed. And now that these have all departed, the streets are so filled with people that they are no emptier. The King was seen to leave with 8 million in silver on 84 carts. That is the very essence of war.[40]

In May 1672, an unprecedented army of 118 000 men set out from Charleroi to Cleves and crossed the Rhine.[41] For that time, the army moved fast. On 13 June it stood at the walls of Arnhem and Nijmegen and on 21 June it marched into Utrecht. One more month, and it seemed that Holland would be taken.

Then General Henri de Turenne and Minister Louvois quarrelled, and Louis XIV made the mistake of taking his minister's side. The general wanted to push on immediately to Amsterdam, but the

---

[39] OC 7, 100   [40] OC 7, 172   [41] Israel (1995) 796–803

minister preferred to wait until the besieged towns had surrendered. While they were waiting, the polders between Gorkum and Weesp were flooded, to stop the advance of the army. In the meantime, troops of the Bishop of Munster occupied the towns in Overijssel, after which they proceeded to Groningen. In July a vast English fleet appeared in the Marsdiep, and was only prevented from landing by a storm.

The day he entered Utrecht, Louis XIV received a delegation from The Hague. They offered the Generaliteitslanden and 10 million pounds, but he demanded the occupied provinces and 24 million. Haggling ended when Willem III, who had been given command over the army by De Witt, broke off the negotiations. He did this as Stadholder, which he was proclaimed on 4 July by the States of Holland, after that of Zeeland, and after rebellions in the towns. He wanted to fight.[42]

The drama of Johan de Witt, who had been too late in seeing the war coming, unfolded rapidly. On 21 June the Grand Pensionary was stabbed, on 4 August he resigned, and on 20 August he and his brother Cornelis were assassinated in the street.[43]

How did Christiaan react to all this? 'The fate of the State Pensioner and his brother is horrifying. I already knew of it on Friday, but not in the details that you furnish me. If such things occur, perhaps the epicurists are right [with their utterance] that government of The Republic is not for the wise. The State Pensioner was indeed extremely careless to expose himself in broad daylight to an excitable crowd, but I grieve deeply for him, for in my opinion he has done nothing for which he deserved to die.' He wrote this to Lodewijk on 4 September.[44]

Earlier, on 15 July, Christiaan had reported:[45]

I am greatly gladdened at the success of the Prince [the Stadholdership] and I hope for favourable consequences for the State and especially for our family. When I last spoke to Colbert, I discussed the possibility of a protective escort [to and from

Zuylichem] if that should prove necessary [read: in case of
plundering], but he assured me repeatedly that it would not come
to that, because the matters would be settled by arrangement. If
we still feel under danger, then in my opinion it would be best to
apply to Mr Turenne, who knows my father. It has rained
continuously this past week, and if that is also the case in
Holland, then that will prove fortunate, and will help to
strengthen the defence.

Earlier, on 1 July:[46]

Rumours are buzzing everywhere; the Dutch losses are greater
every time. Yesterday we heard that the King was to enter
Amsterdam that very day, that De Witt was dead and that Mr Van
Beuningen [the ambassador] was under arrest and to be handed
over to the King, but that he had fled to Spain. How much of this
is true? I shall believe only what you write to me.

And shortly after the start of the invasion, on 17 June:[47]

The last weeks I have been out of town, in Chantilly Liancourt
and after that in Viry, so that is why you have had no letters from
me. When I returned I learned the news of the surrender of four
towns and I assure you that I was greatly surprised. If my friends
here were not so reasonable and so moderate, then these would be
difficult times for me. Everywhere there is rejoicing and talk of
the misfortune and the scandal of the fatherland. For two days
together, people lit bonfires, the first for victory, the second for the
birth of the Duke of Anjou. Around thirty trophies won from the
Dutch were paraded in triumph on the route to Notre Dame,
where they were singing the *Te Deum*. You can imagine what
feelings were kindled within me when I saw and heard all these
things. But I try to set my face as much as possible so as not to
show my feelings, and I do all I that can to stop those terrible
thoughts of what might still occur.

[46] OC 7, 182   [47] OC 7, 176

In spite of these and other similar expressions of feeling, the war did not disturb his equilibrium. Christiaan knew people in France who felt ashamed. In April, when war was declared and all the Dutch were ordered to leave France within six months, Colbert assured him that this did not apply to him.[48] Even in 1669, when Van Beuningen realised that war was inevitable, he found that 'there is nothing in the post I occupy that has anything to do with the war'.[49] For his own family the war was even a matter of good fortune; in February, Constantijn became secretary to the Prince, when Willem III was appointed head of the army, and in September Lodewijk obtained the post of bailiff and dike warden in Gorkum.[50] At the end of July King Louis returned to Paris empty-handed, while troops from Brandenburg and Spain came to the aid of the isolated Republic.

So if he had so little to fear, why did he then write such a dishonourable dedication for *Horologium oscillatorium*? This was all that was still lacking, after the addition of Part IV which he had written in 1671 in response to Roberval's critique. This remarkable piece of work is dated 25 March 1673, but was certainly written earlier. Let us try and be understanding of the fact that he was obliged to surrender his gems of 1659 to a pompous show-off. How much time did he need to polish his words before they achieved that refinement accorded to them by Norbert Elias?[51]

TO LOUIS XIV
Illustrious King of France and Navarra

O Great King, it is thanks to France especially that in this century geometry has been reborn. Indeed, it was here that men arose and by their great and prolonged exertion brought to life once again this lost and almost forgotten science. Treading in their footsteps, men of talent in Europe have made such advances in geometry, and have surpassed those findings of the ancients to such a degree that there seems little that remains to be discovered by future

[48] OC 7, 164   [49] OC 6, 544   [50] Vollgraff 669   [51] OC 18, 74–81

generations. This science, which I have always cherished and admired, has led me to prefer those useful appliances that may add enhancement to life, or that may enable me to understand better the natural phenomena that surround us. And I believe that I have carried out my best work when occupied with matters in which usefulness was combined with subtle reflection and arduous discovery. If I may be permitted to express oneself in such a manner that in no way diminishes your greatness, then I confess to having pursued this double goal with no greater success than in the invention of my clock. For this clock is not only a mechanical invention, but – even more important – it has been assembled according to the very fundamentals of geometry. With regard to the latter, I speak of the very heart of this art, which is no small matter. That is why I place this consideration without doubt above all those other matters that I have struggled with during my studies. Few words are necessary to show you, Almighty King, how useful it is. Since my clocks have already been found worthy enough to be placed in the private rooms of your palace, you know from daily experience how vastly superior they are to all other instruments that show the same hours. For the rest, you are well-informed of the other more specific uses which, from the beginning, I had in mind. They are, to give an example, especially appropriate for the purpose of observing the heavens and for seafarers who wish to determine the longitude in various positions. In accordance with your orders, our clocks have more than once been transported across the sea. Many astronomical clocks, also under your supervision, can be seen in the excellent observatory that you have recently had built, with such remarkable generosity, greater than that of all other kings.

When I consider this, I rejoice that this invention has, by happy coincidence, occurred during your reign. No one who understands the acknowledgement that you deserve for this invention shall ask why I have dedicated this painstaking study to your illustrious name, a study in which I have described in full the theory of my clock. And absolutely no one will in any way be astonished when they learn that it is through your graciousness that I have enjoyed the pleasures of freedom and tranquillity for these and other reflections. In some measure I must justify myself to you for this freedom, and to acknowledge to you that all your perpetual favours oblige me to express my gratitude towards you. Since you are so fully occupied, and must have attention for the most weighty of issues, such as those that may be expected to befit a leader like yourself, these matters will hardly be likely to be worthy of your attention. But I venture to think, illustrious King, that this work is no less welcome [than matters of state]. For the general good is your greatest pleasure, and progress in the sciences through new inventions and discovery, your greatest concern. Your extraordinary generosity towards the advancement of the sciences, and your rewards to those who excel in them are ample proof of this. Your benefaction is limited neither by the costs of war, which are enormous, nor by the frontiers of France. Indeed, it is clear you extend these to those from elsewhere who are worthy of your favours, and grant these in order that they may be happier and wiser in the knowledge that they are better off with such a ruler. Perhaps these lines may contribute something to your truly august glory. And if what we have written should, in future, be regarded as evidence that, in these times, the arts and sciences blossomed, then it shall also be seen as evidence that it was, above all, thanks to your graciousness and your magnanimity.

If it had contained even a trace of irony, then it would, at least, be palatable. But the only sentence that might be ironical (on wars and frontiers) cannot be read as such, because otherwise the following sentence, which explains it, would also have to be ironical. Surely no one can imagine that? The pressure that led to this polished dedication must have been extremely strong, for the subject who wrote it had a brother who, at any moment, could be destroyed by that 'truly august glory': Constantijn would indeed experience, at close hand, the horrors of this pointless war. Or was the king not the only one exerting pressure?

At that time Christiaan had his dining room hung with gilded leather that was raised to look like beautiful Genoese brocade. He gave good parties there, being served by two lackeys and a woman-cook. And of course he had a coachman for his carriage and horses.[52] Romf, whom we have come across before in the letter to Lodewijk of 9 July 1671, wrote to father Huygens, on 13 January 1673, about what people in Paris thought of his learned son: 'Many of those whom I have heard, Sir, in passing, do not hold it against your famous Archimedes that he publishes his great discoveries and secrets in the land of the enemy, for it earns him a pretty salary.'[53]

Horologium oscillatorium was printed by François Muguet, after Colbert had approved the text in the beginning of August 1672. On 4 November printing was so far advanced that the text was registered at the Royal Library, but it did not actually appear until 1 April 1673.[54] The magnum opus numbered 159 pages. In the new edition of the *Oeuvres complètes* there are ten fewer: three for the introduction, sixteen for Part I, twice that many for Part II, twenty-seven for Part III, fifty-nine for Part IV, and three for Part V; the appendix with the thirteen theorems on centrifugal force, which are written not in Latin but in French, consists of two.

As a matter of course he sent it to the king and to Colbert, and for the rest, to members of the Academy (Borelli, Buot, Cassini, Frénicle, the editor of the *Journal des Scavans* Gallois, du Hamel, Mariotte, Niquet, Pecquet, Picard and Roberval), and to acquaintances in Paris such as Arnaut, Baluze, Chapelain, Chevreuse, Conrart, Huet, Justel, La Lovère, Petit and Thévenot. Naturally, his father and both brothers also received a copy. In his homeland, De Witt was on the list, but he, in the meantime, had been assassinated; further copies

[52] Vollgraff 674; OC 7, 101, 113, 211
[53] Brugmans 75 (n4) (OC 22, 104)   [54] OC 18, 84

went to Hudde, his correspondent from Delft Van der Wal, and Sluse, the mathematician in Liège. For the rest, he sent a parcel of books to Oldenburg, requesting that they be delivered to Ball, Boyle, Brereton, Gregory, Hooke, Lord Kincardine, Moray, Neal, Newton, Wallis and Wren.[55]

A new name has appeared: Newton. When this gentleman had read the book, he wrote to Oldenburg, on 23 June, that it was 'full of very subtile and usefull speculations very worthy of ye Author'. But because he missed the most important thing of all, he expressed his wish that the author publish more on centrifugal force, 'wch speculation may prove of good use in natural Philosophy and Astronomy, as well as Mechanicks'.[56] Apparently he knew what he was talking about. He also knew of a cleverer way to determine the comparison of the evolute. 'If he please, I will send it him.'[57] But Christiaan did not respond to this invitation, and nor would he grant Newton his wish.

How did Newton come to be included on Christiaan's list? To understand this, we need to go back a year and delve into his correspondence with the secretary of the Royal Society in London, of which he was, after all, a member. Heinrich wrote him letter after letter. These increased in number from 3(1) in 1671 to 12(4) in 1672 and 13(5) in 1673, fell to 7(1) in 1674, but was back again to 25(10) in 1675 (the number of letters Christiaan wrote back are written in brackets). The exchange of ideas with members of the society went via him. He sent on letters and articles, but if he thought that they might give offence, he would make a summary of their professional contents.

It was on 1 January 1672 that he wrote to Christiaan, 'Let me take this opportunity to give you the description of a new type of telescope by Mr Isaac Newton, professor of mathematics at Cambridge.'[58] And he included, along with his letter, the description of a telescope in which the large (convex) lens is replaced by a (concave) mirror. Because the image was formed in the path of light to the mirror, the

[55] OC 7, 321   [56] OC 7, 326   [57] OC 7, 328   [58] OC 7, 124

inventor had placed a small flat mirror in this path of light, so that it could be viewed with an ordinary lens sideways on. The advantage of this telescope was that the colour flaw of the large lens in no way interfered. It was as elegant as it was efficient.

Christiaan was delighted with the idea. In February, the same month that Newton's description appeared in the *Philosophical Transactions*, the Dutchman pointed out its advantages in the *Journal des Scavans*. Just as promptly, he had a small one made, with the focus at 1 foot, and when he found this to be satisfactory, he had a large one made, this time with the focus at 12 feet and a mirror 11 inches in diameter. This quickly showed up the problems. The metal in front of the mirrors was softer than glass 'and hardly allows it to be ground without spoiling the shape'.[59]

This was not, however, the only reason, or even the most important reason, that Newton had attracted his notice. In 1672, *A New Theory About Light and Colours* had featured in the March edition of *Philosophical Transactions*.[60] It was Newton's first publication in experimental and theoretical research, in this case on the refraction of light in lenses and prisms, and it had earned much admiration. But it also aroused fierce criticism, and once again he disappeared from the public eye, in so far as he was still able to do so. Christiaan found it a clever work, but his reservations were such that, with the passage of time, they did not diminish.

On 9 April he wrote to Oldenburg, 'As far as his new theory on colour is concerned, it strikes me as highly ingenious, but it remains to be seen whether it applies in all instances.'[61] On 1 July he went on to add:[62]

As far as his new hypothesis on colour is concerned, on which you seek my opinion, I admit to you that it strikes me as extremely probable and that the decisive experiment [with a second prism, in which the colours, which had passed through the first prism and were separated, became separated even further] strongly confirms

[59] Vollgraff 673   [60] Westfall 240   [61] OC 7, 165   [62] OC 7, 186

it. But I do not agree with what he says about the refraction in convex lenses, for I find that, according to his own theoretical assumption, to be twice as great as he claims.

And on 27 September:[63]

> What you sent me on Mr Newton in one of your last journals confirms his colour theory, yet again. But the matter could well be otherwise, and it seems to me that he must be content with the fact that he has proposed a highly probable hypothesis. If it is true that, from the start, the light rays exist as red and blue etc., the difficulty remains to account – by means of mechanical hypothesis – for the diversity of colours. What I have said on the refraction in convex lenses [see previous letter] is certainly a misunderstanding.

Christiaan had made a mathematical error . . . Finally, on 14 January 1673:[64]

> I have seen how Mr Newton has exerted himself to defend his theories on colour. It seems to me that the greatest objection, put as a question, is this: are there more than two colours? Speaking for myself, I believe that a hypothesis, which could account for the colours yellow and blue, with motion, also accounts for all the other colours. For these are simply more charged, as appears from the prisms of Mr Hooke, which results in red and dark blue. From these four, all colours may be concocted. Also, I do not see why Mr Newton should not be content with the two colours, yellow and blue, for it would be far easier to make a motion hypothesis that accounts for this one difference, than a hypothesis that accounts for so much diversity of colour. And until he has found that hypothesis, he shall not have taught us what is the nature and difference of colours, but only the fact (which is certainly important) of the difference in their refraction.

[63] OC 7, 228–229   [64] OC 7, 242–243

Newton, who had also become involved in a conflict with Robert Hooke, waited until 13 April before replying. Oldenburg sent this reply on to Christiaan: 'It seems to me that N. takes an improper way of examining the nature of Colours, whilst he proceeds upon compounding those that are already compounded.'[65] 'N.' stood for Nobody, the man he did not wish to know: Christiaan Huygens. He wrote that it was no easier to make a hypothesis for two than it was for more colours, and that it surprised no one that the waves of the sea and the sand on the beach showed an infinite diversity. Why then should the particles of luminous bodies have only two sorts of rays?

> But it is not my aim to explain how colours may be accounted for with a hypothesis. I have never attempted to show what is the nature and the difference of colours, but only that they are *de facto* characteristics of the rays, from beginning to end. And I leave it to others to explain, with mechanical hypotheses, the nature and difference of those characteristics. It doesn't seem terribly difficult to me.[66]

His tone was heated, bordering on the insulting. N. felt this, and wrote to Oldenburg on 10 June: 'Now that I see how passionately he sticks to his viewpoint, I have no wish for further discussion with him (*cela m'oste l'envie de disputer*).'[67] But he had recognised excellence, and put Newton's name on the list for *Horologium oscillatorium*.

This episode with Isaac Newton raises a number of questions. Had he ceased to engage in difficult discussions? (He also heard little more from Gilles de Roberval, who was to die in 1675.) Did he feel that he had already 'made it'? Had he stopped paying attention? In any case, when the great book was published that would re-establish his name, there was much going on of which he was unaware.

The plagiarism of his masterwork of 1652 was among the lesser of these occurences. Edmé Mariotte managed to practically take over

[65] OC 7, 265 [66] OC 7, 266 [67] OC 7, 302

De motu in his *Traité de la percussion ou chocq des corps* (Treaty on the Impact or Shock of Bodies), which he had published in 1673 under his own name.[68] 'Mariotte has cribbed everything from me; he should have cited my name!'[69] Certainly, he made complaints, but he had only himself to blame that it had ever come this far. The text of *De motu* had been lying around the Academy for years, and he had never gone to the trouble of having it published.

More important was that he lost the name he had made for himself in 1655. He had been given two rooms in the observatory, which, in the meantime, had been completed, but he seldom used them.[70] Giovanni Cassini, who lived there and looked through the telescope on every clear night, had him fetched on the night of 23 December 1672.[71] For an hour, he had had the certainty of the discovery of yet another of Saturn's moons. He had also fetched Mariotte-the-plagiarist, Jean Picard, and Ole Rømer, a Dane who had recently become a member of the Academy. They looked through their best lenses, new ones that Eustachio Divini had supplied, and they saw, beside the apparently ring-less planet, a small speck of light that was to be called Rhea. 'I am satisfied enough with my original discoveries; they are more valuable.'[72] Even so, he, the discoverer of Titan, wished to view Rhea independently on 30 December and twice in January. But things went wrong. He confused her with Japetus, the moon that Cassini had discovered the year before, and made a calculation error in the orbiting time.[73]

That he was somewhat absent-minded is confirmed by his notes on the Icelandic crystal, which he had made just after the new year. Erasmus Bartholin (Berthelsen, actually), an uncle of the previously mentioned Rømer, had seen remarkable double images through it, and had already decided in 1670 that, in addition to the usual refraction of light, a '*mira et insolita refractio*' ('curious and unusual refraction') occurred.[74] Picard, who had looked him up in Copenhagen in 1671,

[68] OC 16, 207 (n10)   [69] OC 16, 209   [70] OC 7, 349   [71] OC 15, 115 (n11)
[72] OC 7, 348   [73] OC 15, 116–118   [74] Vollgraff 676

returned to Paris with a few of these crystals (now known as calcite). Christiaan wished to understand more about these than Bartholin had. The secret of the crystals became his own; he wished to discover it on his own.[75]

Although he was apparently now keen to maintain his reputation as a researcher, his notes on what he saw were so confusing that he kept them to himself for years to come. If he caused a light ray to pass through a crystal, so that he obtained two rays (a regular and an irregular), he found that the refraction of the two rays through a second crystal did *not* produce four rays (one regular and one irregular from the regular ray, a regular and an irregular from the irregular ray) each time. He viewed a small object under two flat polished calcite plates and saw it either twice or four times, depending on the '*positu duorum frustrorum*' ('position of the two plates').[76] If he turned one plate vis-à-vis the other, two extra images appeared, and if he turned them further, they disappeared again. We understand this to happen because the light is polarised: the regular ray oscillates perpendicularly to the irregular ray. He himself did not get any further than: 'It seems as though, as it passes the upper plate, the [regular] ray has lost something that is necessary to bring the matter into motion, which is needed for the irregular refraction [in the lower plate]. But to say how that operates – up until now I have discovered nothing that satisfies me.'[77] He rarely wrote about something he did not understand, but this he did publish, seventeen years later.

Does this mean there was nothing more to be discovered that suited his mechanical view of nature? The strange thing is that Jean Richer had supplied him with something of the kind, but, assuming that this man would never learn how to treat a pendulum clock, he never made use of it. After his failed expedition of 1670, Richer set out on another voyage, and found himself, between April 1672 and May 1673, in Cayenne. He wrote that the second-pendulum was 5/4 of a millimetre shorter there than in Paris.[78] If this were true,

[75] OC **19**, 407–415   [76] OC **19**, 413 (§7; n4)   [77] OC **19**, 518   [78] Mahoney 258

then the acceleration must become smaller, and the radius of the
Earth greater, if one travelled from the pole to the equator. Christiaan could not believe it. And yet . . . what are we to make of
his appearance in the following chronicle of 1673? It deals with
the popular Paris salon for ladies, held at the home of Madame de
Chaunes:[79]

> One of our immigrants at the Academy read aloud letters from
> Cayenne that were bursting with figures, curves, tangents and
> triangles. The physicists, who had been sent there a year earlier
> for the purpose of carrying out observations, note that the earth is
> not round, like a ball, as people have thought up until now, but is
> flattened off on two sides.

For the sake of contrast, we cite two sentences from his brother's
journal. Constantijn was staying in the Rhineland, the area to which
the war had shifted: 'During the plundering [of Rheinbach] a dragoon
had robbed a farmer who had been taken prisoner, and also demanded
his shirt. When the farmer had removed his arms from it, with some
difficulty, another dragoon from Kurland came up from behind and
hacked off his head, in one fell swoop.'[80]

In 1673 dragoons also threatened Elizabeth Buat. In an act of
vengeance, she had been placed under house arrest and had dragoons
quartered in her house. 'I have called on her two or three times and,
as usual, these visits are always convivial, but each time there is an
officer on guard and, moreover, she has three dragoons billeted with
her, and all at her expense.'[81] Elizabeth was being given a foretaste of
the 'dragonnades', which would occur later on a large scale to drive
the Protestants out of France.

A short walk through the gardens of the Palais-Royal brought
him to Constantia Caron, but in 1673 the doors of the garden were
bricked up, so he had to walk round. Even then, she was close by.[82]
When war was on the point of breaking out, they spent a day walking

[79] Brugmans 80   [80] Huygens **32**, 15   [81] OC 7, 370   [82] OC 7, 83, 171

in Versailles, from eleven o'clock in the morning until seven in the evening.[83] In his letters to Lodewijk he – understandably – said little about her. He had more to say, though, about her daughter Suzette, who hated the countryside in Normandy, the home of her husband François de la Ferté, and spent the winters with her mother. He found her attractive ('*jolie*', later more forthrightly '*belle*'), described her in tears, and in February 1674 became godfather to her first child, a daughter.[84]

He revealed his weakness for young girls again in 1671, when he flattered the nineteen-year-old Haasje Hooft with his attentions.[85] From Paris he continued to pursue her with gifts, but she understood that it would come to nothing more, and in 1673 married a mayor of Amsterdam. The significance of his attentions in Paris to Constantia le Leu de Wilhem will always remain obscure.[86]

Even more important than his waning interest in astronomy was that he was losing ground in mathematics. He kept working, certainly, but along the same beaten track. He was currently occupied with the proofs of Euclid and checking whether his basic assumptions were either essential or superfluous, matters he discussed in the Academy.[87] In addition, he was looking at positive and negative values in the equations of analytical geometry, but this was hardly new. A student would show him something else.

In the autumn of 1672 Gottfried Leibniz turned up on his doorstep; he claimed to know something of mathematics, but lacked the finer points.[88] If he would teach him? We know little more about their first contact, other than that Christiaan said yes, and that he set him rigorously to work ('*plus credo in me querebat quam erat*' – 'I think that he looked for more in me than I possessed', Leibniz later wrote[89]), and that he gave him a proof of *Horologium oscillatorium*

[83] OC 7, 173   [84] OC 7, 210, 377   [85] OC 7, 113 (n10), 120, 160
[86] OC 7, 87, 105, 107, 113, 123   [87] Vollgraff 675   [88] Aiton 41
[89] Heinekamp 101

when he wished to spend some time in England. We know more about their second contact, which began after Leibniz's return to Paris and would last for the whole of 1674.[90] The teacher let him study the quadratures of Gregorius van Sint-Vincent, Pascal, Roberval, Sluse, as well as his own. On 30 December 1673 Christiaan lent him an edition of his *De circuli magnitudine inventa*.[91]

But Leibniz did not become a mathematician like him. Undoubtedly he was influenced by the fact that he had spoken with Wallis in England and had come to understand better the work of Gregory, but mainly it was his own insights that led him into new avenues. Before the year had ended, the student sent to the master a calculation of π. We give, in full, Christiaan's response of 7 November 1674, the first of sixty-four letters that they would exchange:[92]

> I return your writing on the arithmetical quadrature, which I find most elegant and most fortunate. It is no trifling matter to have discovered a new method in a problem that has preoccupied so many minds, a method that might lead to a real solution. For you have found that the quadrature of a [quarter] circle is in proportion to that of the square that surrounds it as the infinite sum of the fractions
>
> $$1/1 - 1/3 + 1/5 - 1/7 + 1/9 - 1/11 \ \&\text{ct.}$$
>
> It is very possible that this sum can be [precisely] calculated, for you have succeeded with other sequences that resemble it. But even if this is impossible, you have still discovered a most remarkable characteristic of the circle that shall always remain renowned amongst mathematicians. With regard to the nameless curve in your proof, I had wished to give it a name in which the circle and the cissoid occur, for, according to me, it follows from these. But now I have seen that J. Gregory has already propounded this curve, so that he has the right to name it. He has used it to

[90] Aiton 49–55   [91] Vollgraff 688–689   [92] OC 7, 393–395

find the proportion between the squares of the cissoid and the
circle, the latter of which I discovered myself. It also arises in the
treatise of Mr Wallis on the cissoid, and because he discusses it in
his treatise on motion, I enclose my proof of this theorem.
Because it has been found, you can shorten your arithmetical
quadrature. But do as you think best. I wish you a good day and
am entirely at your service.

This, he could still follow, but later Gottfried would journey along
paths wholly obscure to Christiaan. In a following letter, of 30 September 1675, he wrote of Gottfried's calculation of the root of 6 in which
'there is something concealed from us [!] that is not to be understood.'[93] It was right before his eyes, yet he failed to see it: Leibniz
had discovered the *calculus differentialis* (differential calculus).

We can say that Christiaan had yet another student, practically the
same age, who also tended to go his own way: Denis Papin, with a
doctor's degree in medicine (obtained from Angers) but who did not
wish to become a medical doctor. He was a born engineer.[94] Christiaan
had got to know him in 1671 in Versailles, when he was required
there to advise on the capstan wheel mechanism in windmills. These
mills powered the machines that pumped the water in the fountains
upwards, and the engineer worked on these pumps. This frivolous but
troublesome project brought them together so often, and Christiaan
was so impressed with Papin's practical insight, that he decided to take
him on as his assistant in the experiments with pumps and vacuums.
He was twenty-five when he became assistant, in the course of 1672,
at the Rue Vivienne.[95] As soon as Gottfried came by again, they got
to talking. The students struck up a friendship that would last for the
rest of their lives.

From the moment that the teacher gave Papin the task of building a replica of the Kalthof machine, he continued practically on his

[93] OC 7, 504   [94] OC 7, 412 (n11)   [95] Vollgraff 672

own.[96] Christiaan must have picked up the idea while in London, when he called in on Caspar Kalthof, his former instrument-maker from Dordrecht. It concerned a steam engine.[97] This is an engine that is based upon the phenomenon that vapour, if heated and confined within a vessel, rises in pressure, so that a movable wall of this vessel can be thrust upwards. Just what Papin made of it, we do not know. What is important is that, one day, he replaced the steam created by boiling water with hot gas from burning gunpowder. In October 1673, he invented the internal combustion motor.

Just as it would be inappropriate to honour the teacher with the great discovery of his student Leibniz, so is it inappropriate to honour him with the invention of his student Papin. The difference between them is that while he was unaware of the first, he was quite unable to ignore the second. Christiaan would certainly have heard the motor, and smelled it. That is why he was able to write about it, and to create the impression that he had thought of it himself: 'Recently I have shown to the gentlemen of the Academy, and after that to Mr Colbert, an invention that appeals to the imagination [. . .] and is based upon the moving force of canon powder and gas pressure. [. . .] Four or five footmen were lifted by it with ease.'[98] This is what the Dutchman wrote to his brother on 22 September 1673. Papin did not hold it against him – at least, not in public. The following year he dedicated to his boss the *Nouvelles experiences du vuide* (New Experiences with the Vacuum).[99]

Nevertheless, Christiaan had risked a quarrel with his assistant. But he did not manage to avoid a similar quarrel with Isaac Thuret, when he instigated an idea and then left it to someone else to work it out in practice. Thuret was the best clockmaker in Paris.[100] He had a licence for the sea clocks that – apart from the pendulum – were distinguished by a remontoir. In his experience a remontoir was practical, but vulnerable when hoisted up too quickly. Thuret must have shrugged his shoulders when the inventor, who had originally

---

[96] *Ibid.* 680–681   [97] Payen 200   [98] OC 7, 356; OC 9, 79
[99] OC 19, 216–238   [100] OC 18, 505 (n9)

regarded him as a friend, prescribed that it should be done every half minute. Such remontoirs were also unsatisfactory when used at sea.

On 20 January 1675 Christiaan calculated that a spiral-shaped metal spring, which seemed to wind and unwind itself at the same time, acted like a pendulum, so that it could be used to make accurate portable clocks or watches.[101] The next day he went to his clockmaker with a sketch and ordered him to make a model with such a 'balance spring', so that he could apply for a patent on it. The invention was certainly clever, though it hung in the air, and he was highly pleased with it. Two months later he wrote to Lodewijk, after congratulating him on the birth of his first son, 'You have a fine son, and I, a "daughter invention" that is fine in her own way. She shall enjoy a long life alongside her elder sister the pendulum and her brother the ring of Saturn, just like the children of the good Epaminondas.'[102] But although he succeeded in obtaining his patent, Isaac Thuret nearly beat him to it.

Apparently the servant had had enough of playing second fiddle to Christiaan. After delivering the model, he made a second one and took it straight to Colbert, with the claim, or the suggestion, that he had invented it himself. Not entirely proper, but understandable when we see that he had improved on a number of inaccuracies in the sketch. The sketch that Christiaan made a few days later showed a number of significant adaptations that betrayed the craftsman.[103] That second sketch was proof of his own theft; a strong word that arises when we learn how roughly the servant was dealt with. We quote from Christiaan's journal on the affair:[104]

Friday 8 [February 1675] a response from Colbert, who had placed his [consent] in the margin. I discussed the matter with the King. Mr Duverney came to me and said that he understood Thuret to have paid a visit to Colbert and to have submitted to him a

[101] Leopold 227–228   [102] OC 7, 430
[103] Leopold 229–230 (OC 7, 408)   [104] OC 7, 411–412

petition. I said this to Mr [Claude] Perrault, who told me what he knew up until then: that Thuret had shown my invention to Colbert eight days before I did. I believe that it is true, and that it was the model he showed to me on 23 January. And I believe that Mr Gallois will have been present. I recall that Thuret displayed some confusion when I told him that I had showed my model to Mr Colbert just shortly before.

On Saturday the 9th I was at the house of [Charles] Perrault, the controller, who requested that Mr Gallois be present, and at the same time asked that Thuret be fetched. When they arrived, I presented to Thuret my arguments [to then think] that he had, the day after I asked him to make my model, without my knowledge, shown to Colbert just such a model and said that he had invented it. (He denied it.)

On the 11th Thuret came to speak with the controller. He admitted what he the day before so strenuously had denied.

It was on a Sunday, then, that he had been overcome by remorse. Christiaan did not forgive him. In the margin of his journal, beside these events, we read, 'I gave the manufacture of the clock to Gaudron, a cousin of Papin.'

Unfortunately, the other side of the story is unavailable to us. The invention of a mechanism is no invention if the mechanism doesn't work. The hardly insignificant design of the idea was not his alone, and the question is whether the idea, too, belonged only to him.

When he wrote about a constantly winding and unwinding spring in the *Journal des Scavans* on 15 February 1675, it aroused an angry response. To begin with, there was Jean de Hautefeuille from Orléans. In September 1674 this priest had sought support at the Academy for his idea of fitting a clock with an erect spring, a sort of tuning fork. Christiaan must have seen this proposal and – most probably – had a decisive voice in its rejection. Not entirely without reason, Hautefeuille saw this idea as a predecessor of the spiral spring; he had his lawyer note down, '*Difficile est invenire, facile autem inventis addere*' ('To invent something is difficult, to elaborate upon it is easy').[105] He decided to contest Christiaan's patent and went as far as bringing the case to court on 16 May 1675. 'It causes me scarcely any difficulties and I can easily refute such foolish claims.'[106] Hautefeuille lost his case in July.

Bitterness came to the fore in a quarrel with Robert Hooke. A large share of Christiaan's correspondence with Oldenburg in 1675 is devoted to it. After learning of the invention, Hooke notes down for himself, 'Zuylichem's spring not worth a farthing',[107] but he tried to be slightly more diplomatic in his public comments about it. He claimed that he had already put forward the idea of a spring fourteen years earlier, that Lord Kincardine could vouch for him that he had made clocks with springs and that 'Indeed, Mr Hugens hath made use of that part I discovered.'[108] These statements were quite unfounded, although it is true that Hooke wrote about metal springs in 1661; a law of elasticity is named after him.

On 15 October Oldenburg sent Christiaan this commentary, which also contained the slighting words '*facile inventis addere*'. He responded furiously, 'I am greatly astonished that you wished to inform me of the shameful accusations of Mr Hooke. I had already perceived him to be vain and hot-headed, but I did not know that he was so malicious and shameless as I now see him to be.'[109] Seldom did he resort to such language. It shows that he was beginning to come to the end of his reserves. Oldenburg consoled him with the promise that he would write to Hooke, 'as gently as possible', that he had supplied no convincing evidence for his assertions.

As we said, the idea of a spring hung in the air, but he was the one who cleverly seized it and made it real. Gilded clocks with a window, so people could watch Christiaan's spring pendulum in action, were gifts fit for kings. In July, Willem Boreel had even transported one of these clocks through the ranks, as far as Leuven, to present to

---

[105] OC 7, 443   [106] OC 7, 438   [107] Bell 70   [108] OC 7, 519   [109] OC 7, 528

Willem III in the army camp there.[110] But the man from The Hague had taken the criticism too much to heart.

> I have allowed the clockmakers every freedom to take possession of this invention, for I have understood that the patent would oblige me to have it registered with the parliament, and even then still more lawsuits and quarrels could be expected.[111]

He wrote this as winter began and melancholy returned.

[110] Huygens 32, 44   [111] Frankfourt & Frenk 152–153

12 Light

It is thanks to this crisis that we have Christiaan's magnificent piece of work on light. He had resolved to carry it out ever since Newton challenged him to provide a mechanical explanation for the effects of light. In 1673 the essence was clear to him: 'Light spreads in circles and not in an instant (*Lumiere s'estends circulairement et non dans l'instant*).'[1] But up until now he had lacked any tranquillity to patiently elaborate upon it. The theory came into being when, one year after the return of the horrors, and still withdrawn, he managed to pull himself together again. In this work, he was born anew.

Melancholia probably overcame him in January 1676. There are no descriptions of his sickbed, so we are unable to say whether it was less serious than in January 1670. Now that it had repeated itself, the most we can assume is that he experienced it differently. Six months afterwards, according to his father, Christiaan's face still showed 'certain traces of what he had suffered'. He himself said that the illness had returned due to painful experiences too hard to bear ('*trop facheuses experiences*').[2] The little that we do know, in this case, comes from father Huygens.

In a letter of 20 February 1676 to an acquaintance in the principality of Orange, there is talk of 'a melancholy illness suffered for some time by my precious son living in Paris.'[3] And in a letter written on 27 February to De Behringen:[4]

> I do not know what to make of this illness. Since he has no fever, the experts assure me that he has nothing serious to fear. But the illness is tenacious; let us hope that it is not his spleen. A

---

[1] OC 13, 742   [2] OC 8, 10   [3] OC 8, 7   [4] OC 8, 7

brother-in-law intends to visit him who is of a cheerful
disposition, always full of jests and pranks. If one person in the
world can restore to him a zest for life, then it is he.

This time Lodewijk was not sent over, for he had more than
enough problems of his own. Due to 'insufferable arrogance', this
brother, who had been bailiff and dike warden for three years in
Gorkum, had managed to get on the wrong side of just about everyone
there. Moreover, he had been foolish enough to indulge blatantly in
fraud and extortion. There was a court case in progress against him.
The unavoidable verdict, which would be pronounced on 3 July, was
resignation. But in the Republic, the judge did not have the last word.
Lodewijk would play the role of holy innocent and remain at his post.
He lodged a complaint with his influential father, and through him,
got Willem III to gloss over the whole matter. In the end he was able
to buy off his resignation by paying a fine.[5]

Philips Doublet, the brother-in-law who came in his place, may
have been full of jests and pranks, but we cannot help wondering what
kind of jokes he cracked to bring any kind of a smile to Christiaan's
face, when he arrived on an unknown date that spring at the Rue de
Vivienne. In his letters he expresses himself crudely. He would have
been the last person in the world to be able to feel what Christiaan
was going through. But, gossip that he was, he probably would have
understood the symptoms of the hypochondriac, and have accentuated
them. It therefore comes as no surprise that Christiaan gave a name
to his ills. 'The life that I lead here [that is: in Paris], disagrees with
me,' he wrote to Constantijn on 29 July, 'I left as if I would return
[tamquam rediturus]. But I do not believe that I shall ever return to
Paris.'[6]

He was then in The Hague, where he had arrived, 'still not quite
himself,'[7] on 14 July. The brothers-in-law had sailed to Vlissingen
(from Dieppe?), from where they had been invited to board the

[5] Smit 287–288; OC 8, 72 (n2)   [6] OC 8, 10   [7] Vollgraff 697

Stadholder's yacht as far as Brielle. This enabled them to avoid the road through army lines stationed in the Spanish Netherlands, where there was still considerable fighting that summer. Constantijn was with Willem III outside Maasdricht, which still had to be reconquered from the French; the Prince suffered a gunshot wound, his secretary, dysentery. It is possible that the secretary had had his fill when, one relentlessly hot day, he wrote in his journal, 'The enemy killed one of our gunners as he lay sleeping on the battery.'[8] And to what purpose? Negotiators had already agreed in 1675 on the substance of the peace treaty (the French booty would come from Spain, not from the Republic) and on where peace would be signed – in Nijmegen – but the insane thing about this war was that, even then, it did not stop.

It was because of this war that the first letter Christiaan wrote, after remaining powerless for several months, was to Constantijn. He missed him at Het Plein. This melancholic brother of his, to whom he had written for years about the various tricks of the trade in lens-grinding (little has been said of this, for he himself did nothing during these years), was the best one to be able to understand what was wrong with him. The letter expresses a powerful longing, from the very first sentence, 'Do not think that I shall return to France before the end of the campaign. If I should be recovered before that time, then I shall still wait for you here.'[9] To be followed by the sentences that we have already quoted, expressing his aversion to Paris. And then, 'You can imagine my pleasure as I was met on my arrival by four fine little nephews. Your own [Constantijn or Tientje, born on 5 February 1674] has just been to visit me. After he had looked for some time at a new sort of thermometer, with small glass balls that float in tartaric acid, he said, that's pretty, just as his lady mother would have said it.'

It is not apparent in his letters, but he seems to have got on well with 'his lady mother'. From now on, we shall call her 'Suzanna Huygens-Rickaert', sister-in-law, to avoid confusion with his sister

---

[8] Huygens 32, 88   [9] OC 8, 10

Suzanna Doublet-Huygens. She lived with her small son – and Constantijn, when he was not serving in the army – with father Huygens, and kept house at Het Plein.

'That's pretty.' We ask ourselves, in vain, why it would take another year before the light returned to his mind. Did he play with the toddlers? Did he stand and gaze on that piece of ground beside the gatehouse, scene of the assassination of the De Witts, reflecting on his own chances in life's game? This question arises because, during the months of August and September, he once again was solving problems in his notebook on probability.[10] His remarks in the margin of his old notations, on the thirty-one notes in an octave, betray an unadulterated nostalgia, '*Bon*' ('Good'), as if he is Colbert himself, and giving his stamp of approval to a project. We see the stern gaze of Colbert in his homework on static machines,[11] and that of the father in his translation of a letter, in November, from Antoni Leeuwenhoek.[12] This was about the discovery of infusoria, and Christiaan was not inclined to believe it.

We have practically no documentation of that winter and the spring that followed. Most probably he assisted Constantijn, who must then have been living at home, with grinding his telescope lenses. Possibly he also tried to make small lenses, like the glass balls in the tartaric acid that were in Leeuwenhoek's microscopes. We know that he wrote to Leeuwenhoek on 9 February, because he answered him, 'I was glad to hear that my observations were found pleasing in France.'[13] And we also know that he must have visited a picture gallery in Rotterdam in March, together with Constantijn, because the secretary, who had to go to war again, describes it in his journal.[14] But apart from that? A darkness precedes the light of 1677.

Summer came. He was determined not to return to Paris. 'I still have no wish to do so,' he wrote to Constantijn on 5 June, in one of his

[10] OC 14, 15–16 & 151–163   [11] OC 19, 29–33   [12] OC 8, 21 (+n2)
[13] Leeuwenhoek 2, 193; OC 8, 21   [14] Huygens 32, 143–144

scarce letters.[15] 'If I write to these gentlemen, I still use my health as a pretext, but thank God I am in fairly good form.' Probably he only corresponded with Claude Perrault, but we have no record of the letters. We do know that this was the reason he gave to Colbert on 16 September. It is strange that, to his brother, he calls it a pretext, for it provides a clear indication of the nature of his illness:[16]

> People have long considered that I am fully recovered, but I only believe myself to be in good health when I am able to study normally. I find I still lack the strength that is required for this kind of work. I would even do myself an injustice, and you, Sir, would censure my imprudence, were I to expose my fragile state of health [santé peu assurée] to a third dangerous setback, after having suffered two already.

That summer he read Ole Rømer's article in *Philosophical Transactions*, and this must have set him thinking: *A Demonstration Considering the Motion of Light*.[17] It was published in the edition of 25 June, and was a translation of the original article that had appeared in *Journal des Scavans* in December 1676. Christiaan must have missed the French article, for he never referred to it.

Rømer had a brilliant explanation for the course of the Jupiter clock. By this clock we mean the strict regularity with which the moons of Jupiter pass in front of the planet, or disappear behind it. The Dane had compared the Jupiter clock with the pendulum clock of the Paris observatory, and he had discovered that the first ran slower than the second for six months of the year, until the difference ran up to at least ten minutes, and then six months faster, until the difference disappeared. He reasoned that because the distance between Earth and Jupiter changes every six months, with the diameter of the path of Earth around the Sun, the light that the Jupiter clock reveals must move with an equal speed to the ratio of this orbit diameter and this time of at least ten minutes. '1,100 times 100,000 fathoms in one

[15] OC 8, 28–29   [16] OC 8, 29   [17] OC 8, 30

second.'[18] The reasoning was as correct as it was shocking. Light, it seemed, travelled fast, but not infinitely fast, as practically everyone thought: it wasn't there 'in an instant'.

Rømer proposed his explanation to the Academy on 22 November 1676, and it was attacked.[19] Cassini, in particular, would have none of it. But Christiaan, who was busy translating the discoveries of Leeuwenhoek, many miles away, believed immediately what he read in *Philosophical Transactions*. Had he not made the same assumption in 1673? 'Light spreads in circles and not in an instant.'

We now come to the mainspring of this biography, to the need that caused it to evolve. There is a personal element in this, which is why the writer removes, briefly, the mask 'we', to become 'I'. Three centuries after it first came into existence, the theory of light was commemorated at the Teylers Museum in Haarlem. Here Dieuwke Eringa presented there her translation of *Traité de la lumière* in Dutch – the first that was ever made[20] – and Hans Duistermaat spoke about the fantastic insight Huygens had had in the propagation of disturbances when he came to his Principle 'without using calculus, let alone partial differential equations'.[21] On this occasion I was given the opportunity to speak about the maker.

> My introduction to Christiaan Huygens took place at the gymnasium, when the physics teacher explained light. He used an experiment to demonstrate that light had to be a wave phenomenon, which had to be proved – *quod erat demonstrandum*. Therefore, he said, Huygens was right and Newton was wrong! He said it triumphantly, because Huygens was a Dutchman.
>
> I have learned in the meantime that we may not speak in terms of right and wrong, that insights must be understood in the context of their time, and that Huygens' wave theory of light

---

[18] OC 19, 469   [19] OC 8, 30–31 (n2)   [20] Eringa *passim*   [21] Duistermaat 275–281

cannot predict interference. But even so, this in no way erased the deep impression made by this introduction. My interest stems from the fact that he was a Dutchman.

But not only from this. My interest has also become professional. Anyone who wishes to know about propagation of a wave front still uses the Huygens Principle, because, in all its simplicity, it still provides the correct solution to a very complex problem. I, too, have made use of it, to understand how and why the image of a star in a telescope becomes blurred if the refractive index of air fluctuates.

I have the honour of speaking at the occasion of the publication of the first translation of Huygens's *Verhandeling over het licht*. It contains this Principle. The original French text was published 300 years ago, around 1 February 1690, according to a letter of 7 February.[22] The foreword is dated 8 January.

Huygens tells us in this foreword that the text had come into being twelve years earlier, and that he presented it to the *Académie des Sciences* in Paris. It therefore concerns work dating from 1678. The mainspring is even older, from 1677, when Huygens, for reasons of health, was living in The Hague. Various pages with notes on what is indisputably the theory of light date from the summer of that year, and on one of these is written: 'EYPHKA 6 Aug. 1677 *Causam mirae refractionis in Crystallo Islandico* (On 6 August 1677 I discovered [eureka] the cause of the miraculous refraction in the Icelandic crystal).[23]

On 14 October he wrote to the French minister Colbert, who had allowed him leave of absence from his job as scientific director of the Academy, and whom he was expected to keep informed, from time to time, on the state of his health.

'Recently I have had the pleasure of becoming acquainted with the admirable discovery of Mr. Rømer, who shows that light needs time to propagate, and who even measures this time. This

[22] OC 9, 357   [23] OC 19, 427

discovery turns out well for myself, for I have assumed this characteristic of light in what I write on dioptrics. I have used this to explain the refraction of light, and recently also the refraction in the Icelandic crystal, which is a miracle of nature that is hard to fathom. I frequently lament the fact that my physical constitution is so little suited to my inclination to work on this precious knowledge.'[24]

Huygens wrote this in French, of course.

So already in the autumn of 1677, he was busy writing up his theory. After his return to Paris in 1678, he made it public. He was not in any apparent hurry. He didn't read out the text to the academy until sometime between 13 May and 22 July 1679.[25]

It is not my task here to examine this theory. I have been asked to speak about the person who created it: Christiaan Huygens, the sickly scholar who was not in a hurry to publish.

After this, I read part of the biographical sketch that I had written in the summer of 1990, a sketch that went even further back, to notes made over previous years. One person at this gathering asked me if I intended to analyse the character of the scholar, as had been done in the somewhat notorious *Portrait of Isaac Newton*.[26] Although this was not my intention, nor would I be capable of doing so, I nevertheless did want to bring together the life and work of Christiaan Huygens: 'We ask ourselves: what was the interaction that brought about the remarkable work of this son of Constantijn Huygens and Suzanna van Baerle?' Since then this question has haunted me, as if it were my own life that was concerned. For over two years I devoted all my spare time to elaborating on this sketch, and to writing a book that attempts to answer this question. I delved more deeply into the source material, which usually came down to translating, and then I started to write. But in the very process of writing, it became clear to me that Titan, the name for the genius, was not to be derived from Christiaan Huygens, the name of the man, and that my question was

[24] OC 8, 36   [25] OC 19, 441–443   [26] Manuel *passim*

to remain unanswered. The interchange to which I refer may explain
the subjects, but not what is so remarkable about his work. Titan is
autonomous. In describing the year that produced this work, through
which I had become acquainted with him as a pupil at school, I am
once again confronted with a manifestation of this genius. I feel small
in comparison, and quickly retreat behind the mask.

'Light spreads in circles and not in an instant.' This was Titan's point
of departure. When he was strong enough to work again, he completed
the theory surprisingly quickly. The main points must have been ready
between 25 June and 16 September 1677; on the date first mentioned,
Rømer's publication appeared, and on the second he took the time to
write to Colbert.[27] He had certainly worked on the *Traité* earlier on, in
Paris, probably in his reaction to Newton's theory, and possibly also at
home, during that dark spring. But the discovery that his description
of light could offer an explanation for the double refraction in the
Icelandic crystal, that 'EYPHKA of 6 August, gave him the assurance
that he was on the right track. This meant that, in his view, the theory
was already in place, however provisional and incomplete it might still
have been. What remained now was to write it up.

> There is no possible doubt: light consists of the motion of
> particular particles. Indeed, if we look at where it comes from,
> here on Earth, it is chiefly the flame of the fire. This most
> certainly contains particles in a rapid motion, for it dissolves and
> melts other particles of fixed bodies. Or if we look at their effect,
> then we see light that is brought together by hollow mirrors
> operates like fire, which is to say that it causes the particles of
> matter to disintegrate. This strongly indicates motion – at least, in
> the true philosophy in which man explains, with mechanical
> arguments, the origin of all natural phenomena. This, in my view,
> is imperative, or else we must relinquish all hope of ever
> understanding anything of physics.

[27] OC 8, 29

And because it is certain, according to this philosophy, that the stimulus of seeing is only brought about by the motion of particles that impact upon the nerves behind our eyes, we have yet another reason to believe that light consists of a motion of particles between ourselves and the luminous body.

If one considers carefully also the enormous speed with which light spreads in every direction, and the fact that light rays from different, even oppositely situated points, cross one another, then one understands that this – the seeing of a luminous object – cannot be based on a transportation of particles that come to us from the luminous object, as if they were canon balls or arrows. Indeed, that would be in conflict with those two characteristics of light, and especially with the last. Therefore, it propagates in a different manner, and the knowledge of how sound propagates in the atmosphere helps us to understand how it occurs.

We know that sound propagates from the place where it originates by way of air, an invisible and intangible matter. It reproduces itself in a motion that passes from one part of the air to the other. Because this motion proceeds from all directions with the same speed, it must form spherical surfaces that reach our ears. There is no doubt that light from a luminous body, too, reaches us by means of motion in matter that lies between us and that body. We have, after all, already excluded the possibility that it occurs by a transportation of particles. Moreover, if light requires time to propagate – and we shall soon investigate this – then it must follow that the motion imposed upon this matter passes from the one part to the other [*que ce mouvement imprimé à la matière est successif*]. Therefore, as in sound, it must propagate in spherical surfaces and waves. I call them waves due to their resemblance to the waves that one sees in water when a stone is thrown into it, although these have a different cause, and only occur upon the surface.[28]

---

[28] OC **19**, 461 & 463

LIGHT 293

These evocative sentences from the beginning of the *Traité* bring us straight to the heart of the matter. Titan talks of waves, but these are different waves than those meant by Thomas Young and Augustin Fresnel, when they developed a new wave theory at the beginning of the nineteenth century.[29] Those were regular waves in an elastically coherent medium, the ether. Since then, we have come to regard the ether as superfluous, and we have retained just the regular waves, or electromagnetic phenomena. Christiaan, however, saw it in terms of a sea of separate particles, and he had a vague idea about how this sea, in a manner similar to a wave, must convey a pulse when the particles randomly collided with one another.

We are familiar with the idea; apparently, it contains nothing new. He did also refer to Ignace Pardies, Robert Hooke and, of course, to René Descartes, who had worked on this idea earlier. With regard to the spherical shape, he could also have referred to Kepler, who was convinced that light propagates not in rays but in spherules, although this, he felt, must be infinitely fast.[30] He could even have referred to the translation, published in 1572, of the work of al'Hasan, in which we see spherical constructions as well as rays.

And yet, he did have something new. The *Traité*, which touches the limit of what can be explained with particles, passes that limit in his consideration of the wave effects, which, as it were, are free from particles and filled with abstractions. There was something about light that was unrelated to matter. It was intangible. In a similar manner, Fermat had earlier referred to the fact that the law of refraction could be regarded as manifestation of a minimum principle. In the same way, Titan also arrived, by way of abstractions, at results that could be used very much later when it came to working out the theory of Young and Fresnel. It was not the idea, but the mathematical structures underpinning it, that turn Christiaan's theory into something new.

That isn't to say that the idea of particles was unimportant. It enabled him to think about the transfer of a pulse, a transfer in

[29] OC 19, 401 (+n9)   [30] OC 19, 476, 477; OC 13, 741

a sea of particles in which no regularity could be expected. Still, he thought back to straight lines of balls calculated in *De motu*, written twenty-five years earlier. In these lines, a pulse to the first ball is transmitted to the last, without movement in the balls in between, and the speed with which this occurs depends on the elasticity of these balls. 'Motion is perceived to occur with tremendous speed, which is greater in proportion to the hardness of the ball.'[31]

The particles turn out not to be infinitely hard; that is why light requires a certain time to pass through the line. And this hardness does not depend on the direction; that is why light travels equally fast in all directions and propagates in circles.

He saw this as a general truth. It could have nothing to do with the straightness of the line, or with any arrangement of the particles whatsoever. Anyway, if the cosmos were filled with these messengers of light pulses, then straight lines along the rays of a faraway star would be extremely improbable. This brought him to his first and greatest problem: 'Although the particles of the ether are not arranged according to straight lines, as is the case in our line of balls, but haphazardly, so that they touch various others, this does not hinder them to transmit their motion, and always in a forward direction.'[32]

Rather crass, to say the least. We will therefore repeat the French, *'cela n'empesche pas . . . qu'elles ne l'etendent tousjours en avant.'* Of course, it could not be otherwise; everyone could see that light travelled forward in a straight line, but he did not succeed in proving it. Yet it is thanks to his attempts to do so that we have the Huygens Principle: 'Each particle is surrounded by a wave of which this particle is the centre.'

The principle was conceived in a diagram. As he drew, he worked out what haphazardness must bring forth. 'In our line of balls' (from left to right, drawn as small circles on a straight line) he put (where there was otherwise space for a single circle *X*) two balls right beside it (the

31 OC 19, 472   32 OC 19, 473

one above the line, the other below it). A bit of confusion! And he saw what a pulse to X from the left would bring about: a motion to the right, certainly, but also motion upwards and downwards. A bit more confusion! At location X he put three circles beside the line, 'haphazardly, so that they touch upon various others'. And he saw that the number of directions of the motion that a pulse from the left can bring about on X increases. In the case of very great confusion, he thought, all directions of the motion will become likely. For this reason a spherical wave of motion must exist around X. And X is arbitrary. Therefore, this must be true for each particle: each particle is surrounded by a wave, of which this particle is the centre.

In his *Traité* the principle is written as follows: '*Il faut qu'autour de chaque particule il se fasse une onde dont cette particule soit le centre.*'[33] This wave must be spherical, because the only other thing that he says is that it has a centre. But even with this small addition, the principle is incomplete, for if the particles are of equal weight, the spherical wave will not be able to travel backwards ('to the left').

What does he mean by this wave, and what properties does he give it? The *Traité* is extremely vague on this point, 'All this should not create the impression of being too meticulously or too elegantly elucidated...'[34] He perceived strengthening and weakening properties of the wave, but did not articulate these, simply because the words were not yet available to do so. How, then, can we say that he saw them? Because he managed to combine the waves that existed around the infinite number of particles X in one single, new wave.

How he did this, appears once again in a drawing (Figure 4). Countless numbers of particles work constantly in pairs, and how they operate together follows from the way one pair operates. Around two points X_1 and X_2 that stand a short distance from one another, we draw two equal circles, C_1 and C_2, of which the ray is slightly greater than this distance. The aim is to create a sum that prevails

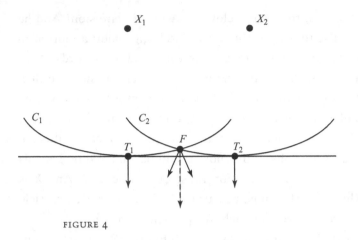

FIGURE 4

for the same moment of time, a fixed time after a spherical wave has emerged in X_1 and X_2. C_1 and C_2 intersect each other at two points, of which we shall call the one F (forwards), and the other, which could be called B (back), and which we lose sight of. After this, we draw the tangent line to the side of F, which touches C_1 at T_1 and C_2 at T_2. Now we have to imagine arrows on the circumference of these circles pointing outwards. These are the impulses that are transmitted at that one particular moment of time. The arrow of C_1 operates in T_1, the arrow of C_2 in T_2, but in F the arrows of C_1 and C_2 operate together: their sum at that point is greater than the length of an arrow. The waves reinforce one another in F. However, F is an arbitrary point. We could have chosen other points between X_1 and X_2. If we draw equal circles around them, then the points of intersection come to lie in an area G between F and the tangent line of T_1 and T_2. Therefore the pulses operate together in this whole area G, a real shambles, and there is something else that is required to find a clear sum.

That 'something else' is a vanishing trick. Titan simply proposed that the waves from the line X_1 to X_2 be added up only on the tangent from T_1 to T_2. On the other hand, he proposed that all waves in the area of G could only be infinitely weak (*'Mais chacune de ces ondes ne peut estre qu'infiniment foible'*).[35] The conjurer simply had them vanish. Did the possibility of interference perhaps occur to him?

Interference does not belong to his time; it is strange to his mechanics. It occurs to us because we know that weakening is possible if the waves recur, if they move in a regular pattern against the movement of other waves. But light pulses, Titan emphasised, are, on the contrary, very irregular. A. E. Shapiro understood why he would have it so.[36]

Where we would have regularity of waves, so that we could possibly interfere them away out of G, he would have the opposite. The summing of arrows (as we have shown above for point F and other points in G), applies only if they are there at the same moment. Since, apparently, no summing occurs, he reasoned, the arrows are not simultaneous, and that is why the waves of X_1 and X_2 could not have departed at the same moment. All the pulses occur quite arbitrarily. But even so.

One insight on his part was extremely clever. It could just have been pure bluff, or else because he never would have been able to explain why light travels forwards in a straight line, but he proposed that these separate circular waves were added up in their tangent. In broad terms: the envelope of the spherical waves around the particles in a wave front forms a new wave front.

The wonderful thing about this kinematic construction is that it is absolutely correct.[37] This is what happens in the case of light. One can therefore also take the view that Titan understood light within a structure that was unrelated to ether particles. Even more wonderful, Titan hereby raised mathematics to the level of physics and the physics of particles – in hindsight they were only will-o'-the-wisps – to metaphysics. His vanishing trick was physically imperative. It is for each one of us to decide where to place the construction on the spectrum from bluff and genius.

[35] OC 19, 475 (lines 37–38)   [36] Shapiro 206–207   [37] De Lang 22–30

If physicists in the eighteenth century ignored his *Traité*, then that is because, in their eyes, he was unable to deliver proof for the rectilinear propagation. They could not consent to the vanishing trick. And the formula for his construction, which in turn was based on his principle, told them nothing about how light was propagated, even though it worked well.[38]

We shall describe briefly how the construction operated, even though it takes up about ninety per cent of the *Traité*.

Let us begin with rectilinear propagation. Here one needs a plane wave that, for example, runs along a line from left to right. We then draw a plane, which stands perpendicular to that line, scatter several points across it, draw around each point a half sphere to the right, in all cases with the same radius, and we find that the envelope of all the spheres forms another surface, parallel to the first and slightly to the right. 'And from this we see the reason why light, unless the rays are reflected or refracted, propagates exclusively along straight lines.'[39]

In the case of reflection the construction proceeds as follows. For the sake of simplicity we shall think in two dimensions, not three. Let the plane wave that runs along the line, from left to right, fall on a mirror. The wave is now a perpendicular line to the direction of movement, and the mirror is the line that, generally speaking, makes an angle to it. Since the wave is not parallel to the mirror, it takes a moment before it strikes it fully. Now we draw two parallel lines at a distance L; the first meets the mirror at point A, the second at point B and in between there is a time moment $t = (L : s)$ if s is the speed of the light.* At point A we draw a circle with a ray L, because this is the distance that light can travel in the time that light from the incoming wave reaches B. Because the mirror and the wave are both a plane surface, we can make do with the one circle. The tangent, which can be drawn from point B to the circle, indicates the reflecting wave, and the line perpendicular to it gives the direction in which it will

---

[38] Dijksterhuis (1999) 267–272   [39] OC 19, 476 (lines 22–23)

propagate. Then it is easy for us to see that the angle at which the light hits the mirror is equal to the angle at which it is reflected.

In the case of refraction the construction proceeds as follows. We can, to some extent, compare the surface between two mediums (such as air and water, or air and glass), where refraction occurs, with the mirror. Up until the asterisk (*) we therefore proceed in the same way as above. However, the wave that we draw around A now has a radius that is not necessarily equal to L. The time interval t, during which the incoming wave proceeds to point B, must be equal to time interval t, during which the broken wave around point A spreads out. But if the speed of the light changes as it passes from one medium $s(1)$ to another $s(2)$, then the distance that exists in that moment of time between the waves will change from $L(1)$ to $L(2)$, accordingly $s(1):L(1) = s(2):L(2)$. This can also be written as $s(1):s(2) = L(1):L(2)$, a ratio that is known as the refraction index. With refraction in a medium with a smaller speed of light, for which $s(2):s(1) = 1:n$ is smaller than 1, it also follows that $L(2):L(1) = 1:n$ is smaller than 1. Therefore, the ray of the circle at A will not be L, but shorter, namely $L:n$.** The tangent of the circle that can be drawn from point B, shows the refracted wave, and the line perpendicular to it shows the direction in which it will propagate (Figure 5). So the refraction law is easily obtained

$$\sin(h1) = n \sin(h2)$$

where $h1$ is the angle between the direction of the incoming light and the perpendicular on the refracting surface, and $h2$ the angle between the direction of the refracted light and this perpendicular.

With double refraction the construction proceeds as follows. We can regard it as a special case of refraction, in which – not until the end – special properties of such a medium as the Icelandic crystal will play a role. Therefore, up until the double asterisks (**) we proceed as above. Then we accept the notion – that golden notion of Titan's – that light in this crystal not only travels at the speed of $s(2)$, but also with another speed that depends upon the orientation of the crystal. If $s(2\uparrow)$ is that other speed in a $\uparrow$ direction, and $s(2\rightarrow)$ in the $\rightarrow$ directions that

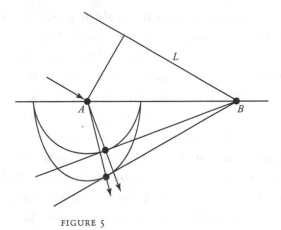

FIGURE 5

are perpendicular to it, then we cannot draw a circle around A with ray $L:n$. Now it is a matter of two refraction indexes, the one given by $s(2\uparrow):s(1) = 1:n\uparrow$, and the other by $s(2\rightarrow):s(1) = 1:n\rightarrow$. Therefore, a wave must be constructed around A that travels in direction $\uparrow$ to $L:n\uparrow$, and in the $\rightarrow$ directions to $L:n\rightarrow$. This produces an ellipse in two dimensions. The tangent that can be drawn from point B to the ordinary circle shows the ordinary refracted wave; the tangent to the extraordinary ellipse shows the extraordinary refracted wave. Perpendiculars show once more the directions in which light propagates: the rays. The beautiful thing about this construction is that it explains the direction of the extraordinary ray by means of the crystal orientation $\uparrow$.

Titan considered this an *'experimentum crucis* [decisive experiment] that confirms my theory of light and refraction'.[40]

The 'ΕΥΡΗΚΑ of 6 August 1677 did not only have to do with his notion of light, but also with his discovery of a new means to obtain knowledge about crystals.

From the *Traité*:[41]

When a piece of crystal remains under water for a day or longer, the surface loses its natural shine. And when covered in nitric

[40] OC **10**, 613   [41] OC **19**, 495 & 499–500

acid, it effervesces – especially, as I have noticed, when the crystal is pulverised. I have also let it become red-hot in a fire, without it changing in any way or becoming less transparent. Yet in a very hot fire it will calcify. It is barely less transparent than water or rock crystal, and it is totally colourless. But light rays pass through it differently, and these produce those superb refraction phenomena of which I shall now attempt to explain the origins . . .

    I wished to verify what elliptical, or rather spheroidal waves would do. I had the idea that the arrangement or the normal structuring of the particles [of the crystal] could somehow contribute to the formation of spheroidal waves. For this it was only necessary that the succession of light pulses travelled more rapidly in the one direction than in the other. I scarcely doubted that there was, in this crystal, such an arrangement of equal and similar particles, because it has a fixed form and its angles have a fixed and invariable measure . . .

He sees the hexagonal shape of the Icelandic crystal and recognises various ratios, such as $8 : 5$, $5 : 3$ and $3 : 2$. Remarkable, too, is that he finds the root of 8 (2.8284), which would be the ratio of the long and short axes of the crystal particles: 'If there existed such a thing as a tetrahedron that was not composed of spherical particles, but of elliptic, spheroidal particles with an axial ratio of the root of 8, then I assert that the solid angle of the obtuse top is equal to that of this crystal . . .'.[42]

He presented more constructions. In Chapter 6 of the *Traité* he gave the solution to the problem of which reflecting or refracting surface plane could return light from a point A to a point B: 'Now I say that the surface need only be made in such a way, that the path of the light from A to the points on this surface, and the path from these points to B, be made in equal times for all the points.'[43]

By phrasing the question in this general manner, the power and elegance of his kinematic construction came clearly to the fore. He

[42] OC 19, 519   [43] OC 19, 525

converted the requirement for equal times into a requirement for equal optical paths (as we have done above by taking not L, but $L:n$, when refraction could occur). He easily found these surfaces and their radii of curvature. Using his lucid methods, he managed to extend this wave optics to surfaces that do not have a focus, but only a focal plane. And this is how he reached the brilliant conclusion that the focal plane is the evolute of a wave surface.

He wrote it down far too cryptically, right at the end of his most remarkable book. As explanation for a complicated figure, we have only, 'All waves impinge on EK . . . Now the line EK is not part of the circle, but a curve that comes from the evolute of another curve ENC . . . originating from the refraction of parallel rays.'[44] ENC is therefore the wave surface, and EK, by definition, the focal plane, since all waves impinge upon it. It is concealed so effectively in these dry sentences that Etienne Louis Malus, William Rowan Hamilton, Joseph Diez Gergonne, Adolphe Quételet and others had to discover it anew in the nineteenth century.[45]

It is fairly clear when Titan wrote all this down. The eleven successive pages in his notebook devoted to the focal plane and the double refraction (which clearly form a whole and cannot be split up in five sections as has been done by the editors of the *Oeuvres complètes*) are dated in the summer of 1677.[46] The informal narrative text from the *Traité* follows on from it, leaving out only minor inaccuracies and, for that matter, discoveries, too. He did not do any more serious work on it; although he did plan to set it out more formally and translate it into Latin, but he never got round to it.[47] We can therefore assume that he wrote the entire text in The Hague, because he only returned to Paris on 24 June 1678.

The question remains whether he was ill, or he was playing truant, so that he could finish work such as this. In September, or earlier,

[44] OC **19**, 535   [45] Shapiro 220 (n30)
[46] Dijksterhuis (1999) 173   [47] OC **19**, 458–470

Denis Dodaert must have paid him a visit, as examining physician, because he wrote to him in October that he had no time to come and visit him again in The Hague. 'I have requested news about your health from the Duke of Roannez.' He was an army physician and had a scholar on his roll, as well as French gunners. He explained why he had been unable to visit: 'The army marches out from here; I hope to be in Leuse on the 25th and then speedily in Paris.'[48]

But the campaign, which was over by then, also made it easy to make a trip in the other direction. Around Christmas, Miss Anne de Longueville came to The Hague, bringing with her a book for him by Antoine Menjot: the fourth part of a *Dissertationum pathologicarum* (Exposition of Pathology). He thanked Menjot on 22 December: 'As if it were not enough that I have managed twice not to succumb to a well-nigh desperate illness, you give me further food for thought, complaints that are all more serious than the one you have seen me endure.'[49] It is apparent from the irony that he had managed to get the upper hand over his melancholia.

Apart from irony, we see something else that surprises us: Christiaan on horseback! Who would ever have thought it? From the 25th to the 29th of April he was in Breda on a visit to Constantijn. Since neither party wished to fight, the brother was permitted, briefly, to leave the camp. According to Constantijn's journal, on the first day he was indisposed with a headache, and on the last: 'we left for The Hague, my brother and I, and we arrived on horseback at the same moment in Moerdijk.'[50] A pathological course of treatment?

Actually the family had understood something of his pathology. People were still concerned about him, certainly now that the war was ending and he lacked any adequate argument to extend his leave. His sister Suzanna had found a Miss La Cour, who was prepared to accompany him to the Rue Vivienne, to keep house for him and, perhaps more important, to provide some homely comfort.[51] Miss

[48] OC 8, 37   [49] OC 8, 52   [50] Huygens 32, 250
[51] Vollgraff 704; Keesing (1983) 188

La Cour would be able to write to her and send parcels, so that she could take action more promptly than in the past, if things were to go wrong. When it came time to leave, he was accompanied, not only by this housekeeper, but also by Nicolaas Hartsoeker.

Nicolaas Hartsoeker was twenty-four years old. He had just completed his studies in Leiden and had been engaged that spring, in his father's vicarage in Rotterdam, with microscopic research on sperm. Exciting stuff. Johan Ham, another student, had started it. Not Nicolaas, but Johan, was the first to observe sperm, in 1677, according to the testimony of Antoni Leeuwenhoek, who had advised the young men on how to make small microscopes from a drop of glass.[52]

It was clear to Christiaan that much could be discovered with this miraculous microscopy. He had entered into contact with Hartsoeker earlier in the year, probably because he saw in the young man a useful servant, while Leeuwenhoek, for all his simplicity, was a contemporary who had made his name. But Hartsoeker's ambition was greater than his talent, and Christiaan would prove mistaken as to his compliance. Even so, he could learn a thing or two from him, as is apparent from Hartsoeker's letter of 14 March 1678:[53]

> I take a piece of the clearest glass with the smallest quantity of salt, for this mixes ill with the glass when melted with the weak flame of the candle, and is drawn to the surface – for we know from physics that all . . . fluids cause what is most dense and least agile to rise to the surface to form a crust there – and in this manner cause the formation of different surfaces.
>
> From this glass, I pull either a thick or a thin strand through the flame, according to whether I desire to make my lenses large or small. But when pulling the strand it is essential to avoid too much twisting or moving of the melted glass, because the air that enters the glass forms small blisters and thus renders the glass unusable for its purpose.

[52] OC 8, 58 (n1)   [53] OC 8, 59–60

The pulled strand (which has the added benefit that, in specific places, few or no flaws or blisters occur) I hold in the fanned flame, and thus I obtain accurate round lenses, large or small, according to how I wish to have them. The best I choose are those that have the clearest shine.

On setting these between the copper [I take care] that the aperture towards the object should not be larger, so that when one holds the microscope in stationary fashion to the eye, all sides of the aperture can be viewed with ease, this being to prevent the spreading of light rays and darkening, which would be caused by a larger aperture. Concerning the aperture towards the eye, this is of less importance.

Since you encountered problems in making the lenses round, I hereby send you a pair that are set and ungreased, which I think you will find round enough. I hope soon to hear your opinion of them.

Instead of observing the little creatures with the aid of a candle, by which they do not show up as shadow spots, I now view them against a clear blue sky. But in such a manner that I place before the object a tube of one or one-and-a-half feet that is black on the inside, without which I would see nothing. In this way they reveal to me the following appearance . . .

We cannot resist quoting from Hartsoeker's letter of 4 April 1678, in which he reveals skilfulness of another kind:[54]

I have seen the little creatures in the seed of a bull, but owing to the dark colour of the substance and the little life that remained in them – for they were almost a day old before the farmer brought them to me – I was not able to observe their appearance to such a degree that I would dare to draw them. I already went to much trouble to observe them once again, but because the time has not

[54] OC 8, 68

yet come for the cows to be mounted, my efforts were without
success.

In that of a horse, I found no life. I believe, however, after a
farmer brought me a filled tube, that this proceeded in the same
manner as often occurred for me with that of a human being. For
in the one quantity I find plenty of life, in another very little, and
in yet another, none at all, namely where the white is lacking.

In the seed of a rooster I found an uncountable number of
little eels, about five or six times as thick as the tails of the little
creatures in human seed. They manifest themselves in similar
fashion in that of a drake; but since I could only obtain this by
cutting open the duck – and only with the fourth did I succeed –
I observed it in a dead and defective state. The seed of a dove I
expect any day, but I do not wish to delay any longer my answer to
your agreeable letter. I have tried to obtain that of a tomcat, but in
vain, for those are animals that are not to be seized without gloves.

We do not know what Christiaan wrote to him, because he kept no
transcript or note of letters to subordinates. In his notes he records
observing spermatozoa on 28 February, and also on 13, 14 and 20
March, when he visited Leeuwenhoek in Delft.[55] Shortly prior to,
or on 26 March, when he visited Hartsoeker in Rotterdam, he saw
the sperm of a dog. But at least as important was that he saw the little microscopes. This meant that it was not too difficult for him to
'invent' a new microscope. At least this is how he puts it in one of his
manuscripts, next to a drawing of the glass drop microscope, 'Invented
[*Invente*] in The Hague on . . . May 1678.'[56] He would, indeed, have
made some improvements on it. It was in his own self-interest that
he neglected to fill in the date.

So Nicolaas Hartsoeker accompanied him to Paris, probably as
a reward for the lenses and the help, and he would remain there for

---

[55] Leeuwenhoek 2, 325; OC 8, 27; OC 13, 733–736
[56] OC 13, 680 (OC 8, 112–114)

a year. Less than two months later, on 15 August, they quarrelled. It was one day after Willem III had had a final go at the French at Saint-Denis, after peace had been signed. Frustrated at not having engaged in any real fighting during the war, he wanted, at least once, to see some bloodshed.[57] That day Christiaan's article on 'the microscope that has been brought from Holland' appeared in *Journal des Scavans*.[58]

Nicolaas was furious, for nowhere was his name mentioned. The microscope had caused a great sensation. A lyrical Jean du Hamel, who took down the minutes on 16 July of the meeting at the Academy – the meeting where 'H., returned from his journey to Holland, presented a microscope that enlarges objects in an unbelievable manner' – speaks of 'extraordinarily small things, such as a grain of sand'. In his minutes of the meeting on 23 July he reports that members observed 'small creatures in water'. On 30 July it was still no less a source of fascination. It was all 'the new microscope' and 'his microscope'.[59] Nicolaas was not present at the time, but if he had ever suspected that his contribution had been suppressed, the publication afforded him definite proof.

He wrote a letter. On 29 August an 'Extract from a letter of Mr Nicolaas Hartsoeker to the editor of the *Journal* (*des Scavans*) on the manner to construct the new microscopes referred to in the *Journal* a few days ago' was published.[60] Not the letter itself. Chief editor Gallois was in a quandary, for one does not readily accuse a director of theft. It is almost certain that he discussed the matter with Christiaan, who was against the publication and, to placate the incensed servant, wrote the extract, including the finer points as they stood in Hartsoeker's letter of 14 March. The letter itself, and Hartsoeker's account of the affair, did not come to light until 1725.[61]

It was a dirty trick. Nicolaas was no saint either, but Christiaan had certainly shown himself in an unfavourable light. Why did he have to claim another man's glory for himself, glory that did not even belong

[57] OC 8, 94; Huygens 32, 270–273; Israel (1995) 825   [58] Vollgraff 705
[59] OC 8, 96 (n1); OC 22, 256   [60] OC 8, 98–99   [61] OC 8, 110–113

to Nicolaas, but to Antoni Leeuwenhoek? Was there not sufficient glory to be had in his *Traité*? Or did he feel that the ethereal structure underpinning his explanation for light was no match for the direct glimpse into the world in miniature?

Hartsoeker, who, according to his posthumous report considered himself the inventor of these microscopes ('he had them from me'), sought out Christiaan's enemies. He quickly took up with Jean de Hautefeuille, who had contested Christiaan's invention of the spiral pendulum for clocks. He gossiped with Abraham Cyprianus, Peter Guenellon, Nicolas Lémery and others.[62] Isaac Thuret probably also joined in. And it would not surprise us if, in his bitter ramblings, he did not come across François Catelan, a name we shall meet again in the next chapter.

Christiaan, who was now back in favour with the attention he had received, had a marble medallion portrait of himself made (Plate 6). It was a profile of a powerful face. Very fine indeed. He posed for 'my man', Jean-Jaques Clérion, but when it was finished, he claimed he could make a better one himself – that is, if it could have been bigger and made from clay . . .[63] This make-believe world, where he now seemingly wished to belong, revealed itself even more clearly in his criticism of an Amsterdam stockbroker with whom he visited Paris art collections. This down-to-earth trader, who was incapable of admiring anything, was unable even to pretend to admire anything.[64] It revealed itself in his criticism of Charles Perrault, the friend who would become immortal with his fairy tales of Mother Goose, and who swore by the literature of his time. With little knowledge of such matters, the man from The Hague defended the unparalleled greatness of ancient writers . . .[65] He was more concerned with things than with people, with a lucid articulation of the facts, uncluttered by judgements, and a spelling without fancy accents. But he was fooling himself. At the same time he was, after all, working on experiments

[62] OC 8, 101 (nn7–9)   [63] OC 8, 202, 212–213
[64] OC 8, 181   [65] Vollgraff 708; OC 8, 177

for 'fountains of liquid silver' in Versailles.[66] Also noteworthy: he delivered a boastful text about the Academy, which was to be given a place in Paul Pellisson's exalting piece of writing on Louis XIV.[67]

All false.

He remained preoccupied at the Academy with that glimpse of the world in miniature. Up until well into the following winter they – he, Rømer and Phillipe de la Hire (a new member), but also Hartsoeker and even Thuret – continued to look at sperm, infusoria, bacteria and the like.[68] And it was as if, with this stolen glimpse of floundering little shadows, he was closing his eyes to a shade that could turn out to be frighteningly real.

On 14 April he turned fifty.

On 11 May Suzanna wrote, 'I need not tell you, dear brother, how happy I was to receive your agreeable letter of the 5th, from which it is clear that you are fully recovered. You know well how it grieves me, and how much I worry when you are unwell. Your good health reassures me. I am well able to imagine that Miss La Cour is, as you write, of great service on such occasions in attending to you and to your housekeeping. Now you perceive, as I do, just how much she is worth.'[69] We do not have Christiaan's letter of 5 May.

On 24 May Samuel du Clos informed the Academy that he had cured a servant of H. who spewed blood, by administering a grain of laudanum and Granada syrup.[70] We do not know whether it was Miss La Cour.

On 28 June Christiaan wrote a letter to Samuel de Fermat, son of Pierre, the mathematician who died in 1665:[71]

> When I received your obliging letter [of 15 March] that you were so good to write to me, I was deeply despondent due to a fateful illness, for the third time since I have taken up residence in

[66] OC 22, 267   [67] OC 8, 196–199   [68] OC 13, 702–707
[69] OC 8, 163   [70] Registres T.VII   [71] OC 8, 186

France, and I still find it difficult to emerge from it. Were it not for this, I could never have been forgiven for delaying for so long my expressions of recognition for your esteem and affection. I am happy, Sir, that the feelings that your father cherished for me have passed on to yourself, and you may rest assured that I shall leave no stone unturned to safeguard such treasured kindness. Not without some confusion did I read the verses that you added to your letter, for you compare me to Mr Descartes. I know well that I do not deserve this honour, and that those verses, if they – as they deserve to be – are one day published, will cause others to become jealous of me. I am one of those who has benefited from the lucid insights of this excellent man, and also from those of your esteemed father, who gave me reason to admire all the more the great science of geometry when I considered in what little time he already possessed such consummate knowledge. You therefore need have no doubts that my feelings for his work, which you have seen in my letter to the Abbot of Miramion, are genuine and sincere, as I am also disposed to honour yourself and to remain all my life at your service.

Clearly, then, he celebrated his fiftieth birthday in deepest melancholy. As a gift he had the mediocre verses of Samuel de Fermat from 1666:[72]

> *Huggenium Gallis sua nunc dat patria, quondam*
> *Cartesium Batavis Gallia amica dedit –*

> The Fatherland gave Huygens to France,
> Just as France, in kind benevolence,
> Once gave Descartes to the Batavians –

But once more, he managed to resume his activities. On 7 January 1679 he spoke for the last time at the Academy. On 10 May, according

[72] OC 8, 156

to Jean du Hamel, he read aloud a letter from 'Leorenooke' (Leeuwenhoek).[73] Three days later he began discussion of his *Traité*; the nine or ten lectures that followed would continue until well into July.

He did not restrict himself to reading aloud, he also continued his experiments with the Icelandic crystal. On 3 November he wrote to Constantijn that he had discovered a manner to cut and polish it, 'which had been considered impossible'.[74] The Academy certainly attached importance to the subject, because even before he had said all that he wanted, Edmé Mariotte began a series of lectures on colour, but nothing is known of any response to the *Traité*.[75]

He brought his work on optics up for discussion in a letter of 22 November to Gottfried Leibniz. We shall quote almost all of it, as introduction to the renewed contact with his pupil. The letter is in answer to Gottfried's very detailed exposition of 8 September:[76]

I have examined closely your new characteristic [a concealed form of the differential calculus], but to tell you the truth, I fail to understand why you expect so much of it. For your examples are self evident, and you provide unintelligible proof for the proposition that the intersection of a sphere with a surface is a circle. Also, I fail to see how you could apply your characteristic to such diverse matters as quadratures, finding curves from their tangents, the roots of negative figures, the problems of Diophantus and the quickest and most beautiful constructions in geometrical problems. And what surprises me most is your application to the invention and the explanation of computing tricks. I tell you plainly that these are pious hopes, and I wish first to see proofs before I can believe your claims. But I am cautious in saying that you are mistaken, for I know that you are astute and discerning. I only ask that, besides this, you also occupy yourself with the matters that you have already given us . . .

[73] OC 8, 160 (+n2)   [74] OC 8, 241   [75] OC 19, 439   [76] OC 8, 243

This whole summer I have worked on my refraction, especially those in the Icelandic crystal, in which such strange phenomena occur that I have not yet managed to fully comprehend them all. But what I have been able to comprehend, confirms my theory of light and ordinary refraction in splendid fashion. Amongst other things, I have found the solution to Descartes' problem: if the shape of a lens surface is known, what is the shape of the other surface so that, together, they make a parallel 'pencil' departing from a given point? I have solved it even more broadly than he, for he takes only a sphere or a cone. If my health permits it, I shall try and have this *Traité* printed in the coming winter. It would please me to follow your advice to present my considerations in shortened form, without proofs, but this I cannot manage, for in these matters people would not believe me on my word.

For the rest I have nothing else to report other than my invention of a very practical spirit level, which corrects and checks the image both at once, so that always one knows that there is no error. In not one of the current spirit levels is that the case, at least in those specimens which, like mine, use a telescope. I shall publish it in the *Journal* [*des Scavans*] and send you the article at the first opportunity.

Here speaks a peevish teacher with an unmistakable soft spot for his pupil. When the *Traité* – much later than he predicted here – appeared, he would expect the most favourable judgement to come from Gottfried. This profound young man had acted as librarian and court advisor to the Duke of Hannover since 1676.[77] He had not succeeded in being appointed a member of the Paris Academy, but when he knew that Christiaan was back in Paris, he turned to his teacher to see whether he might still succeed.

Leibniz's letter of 8 September 1679 was no more than an opening move. The letter of 20 October was more to the point. He

---

[77] Aiton 71–99

drew his attention to a phosphor, which might intrigue not so much Christiaan, as Colbert. Then he wrote once again, at the beginning of December, and made clear what he expected from Christiaan. In a highly remarkable supplement to his letter he went so far as to explain to him what he was to do:[78]

> In order to be better able to succeed with M.C. [Mr Colbert] it would be advisable, in my opinion, to say that a German possessed of great curiosity has sent this phosphor and that he can describe its composition, that he is competent in physics and in mathematics, that he offers to write letters from time to time about discoveries in Germany, where he has quite a few learned acquaintances, and that he himself also has much to offer. He could perhaps, in one way or another, be of use to the Academy as member for the correspondence and the appointments. As regards his name, this may not be mentioned. Not even Gottfredus Wilhelmi, for even without calling him Leibniz, it is still him. M.C. has frequently heard his name [*ayant eu souvent les oreilles battues*] when the moment was not right, and he will dismiss him if he should remember. For once an important man has made a problem of something, it is not soon forgotten. One is more assured of success if one starts anew.

He probably saw later that Christiaan might find this advice to be inappropriate, because in his letter of 10 December he wrote:[79]

> With regard to what I mentioned in the supplement of my last letter, you can better proceed according to your own discretion. I thought that a fresh application would be better than an old one, and that in this way people could more easily seek information on its intentions, the more so since important men can hardly be said to take pleasure in asking people's names. If the name could be dropped in passing, it would be good to do so. But if this is

[78] OC 8, 251–252   [79] OC 8, 252

difficult, then it can be better said openly, if requested. Be so good as to mention this small piece of advice to no one. My confidence in your goodwill enables me to venture to broach the subject.

Christiaan's answer on 11 January 1680 was clear:[80]

Since my last letter I have been ill for a full month and have had to keep to my room. During this time, however, Mr Gallois has visited me. I have commended your concerns to him and he was most favourably disposed to send to you a number of things. At that moment I had not yet received your letter before last with the supplement. That is why I have not yet approached him on the manner that you have in mind to apply anonymously. But I do not wish to speak to him about it, for I know very well how badly it would be received by the boss if he were to hear of it.

This embarrassing exchange of letters came to an abrupt end. 'Tear up the supplement,'[81] Leibniz wrote on 26 January, and practically nothing was heard from him until many years later, when Christiaan sent him his *Traité*.

He continued to be troubled with his health and, more significantly, his science. One example is his unrealistic plan of June 1679 to make air condense. Three weeks after the Academy had challenged him to say just how he intended to go about it, he was so free as to state that, for this, a pump would be required, which could compress between eight and nine hundred times. The latter figure comes fairly close to the truth, for the density of particles in solid or liquid matter is around a thousand times as high as in gases. Christiaan had known this since his early experiments with air pumps, but it was impossible to construct such a compressor in those days.[82]

Another far-fetched idea, which he tried to put into effect in the spring of 1680, concerned magnetism. Apparently there had been a

---

[80] OC 8, 256–257   [81] OC 8, 268   [82] OC 19, 340–341; OC 22, 269

reconciliation between him and Carcavy (although there is no letter, or any other evidence, to support this), because he was permitted to borrow the large iron magnet belonging to his neighbour. Experiments with iron shavings gave him the impression of strong influences, stronger than those caused by gravity, and this led him to fantasise that the whirl brought about by magnetism must consist of cruder particles than those of gravity. He read out his short explanation on 25 May, to return to it on 1 June. But it came to nothing, because polarity eluded even him – and everyone else.[83]

It is easy to understand how he failed to understand a phenomenon that could only be grasped in its relation to electricity. It is harder to see how he went wide of the mark in the explanation of comets of 1680. The question was whether the comet, which suddenly appeared above Paris in December, had also been observed briefly in November. He thought not.[84] Just like Giovanni Cassini and so many others, he had made up his mind that comets only travel in straight lines. But the comet in November had changed direction as it skimmed past the Sun, and turned up again, one month later, close to the Earth. Ole Rømer had understood, and so had a man in Cambridge who thought about the effects of gravity.[85] But his unbiased judgement, which had once earned him such high esteem, had now deserted him.

'I thank God that I am in good health,' he wrote to his father on 27 December, 'but for the last two days I have had a cold. For some time there has been talk of a comet, but up until yesterday, there was nothing to be seen. Around 5 o'clock, when the heavens cleared, there it was, surprisingly clear, and the long tail (across practically half of the sky) was remarkable. In all my life I have not seen such a strong comet.'[86] A sign, according to the father.

That autumn Christiaan went to Chantilly to hear the echo from a high stone staircase. It resembled the sound of an organ pipe,

[83] Vollgraff 713; OC 19, 574–581   [84] OC 19, 283–302
[85] Westfall 391–395   [86] OC 8, 312

316 HUYGENS: THE MAN BEHIND THE PRINCIPLE

he thought.[87] And in the latter part of the summer he spent much time in the countryside, where he examined insects and flower petals. 'It has already been warm weather here for more than a month, and the heat is more tolerable than in mid-summer. The surroundings are delightful, and due to the holiday and the grape-picking there is plenty of company to be had. Today I have been for a picnic in one of the prettiest spots, together with the lady who occupies the best rooms in this house. She holds me in high esteem and is most kind to me.'[88] He wrote this to Lodewijk from Viry. No need for Miss La Cour.

After the comet, however, he quickly reached the end of his tether. His incapacity must have begun sometime between 6 February and 3 April 1681. We have this information from letters written by Suzanna Doublet, who kept a finger on the pulse. 'Let God preserve your health; in our family all goes well. Adieu. Convey my friendly greetings to Miss La Cour.' This is how she concluded the first letter, but the second begins as follows, 'It grieves me deeply to read in the letter from Miss la Cour that your illness continues.' She wanted to come and see for herself.[89]

She left on 12 or 13 May, with Philips and her three children. She had never been to Paris, which explains why it was not until August that the small party set forth with him and Miss La Cour – we presume – on their return journey to The Hague.

[87] OC 19, 374   [88] OC 8, 299   [89] OC 8, 321, 325

13 Dismissal

Meanwhile, father Huygens was feeling like a hermit. His shared living arrangements with Constantijn's wife Suzanna and Tientje, her son, came to an end when they moved out on 29 June 1680.[1] Constantijn himself had seldom been at home with them. Most of the time he stayed at Dieren, where he had to endure the dissipated company and endless hunting parties of Willem III.[2] Thanks to a legacy left to him by his mother-in-law, he could now rent a house of his own on the Korte Vijverberg. The eighty-five-year-old patriarch accepted their wish for a home of their own, but missed them dearly, especially the precocious Tientje, whom he had thoroughly spoilt.

In the silence of his residence on Het Plein he added the finishing touches to his last great poem, *Cluys-werck* (Hermitage Work). In it he recounts how he began and ended the day with prayer, read through or discussed letters of petition, presided in the princely auditor's office between ten and two. How he arrived home to find no one to talk to, visited his library and looked at drawings, had a servant read aloud to him or allowed himself to be driven around, was satisfied with just a couple of pieces of bread and fish or cheese and some beer. Not exactly the life of a hermit, but the poet felt it to be so. The mood of *Cluys-werck* is contentment, loneliness the undertone.[3]

Certainly he was glad when the learned son moved back in with him on 11 September 1681. Once again there was conversation to be had. Christiaan was not his dearest child, for – according to the standard formula that characterised his correspondence – all his children were equally dear to him, but nevertheless, he was the most precious. In his autobiographical poem from 1679, *De vita propria* (Upon One's

---

[1] Strengholt 123; OC 8, 272, 293   [2] OC 8, 338   [3] Strengholt 124; Smit 291–295

Own Life), he expressed this preciousness in several places. He had Christiaan say of himself:[4]

> *Jamque per abstrusas rerum caelique solique*
> *Ducebar causas: quas ut de fonte Stagirae*
> *Purius haurirem, (nondum Cartesius auctor*
> *Luce nova sucum vero detraxerat), alter*
> *Ecce labor sudorque, novum maris aequor arandum.*

> Sun and heaven's hidden causes
> Held me captive instantly; drawing from the source
> Of Aristotle (for Descartes' light could not reveal the truth)
> Sham truths once more did I relinquish,
> Laboured to encompass time upon the sea.

Here is another example of how he puts into words how precious the son is to him:[5]

> Son, who by God's discretion,
> Of this steady yet unyielding motion
> The bold inventor be;
> Whether with the tide
> Of world's oscillations, or against it,
> Always keep in sight its constancy.
> If this frail work midst heaving oceans
> Dispirit thee in all that it endured,
> Remember all your cognisance
> Comes from God alone,
> Who sets invention and inventor,
> mind and moving wheel as one.
> Let not the work betray the Master.

This poem from 1671 describes the steady motion of the marine clock, for which the inventor ought better to have realised that all this had been devised by God.

---

[4] Worp **8**, 183–184 (*De vita propria* 129–133); OC **22**, 720   [5] Worp **8**, 40

They attended the Walloon church, and spoke to the organist, Quirinus van Blankenburg, about remedies against false psalm singing.[6] They played and sang from the *Pathodia sacra et profana occupati* (Sacred and Secular Compositions of a Busy Man) which, in contrast to the rather ponderous title, are flowing psalms and songs by the father, quite simple, without counterpoint or modulation, and without drama.[7]

They conversed at table, undoubtedly, but what about? Whether Christiaan still felt ill? It can hardly have been otherwise. What it meant to be melancholic? This is unlikely, for if they allowed little drama to enter their music, then even less into their words. Moreover, a religious interpretation, readily given by the father, would have been unacceptable to the son. We take it that they confined themselves to small talk and the weather. We may even wonder whether Christiaan ever mentioned Catelan's attack.

François Catelan, of whom we know little, other than that he was an abbot and that he later also quarrelled with Leibniz,[8] had published a criticism of *Horologium oscillatorium* in the *Journal des Scavans*. This criticism began:[9]

> Huygens did not wish to leave out anything in his book about the clock, in which he discusses, exhaustively, the problem of the centre of oscillation. But because it is somewhat difficult for the mind to be always attentive, certainly where mathematical abstractions are concerned, it need not surprise us that he is not entirely successful in his solution to this problem.

It was insulting. Rather curiously, the criticism did not appear in the official Paris edition. Possibly Hartsoeker was behind it; Christiaan got to see it at his book dealer when he bought the September edition of the *Journal* in Amsterdam in the autumn of 1681. Not until 16 April did he approach the Paris editor, Jean la Roque. After justifying his delayed response by referring to his illness, he wrote:[10]

---

[6] OC 20, 129   [7] Worp 4, 341–343; Rasch (1987) 106–109
[8] Aiton 130   [9] OC 8, 353   [10] OC 8, 350, 352

I am astonished at his attack on my theory for the centre of oscillation, which no one has remarked upon in the nine years since its publication. Now that I have examined his so-called refutation of my theory, I am surprised that the author has not withdrawn it in the seven months since its publication. For, to put it briefly: he finds that the sum of two line segments cannot be equal to the sum of two other line segments, if the ratios of these segments differ. Imagine that the first two are 4 and 8 feet long, and the other 3 and 9 feet, and then see how you can obtain a sum for either the one or the other that is anything other than 12 . . . It would please greatly me if this could be published, so that those not familiar with my proof [*Horologium oscillatorium*], do not think that the remarks of Catelan are of any significance. Should he still return to the subject, then you will oblige me by submitting his answer to a professional before having it published. Surely that is to the advantage of his honour. And if truth be told, I find it distasteful to be attacked by such a blockhead.

For Christiaan this was the end of the matter. But not for La Roque and Catelan. The former replied that he had known nothing of the Amsterdam publication in the *Journal des Scavans*, called it a piece of villainy, announced steps to be taken by the French ambassador to the States-General because publication rights had been violated, and published a declaration to this effect in June 1682.[11] In spite of this, Catelan managed to get further criticism printed in the Amsterdam version in July and September:[12]

Huygens should not seize upon a result of his principle [that the centre of gravity of a system of moving bodies cannot rise of its own accord] in order to change the tenor of my article. He defends himself only upon the pretext that I must be unable to count, that four unequal quantities cannot have two equal sums, which I never claimed to be the case . . .

[11] OC 8, 370   [12] OC 8, 372

And more such scraping of the barrel. Christiaan ignored it until Paris friends at the Academy and La Roque made it clear to him that this was unwise. In the spring of 1683 he wrote to the editor that Catelan had made such fine blunders that a reply was called for after all. He avoided using such words, however, in his second defence, which appeared in July 1684.[13] After repeating his first defence in greater detail, he ended with: '[Catelan] has therefore twice guessed wrong, for he guesses only, and he would be correct in his assertion that the problem of the centre of oscillation is easy to solve, if he had first decided upon the real significance of this problem.' But no amount of argument would convince the abbot.

Just over a year later he published another article on the centre of oscillation, this time in a discussion with Jacob Bernoulli: 'And I conclude that Huygens' argumentation fails to prove his proposition. It makes no sense at all.'[14] Catelan published this not only in the *Journal des Scavants*, but also in a new kind of journal, the *Nouvelles de la République des Lettres* (News Items from the Republic of the Arts).

These *Nouvelles* may be considered the first popular scientific journal, and its publisher and editor, Pierre Bayle, the first journalist.[15] An extra reason to dwell on this publication is the fact that Christiaan was involved in the setting up of the *Nouvelles*.

Bayle left France and came to Holland in 1681. After losing his post as teacher of philosophy at a school in Sedan, he managed to acquire a similar post at Rotterdam, thanks to the intervention of the broad-minded city magistrate Adriaan Paets. So, at thirty-three years of age, he became teacher to Lodewijk's children.[16] He tried, not only through this brother, but also through cousin Maurits de Wilhem, to get in touch with Christiaan. This man of learning could be of great support to him in his new environment, in his fight against prejudice and superstition.

[13] OC 8, 500   [14] OC 8, 538 (OC 18, 459–461)
[15] Bost 122, 286   [16] Vollgraff 753; OC 8, 454 (n1)

Bayle's opening move came on 14 September 1683: 'Since you are so widely cultivated, I believe, after long reflection, that any book would please you.' He accompanied this letter with his *Lettre sur les comètes* – a piece of work with the highly explosive content that comets are not divine messages of doom, but natural phenomena – and excused himself that it contained 'very little physics, and no mathematics or astronomy'.[17] Years later this heresy would land him in grave difficulties. As far as we know Christiaan did not respond.

His next move came in 1684, when he sent him the first edition of the *Nouvelles*. Now Bayle excused himself for daring to set up this journal without Christiaan's advice. A printed edition, however, would show more clearly what was lacking, and he would be willing to travel from Rotterdam to The Hague to hear his critical comments.[18] Christiaan was unable to withstand such diplomatic persistence.

In June of that year he wrote back to Bayle, outlined a plan for the journal and promised his cooperation, providing that the subjects were chosen with care. He found that, first and foremost, superstition must be opposed, that is to say, everything that was unproven but which time and custom had rendered sacred: 'You are well suited for this.'[19] So as not to rush any further ahead of events, we shall return to the first tasks to which Christiaan applied himself, upon his return.

Which was it, lens-grinding or finishing the planetarium? Constantijn, who was closer than Colbert and sought diversion from life in the barracks, quickly involved him in the grinding. He asked his advice, and was constantly urging him on. They wrote to one another so often that their letters crossed.

Within a short time the brother from The Hague could report to the brother from the Veluwe that 'he [had] just a moment ago set up the previous mould of potter's clay upon a wheel and [had] moved the little blocks over the grinding disc.'[20] Later, in his *Considerations upon the Grinding of Glasses [lenses] for Telescopes*, he said: 'In 1682 we still used grey glass powder [vert-de-gris] for polishing, instead of

[17] OC 8, 455   [18] OC 8, 490   [19] OC 8, 491   [20] OC 8, 346

vitriol, which is infinitely better, [for then] what we call the heating of the dish is no longer necessary.'[21]

The work was hardly healthy: it gave Christiaan migraines. Nor was it light: even with his special grinding table, it took many hours before the glass took shape, and all the time he had to exert considerable force on the lever by which the glass – he came to curse the glass – was pushed down on the turning wheel.[22]

Thirty years after their first grinding activities the brothers set themselves the task of making large objectives with a long focal distance, up to 210 feet (66 metres; a Rhineland foot was 31.39 centimetres). In this way they attempted to reduce the unacceptable degree of spherical and chromatic aberrations. It took them more than a year before their first lenses were ready. The following objective lenses made by Christiaan have been preserved: 12 feet (? March 1683), 35 feet (10 May 1683), 13 feet (30 May 1683) and 34 feet (19 November 1683). And by Constantijn: 34 feet (25 October 1683), 43 feet and 7 inches (7 February 1685), 85 feet (21 May 1685), 43 feet (21 July 1685), 84 feet (19 June 1686), 122 feet (10 May 1686) and 122 feet (15 May 1686). They engraved this information and their names, with a diamond, on the side of the glass. As a rule, they ground two lenses to obtain the desired focal distance.[23]

Obviously, these were intended for observations. In March 1683 Christiaan constructed a container for the 12-foot objective, consisting of five tubes that fitted into one another, which measured 1.32 metres when pushed in and 5.3 metres when pulled out. On 25 April he tested the whole thing mounted with a triple eyepiece . . . Titan. Straightaway he contemplated a container for lenses with a larger focal distance, for which such a construction could prove too weak.

In the meantime, Christiaan had the planetarium made that Colbert had ordered from him.[24] In the autumn of 1681 he went to his clockmaker in The Hague, Johannes van Ceulen, with the design;

[21] OC 21, 287 (297)   [22] OC 8, 433, 428   [23] Crommelin 20–26   [24] OC 21, 163

Van Ceulen finished it by the following summer.[25] It was an elaborate version of the timepiece with a balance and spiral, a prototype of one that had once been made by Isaac Thuret. The elaboration consisted of a speed variation of the planet hand, as happens in reality when the distance of the planet to the Sun changes, and for this he had calculated the necessary cog distance on the wheels by using continued fractions. But above all, this didactic timepiece was wonderful to behold. On 27 August 1682 he described it to Colbert:[26]

> The octagonal case housing the mechanism is 2 feet high and wide and six inches deep. The gilded face, on which the planet system may be observed, is covered with a glass pane edged in gilded copper that opens with a hinge.
>
> The entire circuit of the planets' orbits is carved. The planets protrude above the face in silver semicircles that turn around the centre and are mounted, one by one, on a small gold plate that represents the planet whirl and also makes the planet more visible. Moreover, the small plates for Saturn, Jupiter and the Earth are used to support their moons. The one for our moon rotates regularly around the Earth and shows the phases of new and full moon by its position.
>
> The year and the day of the month can be viewed in two apertures below, between the orbits of Saturn and Jupiter. The hour and the minute can be viewed in the semicircular aperture between the orbits of Jupiter and Mars, in which the disc with the hour rotates from left to right, and which also shows the minute that is engraved on the perimeter of the aperture. When this hour disc disappears to the right, the next one appears from the left, and so on.
>
> In the case is a clock mechanism that must be wound every eight days and sets in motion the hours, days, years and all the planets. It is highly accurate, both in the total circuit as in its

---

[25] Leopold 228–230; OC 8, 342 (+n5)   [26] OC 8, 376–378

parts, which are unequal because the planets move more slowly when they are further from the sun. In this I have represented Kepler's Law.

If one wishes to see quickly how the planets move over many years, or their position on a particular day in the past or in the future, one turns the handle on the right hand side (this moves quite easily) until the desired moment in time appears in the two previously mentioned apertures. Then all the planets for that moment in time are in their correct position. And to return them back again, one turns the handle in the opposite direction, until the present moment. In this manner one can find out when the conjunctions and the oppositions of the planets will occur, and whether they are visible or are concealed by the sun. But before one turns the handle, the pin, which sets in motion the mechanism that governs the planets, must be released inside. The clockwork keeps running. When the handle has been used, the pen must be fastened once more, so that the motion can resume normally.

The entire case hangs on two spills in an iron frame, so that the inner mechanism can be seen. This frame is largely covered by the case. By turning the case on the spills, one can reach the back side, which normally stands against the wall or the carpet. And by removing the lid, one can view the entire gearing. The principal component that appears is a large axis that is hung across the full length of the back panel. This axis carries the cogs that catch onto the wheels for the planets, days and years, which are entirely encased between the two front panels, at a distance of one inch.

The advantages of my clockwork above that of Rømer's are:

My clockwork shows the orbits in their true proportion to one another, while Rømer's makes the orbits of Mercury, Venus, the Earth and Mars, in proportion, much greater than those of Jupiter and Saturn. This fails to convey a proper idea of the solar system, and shows neither the position of Saturn and Jupiter in the zodiac, nor the conjunctions of the three planets Mercury, Venus and Mars, nor of the moon with Jupiter and Saturn.

Mine represents the motion of all the planets much more accurately than that of Rømer, because I have a better way [with continued fractions] of finding the number of cogs of the wheels.

My planets rotate above the dial, his rotate behind it, and can only be seen through small holes, so that the dial can be made from a single piece, but between the holes they remain invisible. I have two further advantages. Firstly, Jupiter and Saturn carry their moons, and secondly, by making the Earth slightly larger than it is in reality, the seasons can be seen, as well as the rising and the setting of sun and planets above our horizon. By making Saturn slightly larger than it is in reality, I also show all the manifestations of the ring that surrounds it.

My mechanism has a built-in clock that shows hours and minutes, while the other only runs if set in motion by hand. And because it runs rather stiffly, it would be impossible to operate it by means of clockwork. If one wishes to follow the motion of the planets with the naked eye, because it moves so stiffly, it cannot be set in motion using a wooden handle, but requires a key, which creates interruption and jolting in the motion. The handle in my mechanism, however, assures one of an uninterrupted and steady view of the motion of all the planets.

Unlike mine, Rømer's timepiece cannot be hung upon the wall, but must be placed upon a table or on a pedestal, so that one can get behind it to turn the key and to see the day of the year.

Mine can be opened like a clock while it hangs on the wall, so that one may look inside, and adjust it, if necessary. This is not so with Rømer's, where it can only be opened on one side.

On mine one can see the day of the month on the front, on Rømer's, on the back.

On mine there is a thread between the Earth and the sun, with which one can establish the position of the planets in the zodiac, which cannot be done on Rømer's because of the joins in the panel.

He asked 620 pounds for it: 520 for material and the work of Van Ceulen, 100 for the design and the description.[27] That was, as he wrote to his patron, considerably less than Rømer's planetarium had cost. Gallois, who replied to him on 27 October on behalf of Colbert, called it a reasonable figure and did not expect that payment would present any problem.[28] He would, however, not receive one cent for it.

Thomas Molyneux, a young Englishman, saw the planetarium hanging in Christiaan's study, in all its golden glory. He wrote the following report of his visit – one and a half hours – on 15 August, 'I was most graciously received by him. When, after conversing for a while, he understood that I was an Englishman, he spoke to me, beyond my expectation, in my own language, and with a pleasing competence. He brought me to his study upstairs, where he showed to me a most curious mechanical movement of his own contrivance.'[29] In his opinion, 'Joncker Christiaan' looked younger than his fifty-five years.

A rare document, when we consider how accurately this Molyneux described other learned figures whom he visited on his travels. For instance, John Flamsteed: unconventional but simple, stutters rather and speaks slowly, expansive in his use of words, just as he writes. Or Robert Hooke: ill-humoured and self-seeking, hated and despised by most others of the Royal Society, claims to have invented everything . . .[30]

From the numerous letters that the brothers wrote to each other, we quote the one that Christiaan wrote to Constantijn on 13 September 1682, because of the tone as well as the content:[31]

> Berkhout has disclosed to me what people are saying to you about my travel plans. I am by no means about to set forth, as my sister would have you believe, but I shall have to one day. The moment

[27] OC 8, 375   [28] OC 8, 400   [29] OC 8, 528–529
[30] OC 8, 529 (n)   [31] OC 8, 388–391

of my departure depends on Colbert's reply to my last letter, in which I informed him that my planetarium was ready, and that I am prepared to present it to him and, if he so pleases, to the King. But, in any case, I do not wish to remain in France, for I have fallen ill on three occasions there and am afraid of doing so again. I also have further reasons. I shall see if it is not possible to retain a part of my salary without being obliged to live in that country. I am not telling any one of this, and you must speak of it to no one. Besides this, I should regret it deeply if I should depart before seeing you. Most certainly you must see my planetarium before I take it with me, for it is so successful and runs so well that I shall find it hard to part with it.

Naturally I have foreseen the superior conjunction – not that of all the constellations, as our ambassador writes, but that of the three outer planets Saturn, Jupiter and Mars. I purposely arose on four mornings at four o'clock, to see how the heavens accord with my mechanism, and I saw these three planets quite close to one another, exactly as the machine foretold. But within a couple of days they must come even closer together. I expect the conjunction of Mars and Jupiter between the 16th and the 17th of this month, with Saturn at a distance of around 2 degrees. Between the 21st and the 22nd of the same month, the conjunction of Mars and Saturn, with Jupiter at a distance of around 1.5 degrees. But the conjunction of Jupiter and Saturn will only occur around 13 October, and the other will be quite far away, 8 to 10 degrees. These triple conjunctions are extremely rare, and there will not be such a remarkable occurrence for many hundreds of years.

I have observed the comet [later named after Halley] one more time [that was on 5 or 6 September] with my twelve-footer. Greater focal length is of no use for these [extended] phenomena. The thirty-footer was difficult to align; moreover, it was unsteady because the stand is not as heavy and robust as that of our viewers in Paris. The head of the comet became a luminescent speck, just as I had seen in the comets of 1664 and 1665. The vapour of the tail was thickest in the direction of the sun, in spite of the tail in this direction being short, and it seemed as though the vapour were being driven towards the other side. The next time I wished to observe this comet was disastrous for my good glass of twelve feet. I had hastily placed it to the fore in a large tube, and when I slid the tube through the garret window on the garden side, the glass fell down into the courtyard. The supporting frame, made of white iron, was quite bent and misshapen, but fortunately the glass was still intact, since it was set in a cardboard mounting.

Mrs Zeelhem [his sister-in-law] has given me the English glass, a round piece that is nice and large and thick, but also very dark. It will take away so much light that I cannot imagine that we shall achieve a good result with it. For what do you gain with a greater length [focal distance] if, due to the blackness of this glass, you lose what you win with the larger aperture? Even so, in my opinion it is good glass, with few marks [irregularities], so that it can always serve to test the veracity of the grinding [method]. But I wouldn't have begun with such large pieces if you had not planned to do so. I have assigned the glassmaker of 'The Back Alleyway' [Master Dirck] to make two glasses according to your design. They are for a spirit level that I have just adjusted for brother Drossart. He brought it yesterday from Gorkum, where he wishes to use it for dike measurements. It is encased in a triangular mount instead of a cross-shaped one, like my spirit level, and I think that this one is better, sturdier, and easier to adjust.

My sister has sent to me a parcel of books from Amsterdam that, according to her, were intended for you, and she asked me to write down what they contained. It regards only the work of [Marcus] Welser, but it was addressed to my father, who said that it had been promised to him in exchange for copies of letters from [the humanist Justus] Lipsius. And as you may imagine, he has succeeded once again. Farewell, and I wish you bad weather that forces you to leave where you are.

We now come to the main topic of this chapter. On 19 November 1682 Christiaan thanked the editor Gallois and, by the same gesture, Colbert, for his letter of 27 October. He mentioned his wish to return to Paris, now that his health was recovered, but found that the best travelling season was over.[32] On 7 January Gallois let him know that he could stay away for another three or four months to continue working on his mechanical constructions, but added that it would please Colbert if he had these done in France.[33] This gave him plenty of time to work, up in his garret on Het Plein, on pendulums that were impervious to a ship's movement.[34] And because he had such ample time, he built, for the first time in his life, a complete clock.[35]

In the meantime he was able to enjoy the wedding of Constantia, daughter of his sister Suzanna en Philips Doublet, and Mattheus d'Oyen in the spring. He also attempted to shed some light on a number of thefts that occurred in the paternal residence.[36] Guiltily, he wrote to Gallois on 22 July, 'I should have come to Paris one or two months ago, and possibly you are surprised that I have delayed my journey.'[37] He offered as an excuse a few unforeseen setbacks in the construction of his marine clocks; but now he had two clocks that ran well. He promised to be in Paris in September. That month, on the 5th, Colbert died.

François Michel Le Tellier, Marquis of Louvois, became prime minister. How was Christiaan to react to this appointment? Was he now expected straightaway in Paris, after all his procrastination? On 16 September he wrote a letter to the man who must now be his new patron:[38]

> My absence does not release me from my obligation to offer to you my respects and my obedience, now that I have heard that the King has entrusted you to be in charge of public works, including the Academy of Sciences of which I am a member. I was on the

---

[32] OC 8, 401–402   [33] OC 8, 406   [34] OC 18, 509–510, 532
[35] Leopold 230; OC 18, 527   [36] OC 8, 422   [37] OC 8, 429   [38] OC 8, 456–458

point of leaving for France, after a period spent in this country for reasons of health, when news came of the death of Mr Colbert, which induced me to delay my departure. I do not know what is to happen with this institution. But because I know its care to be laid in your hands, Sir, you, who know how to appreciate the arts and useful inventions and even delight in the perusal of these matters, in so far as your infinite pursuits permit you to do so, I am convinced that our affairs shall proceed even better than in the past. I even hope, knowing your goodness and generosity, . . . , that my conditions of employment may be better than in the seventeen years that I have served His Majesty. I have seen a number of colleagues receive more salary, while my own has been reduced in the time that I, with permission, was absent for reasons of health but still continued to work and to study. I think, Sir, that this does not show great propriety towards a man who leaves his country of birth, and all the advantages that are attached, in order to serve such a great king. His goodness and generosity come in the place of what has been left behind. But all shall depend on your favour and benevolence. Therefore I request that you will be kind enough only to send me your commands.

The answer, however (a few lines? – the letter has been lost), left him in uncertainty; he would receive further commands.

Christiaan waited several months for them, but in vain. On 9 March 1684, the father, now seriously apprehensive, called upon the assistance of his Paris friend, De Behringen, the royal equerry:[39]

My Archimedes continues to live here, and he waits for the Marquis of Louvois to take the trouble to reply to him. Patience is a virtue. For myself, he need make no haste, for I delight in the conversation of such a precious child, as do all others who wish to understand the quintessence of matters. But now that my days are numbered, I wish to know what is expected of us. If you are able

[39] OC 8, 483–484

> to find out, without any problem, then you will do me a great
> pleasure in telling me of it, so that I may advise this beloved
> Archimedes.

And after praising his child's most recent work to the skies, he
concluded:

> If God does not choose to grant this boy a long life of good fortune,
> which I am often of the opinion, people may lament such a loss, as
> they lament the death of the goose that lays the golden eggs. For
> many astonishing things are brewing in that brain of his.

Father Huygens received no answer. Subsequently, on 18 May,
Christiaan sent Louvois his *Astroscopia* (brochure on the tubeless
telescope, still to be discussed) with an explanatory letter in which
only the closing words were of significance: 'recommending myself
to your attention and still awaiting the honour of your commands'.[40]
He received no answer. They practised patience, in the knowledge that the longer the uncertainty continued, the worse the situation
became. In the end, the waiting proved too much for the father. On
2 November he wrote indignantly to De Behringen, 'If people perhaps
wish to be rid of him, after all that such a deserving colleague has
contributed to the brilliance of the Royal Academy, then it should
not be performed so underhandedly, in a fashion that is unworthy of
a foreigner of his calibre.'[41]
His suspicion proved to be all too well founded. It was confirmed by the reply from De Behringen, which has been lost, but can
be reconstructed on the basis of other letters. Although he was not
the minister's official spokesman, the royal equerry made it clear that
a return would not find favour and that this decision was irrevocable.
That is why, on 4 January 1685, Christiaan asked him for the least
that he could still expect:[42]

[40] OC 8, 489   [41] OC 8, 550   [42] OC 9, 1

'Now that it is all over . . . , it only remains for me to request permission to come and fetch my books and my furniture that I have left behind and, if possible, that I may be granted some form of honourable discharge.

This too was refused. No trace of a letter of dismissal has been found. Then on 5 April, Christiaan appealed to Louvois for the third and last time, and his tone was bitter:[43]

Now that I have understood from the royal equerry, who honours me with his friendship, that there is little chance that I shall be recalled, I believe that I have received your orders. The reason for this decision is unknown to me, but I must resign myself to it. Therefore I have requested that one of my friends send to me my books and furniture that remain in part of the living quarters of the royal library.

Most likely it was Friquet who took this task upon himself.

It did not go by unnoticed by the Academy. The letter that Jean du Hamel wrote to him on 23 May on behalf of the members does indeed express regret, 'for you have been most useful', but concludes with the instruction that in the event of future correspondence, this must be sent without an envelope.[44] In a letter of 8 November 1685 that father Huygens wrote to Jean Antoine, Duke of Avaux, he returned to the matter once more, to vent his anger: 'The only concession that has been made to us lay in those fine words, that he was welcome to come and fetch his possessions. I think, Sir, that a groom would hardly have been gratified, had he been addressed in this manner.'[45]

To what did he owe such treatment? Christiaan himself blamed it on the intrigues of a few jealous colleagues, who were happy to seize on his leave of illness as an opportunity to get rid of him. In one of

[43] OC 9, 4–5   [44] OC 9, 9   [45] OC 9, 39

his letters he complains at their lack of cordiality and regard. This would have been intended for Cassini in particular, who had assumed the role of discoverer of the moons of Saturn, and in this capacity had named them after Louis XIV, thereby suppressing Christiaan's role in their discovery. But these are vague clues.

What is clear is the following.[46] Ehrenfried Walter von Tschirnhaus (a Saxon nobleman who would become famous for the furnace technique of very hot fire with which strong (Meissner) porcelain could be made) pointed to La Hire as the culprit, in his letter to Christiaan on 30 August 1683. He was the one who had opposed his, Von Tschirnhaus', nomination as member of the Academy, because La Hire was opposed to the nomination of foreigners on principle, and in particular of friends of Christiaan; he considered that the man from The Hague had damaged the good name of the Academy with his criticism. If that is true, then clearly Christiaan had an opponent. But he also had an ally at the Academy. The abbot of Lannion wanted to plead with Louvois to have Christiaan's membership continued, but he came up against the objections of Henri de la Chapelle, Louvois's stooge in the Academy. Moreover, in spite of his highly esteemed mathematical contributions, the abbot was under a cloud for his unorthodox opinions and critical remarks about the expulsion of Protestants. But there is nothing to prove that all this envy and intrigue had any real influence on Louvois's decision.

The dismissal undoubtedly arose as a result of the political changes that followed the death of Colbert, his patron. The appointment of General Louvois as minister, on the same day that Colbert died, had shown that Louis XIV intended to make a clean sweep. After all, he was almost constantly at war with the Republic and he must have come to detest Willem III and all who served him – the Huygens family included. In addition, the Roman Catholic bigot, the Marquise de Maintenon, whom he had set up as his mistress, had incited him to tackle in earnest the suppression of all that was Protestant in his

----

[46] Brugmans 95–96

country. Louvois did not go in for half measures. The French Protestants rapidly began to lose their freedom rights, and were deprived of the only one that remained when the Edict of Nantes was revoked in October 1685.[47]

By that time Christiaan had also realised the truth. In April 1685 he wrote to Constantijn: 'As you see, the behaviour of the French towards me is somewhat barbaric, and I have to ascribe it to their policy to give no one a job who does not subscribe to their religion.'[48]

An excerpt from a letter, written at that time. We may infer, from a note, that the epistle was mainly about music. To whom Christiaan addressed it, we do not know, but apparently the gentleman lived in Paris:[49]

> If my correspondent M. conveys to you my humble greetings each time that I ask it of her, then you surely will have heard news about me. I have so entrusted myself to her good care, that I did not feel it necessary to trouble you with my letters, for I know of your busy engagements, imagine often to myself the room filled with people, and a beating heart on entering, even as far as the staircase. What I have so far told her of my health . . .

Who is M.? Why has the correspondence disappeared?

Music from that time. Christiaan wrote down the tones on his flute obtained by opening and closing the holes, which he had numbered from 1 to 8 (if a number is in brackets, then this hole must be half-closed):[50]

> First row: doh = closed, re = 8 open, mi = 87 open, fa = 86 open, sol = 8765 open, la = 123 closed, si flat = 12 closed, si = 1245 closed; doh = 13 closed, re = 3 closed, mi = 187 open, fa = 167 open (blow hard), sol = (1)23(4) closed, la = (1)23(4)678 closed, si flat = (1)236 closed; (1)25 closed, re = 12347 closed.

[47] Israel (1995) 646   [48] OC 9, 6   [49] OC 9, 41   [50] OC 20, 104

Second row: doh = 8 open, re = 87 open, fa = 5 open, sol = 48 open; doh = 23 closed, re = 18 open, mi flat = 34567 closed, fa = (1)2346 closed, sol = (1)23(4) closed, la = 145 open or (1)2357 closed, si = (1)236 closed; doh = (1)25 closed, re = 12347 closed.

Now in Christiaan's handwriting, written in pencil, a song from those times:[51]

doh sol mi mi fa sol la si doh re si
doh si sol fa mi fa sol la fa si
re si sol la mi fa fa re la sol si sol
doh sol la sol fa mi la sol la si mi si
sol doh sol la sol la sol fa mi re do re doh

L'on se trompe aisement lors que l'on aime bien
Je croios vostre amour de la force du mien
J'aurois moins repondu de mon coeur que du vostre
Helas de bonne fois je disois mon Iris
Mais cette Iris estoit l'Iris d'un autre

One who loves so deep is easily deceived,
Your love as strong as mine I readily believed,
I listened ill to that heart of yours,
Calling you mine Iris,
Alas, this Iris to another is betrothed

Who is Iris? A name with S? Such secrecy is uncalled for. He wrote the song when the news came that the Parisian Suzette Caron, married to François de la Ferté, had arrived in The Hague, one of the 400 000 French Protestants who left after the revocation of the Edict of Nantes.[52]

Several more poetic lines of that time. These are written by the father who, eighty-eight years old, composed the following lines on

[51] OC 9, 103 (n)   [52] Vollgraff 386 (n10)

his way to renting out the princely fishing waters near Geertruid-enberg:[53]

> But for twelve, one hundred years!
> This cannot be forgiven;
> I let a world perceive
> My tenacious life survive
> Many after me who entered it.
> And children, what think you
> whom I as children keep; you as parents,
> grandparents even? Will all grow older
> But for one? Take comfort, but a little patience;
> The day is close, that fatherless and free
> My help to you extended
> In hope your lives unroll as mine,
> So that, white-haired, and whiter than I wore,
> May say, like me: Come Lord, it is enough.

The old man wished, gradually, to relinquish office. Now that his Archimedes had been relieved of his duties in Paris in 1685 and had not been paid for four years, he offered him this stewardship.[54] The son refused, in spite of the princely income that went with it. It was no sinecure: the job would entail a great deal of time and was not suited to him. Yet he needed an income, even if only to pay off his debt to Van Ceulen.

Then the father offered him a fixed portion of his annual income, the gratuity that he had received since 1630 from the House of Orange. This Christiaan readily accepted, for it required his doing nothing in return.[55] It appears to have been no mean sum (in 1630 it came to 1000 guilders per year), appropriate to the wealth of his home-land 'and all the advantages that are attached', as he had pointed out to Louvois. Constantijn discussed the assignment of the gratuity

---

[53] Worp 8, 344   [54] Keesing (1987) 110   [55] OC 9, 103

with Willem III, who gave his consent. This meant that, starting in 1686, he received a private annual allowance that enabled him to live comfortably.

In 1686 a fleet of the East India Company sailed out. One of the galleons, the *Alkmaar*, had two clocks on board that were to determine geographic longitude at sea.[56] These were remontoirs made by Van Ceulen.

They had a new kind of pendulum, a '*pendulum cylindricum trichordon*', of which the length (in other words: the time of oscillation) would change very little with a change in temperature.[57] This was the pendulum that Christiaan had used in the clock he had made with his own hands in 1683. It consisted of three threads of equal length that hung at an equal distance from one another from a horizontal mounting, and were fixed at the bottom to a heavy ring. This ring could swing freely backwards and forwards. A brilliant invention that quickly led to a practical result, but which also sparked off a challenging theoretical problem.[58]

Christiaan had written an instruction manual for Thomas Helder, which would convert the timekeeping of the clock into geographic longitude, with instructions and rules about how it should be hung, wound, regulated and, last but not least, cared for. This *Kort onderwijs aengaende het gebruijck der horologiën tot het vinden der lenghten van Oost en West* (Brief Instructions concerning the Use of Clocks for Finding the Longitude from East and West), dated on 23 April 1686, was far from brief.[59] It contained no fewer than thirty-four extensive articles, and one wonders whether the good seaman understood them all or even read them. Permission for the endurance test (the *Alkmaar* would stay away for a year) came from East India Company director Johannes Hudde, after a short trial at sea had been carried out with success.[60]

---

[56] Mahoney 255–256   [57] OC 8, 475   [58] OC 18, 508–511
[59] OC 9, 55–76   [60] OC 18, 533–535

The inventor had seen it for himself. On 13 August 1685, he travelled with the two remontoirs to Amsterdam to hang them on board a flatboat of the East India Company, and to regulate them. As Hudde was not there, he returned home regretfully after a week, his task unaccomplished, 'for the wind was right for a trial'.[61] On 3 September Hudde wrote to him in haste that he would send a flatboat to Scheveningen, but Christiaan gave preference to a trial on the Zuiderzee, which he considered calmer than the North Sea. Back in Amsterdam on 9 September, he met with an extremely courteous shipmaster who had, however, received instructions to sail out from Texel.[62]

Two days later they set forth, sailed into rough waters and a storm that tugged at the sails, put in at Enkhuizen Harbour, where they waited until the storm had passed over, and then set course for Texel. One of the clocks continued to run smoothly, the other stopped now and again. Christiaan knew enough. Lack of sleep, due to the rumpus on board, and a physical exertion to which he was unaccustomed, had made him ill. He was feverish, and went ashore at Texel. He returned to The Hague at the beginning of October, where he let himself recover.[63] On 26 October he wrote to Hudde that the marine clocks were seaworthy, and that Helder was prepared and able to operate them on a voyage to the Cape of Good Hope.[64]

Was it because of Christiaan's success story of the calmly ticking clock on the heaving ship that his father wrote that angry letter of 8 November 1685 (about the groom who had done his best) to the French ambassador? The son restricted himself to the taunt that Sire Louis really did need a *Medicina mentis et corporis* (Remedies for Mind and Body), the title of a mathematical work that Von Tschirnhaus had dedicated to the Sun King.[65] But the father felt deeply insulted. He was even angered that Louvois, until then a general, had called his son a mathematician in the (lost) letter of 1683. This was

[61] OC 9, 20   [62] OC 9, 25–27   [63] OC 9, 30–31
[64] OC 9, 37   [65] OC 9, 113

a term usually reserved by generals when referring to the engineers for their armaments. Moreover, this general had occupied the country of Orange with unnecessary violence, the principality for which he had once pleaded for, and won back, the rights of Orange. He was wounded in his honour, or, if the word honour has lost its meaning, his pride. However poor his eyesight, he took in everything, and however arthritic, he weathered the damp night, but he was certainly proud when his Archimedes tried out his tubeless telescope in the garden.[66]

As we saw, Constantijn and Christiaan had ground objectives with such a long focal distance that a tube construction would no longer suffice. In the seventeenth century, there was no question of making a tube 66 metres long (210 feet, their longest focus). So Christiaan resorted to an open construction, whereby the objective was attached to a high mast, and the eyepiece to a low tripod for the viewer. To align both lenses and to focus on a star he invented a simple but effective mechanism.[67] But to use it, he needed a servant at the mast.

Constructing and positioning the mast would have been the hardest part of all, because even minute vibrations of the objective had to be avoided. Would it not have swung in the wind, or sung in the wind when it was more tightly secured? It was, however, not necessary to make the mast as high as the telescope was long. If observations within 30 degrees of the apex were excluded, a lens of 66 metres would require a 57-metre-high mast. Did he ever use this lens, which has not been preserved? What is certain is that in 1686 Christiaan had a 105-foot-high mast (33 metres) that indicates use of his 122-foot (38 metres) long lenses.[68]

He had already described this telescope in *Astroscopia compendiaria tubi optici molimine liberata* (Compound Stargazers without Light Tube) in 1684, in a quarto brochure printed in The Hague which he sent to Louvois, amongst others.[69] On 6 April 1686 he wrote to

Constantijn that he had made some fine observations, but if we read this properly, we realise that he was just testing the glasses.[70] There was no question of star observations, for the mast vibrated too much. It never became a success. The professional astronomers were not enthusiastic about it either; Cassini saw something in it, Auzout did not.[71]

However humiliating the dismissal may have been, Christiaan's self-esteem was undamaged; he was now liberated from the whole rigmarole in Paris. In fact it increased his scientific activity, which, in 1685 and 1686, was considerable. If we put aside the practical work on clocks and telescopes, we are left with his diverse astronomical, optical and mathematical studies.

Undoubtedly it was his renewed interest in lenses that prompted him to return to his manuscript of the *Dioptrica*. To this work, begun in 1652, he added chapters on the passage of rays in a lens system, of both telescopes and microscopes.[72] And in spite of the elegance of his argumentation and the soundness of his calculations, even then he could not bring himself to conclude it with a publication.

Although the imaging rules of lens systems were gradually becoming generally accepted, this work, completed in 1685, offered so much that was still new that its publication would have been widely applauded. It plays an important role, for example, in the correspondence that Bernard Fullenius, professor in Franeker, entered into with him (and which, curiously, was conducted entirely in Latin, while they both used Dutch in their letters to a go-between in Leeuwarden).[73] To give an idea of the work, we shall restrict ourselves to a description of his introduction on telescopes, and of his consideration of the magnification and the field of view of telescopes with two lenses.

Christiaan begins with the invention of the telescope, which he attributes to Johannes Lippershey or Zacharias Janssen. He rules out

[70] OC 9, 51   [71] OC 21, 197–198
[72] Dijksterhuis (1999) 238–240   [73] OC 8, 443, 489, 533; OC 9, 352

Adriaan Metius as the inventor, but Giambattista della Porta deserves mention, for he had provided the impetus.[74]

The importance of telescopes for astronomy lies in their capacity to enlarge, he explains, and this is why he has great expectations of his invention of the long, tubeless telescope. Indeed, greater magnification will bring to light 'countless new phenomena'.[75] He focuses his attention especially on the Dutch telescope, in which the large front lens (the objective) is convex and the small rear lens (the eyepiece) is concave, with the vanishing point of the eyepiece in the focus of the objective.

When he moves on to the operation and characteristics of this artificial eye, he deduces the formula for the magnification no less than four times. This astonishing excess can only be explained by his desire to excel:[76]

> Up until now, no one has penetrated to the heart of the matter, to know the nature and extent of the magnification of the observed object, when the form and distance of the lenses is given. Kepler did not do so, although he is to be praised for the first explanation of dioptric phenomena. Nor did Descartes, and he even went astray when explaining the effect of the telescope.

After the fourth deduction of the formula for magnification, the French philosopher was given another lashing: 'This is how Descartes should have determined the theory of the telescope, not by the cross section of rays on the surface of the front lens.'[77] And then he is dealt one more swipe for his exaggeration of the magnification that could be achieved.

For his first deduction, Christiaan assumes the eye to be directly behind the eyepiece. According to one proposition, which had been proved earlier on in his *Dioptrica*, the object (imagined as being infinitely far away) is seen at the same angle as if, instead of the eyepiece, there were only a minute hole. It can be said that the eyepiece

---

[74] OC 13, 435–437   [75] OC 13, 439   [76] OC 13, 441   [77] OC 13, 451

then exerts no influence on the apparent size of the image, and that it only restricts the field of vision. It then follows that magnification is determined by the ratio of the distances of the objective and the eye-piece at the point of focus of the objective, or, to be exact: at the place where the image of the object would be if there were no eyepiece.

In whichever manner the eye focuses (accommodates), so long as it is placed close to the eyepiece, these rules must always apply. This restriction no longer applies if one assumes that the eye has focused at infinity. In this approach the vanishing point of the eyepiece must coincide with the focal point of the objective, so that parallel rays that enter the objective from the front also leave the eyepiece at the rear in parallel fashion. Because, according to a proposition that was also proven earlier in his *Dioptrica*, the apparent size of the object does not depend on where the eye is placed, the magnification can always be written as the ratio of the focal distance of the objective and the distance of the vanishing point of the eyepiece. This is the result that is now very well known.

To find this formula one can also follow a ray that meets the axis at an angle through the objective and the eyepiece. Christiaan chooses, successively, a ray through the middle of the eyepiece, a ray through the focal point of the objective for the viewer (which, after the objective, then passes parallel to the axis as it goes through the telescope) and a ray through the centre of the objective. This enables him to obtain the three subsequent deductions of the formula.[78]

After this he examines the field of view.[79] He has seen that this is greatest if the eye is placed immediately before the eyepiece. Normally, then, it depends on the pupil of the eye, and is determined by the cone with the pupil as base and the centre of the objective as top. It will not, however, become limitlessly small if one holds a plate with a smaller hole in front of the eye each time. In the end, he notices, it covers the solid angle of a cone with the objective as base and the eye as top.

[78] OC 13, 453, 461, 467   [79] OC 13, 481–511

He then accounts for the reduction in clarity towards the edge of the field of view. He carried out an interesting experiment. After he had allowed his eye to become accustomed to the dark, so that the pupil stood wide open, he quickly placed it before his telescope, and then saw that the field of view became smaller because his pupil contracted in the light.[80] This effect can only occur in the Dutch telescope with its rather large exit pupil, not in Kepler's telescope with its much smaller exit pupil. In this telescope the eyepiece is not a small concave lens located in front of the focal point of the objective, but a small convex lens behind that focal point. Christiaan also deducted the formula for the magnification for Kepler's telescope, this time only three times.

He emphasises that Kepler's telescope has a smaller field of view than the Dutch one, and estimates its solid angle as equal to that of a cone, with the eyepiece as base and the centre of the objective, as top. His calculation of the field of view is, on the whole, rather careless, and comes nowhere near that of the spherical aberration, for which related mathematical problems need to be solved.

In this case, it involves constructions of straight lines through three convex surfaces, that is to say of the eyepiece, the eye and the exit pupil, which are certainly not simple. They do become so, if the eye coincides with the eyepiece, something which, as we saw, Christiaan assumed. But in an addition of 1686, he seeks the position that a point-eye, as conceived in Kepler's telescope, must occupy in order to obtain the largest possible field of view.[81]

We have pointed out that the publication of these results would have been widely applauded. So why did he keep them hidden away in his room? Why didn't he put them together with his early studies and publish a fully fledged treatise on optical systems? F. J. Dijksterhuis has recently explored this question.[82] He noted that Christiaan's work of 1653 and 1665, which we have mentioned above (see pp. 126–127),

---

[80] OC **13**, 505   [81] OC **13**, 611–612   [82] Dijksterhuis (1999) 1–7

did not go beyond applications of Snel's sine law of refraction to spherical surfaces. This meant that the aberration that he had calculated in 1665 – and how difficult it had been! – could not explain the image distortion and colours that he, and everyone else, could observe at some distance from the optical axis. Around 1670, after unsatisfactory attempts to correct for spherical aberration by a slightly curved lens, he knew that something important was missing from his calculations. He even crossed most of them out when Newton confronted him with his colour hypothesis (see above, pp. 269–270). Disheartened, he gave up. What he had in mind was a *Dioptrica* in which first the sine law had to be explained by an appropriate theory of light propagation and then, based on this law, a thorough treatment of the image formation by lenses and lens systems. The work on telescopes that we have just described – work of the 1680s – confirms this. It has traits of a triumphant finale. Is this how he really intended to complete it? Let us not forget that by this time Christiaan had an appropriate theory of light propagation as well, with all the appearances of an introduction that even addressed the curious possibility of double diffraction to justify its mathematics. Yet he left the work unfinished. But could he have done otherwise, being the scrupulous and painstaking person that he was, if he lacked the proper explanation for those well-known colourful distortions of images off-axis? Dijksterhuis concludes:[83]

> He never abandoned the original plan of 1672, in which his light theory would be a preparatory part of his theory of dioptrics. He published his wave theory in 1690 as *Traité de la lumière*, a title he had chosen at the last moment. He did so after many hesitations over the best way to present his wave theory, which suggests that Huygens himself was not sure about its exact status.

Just as the Frisian Fullenius, at that time, became his pupil in optics, so the Swiss Nicolas Fatio de Duillier became his pupil in mathematics. He never actually met the former, but he would entrust

---

[83] *Ibid.* 231

him with the publication of his *Dioptrica* and other uncompleted work. The latter he did meet, for instead of writing Latin letters, Fatio simply turned up on his doorstep.[84] The young mathematician, who led the penurious life of a wanderer, would affect him more deeply.

He had already announced in 1684 that he would come to the Netherlands to have a dissertation printed on the geometry of snow crystals ('ice stars') which he had observed in the Jura. Sometime in 1686 – we do not know exactly when – he paid his first visit, and left behind a copy of the dissertation. The mathematical quality was so high that Christiaan went into it at some length in a letter he wrote to Fullenius on 24 October.[85]

Then Fatio involved him in his attempt to construct tangents to the complicated rosettes that Von Tschirnhaus had discussed in his previously mentioned *Medicina mentis et corporis*. Von Tschirnhaus had also paid him a visit, probably in September 1682, but his mathematical work did not appeal to Christiaan.[86] Fatio, however, managed to rekindle the fire of pure mathematics that had long smouldered beneath practical applications. Flaring up higher than ever, it would never again be extinguished.

Towards the end of 1686, Fatio must have rented a room in The Hague, for he was repeatedly to be found in Christiaan's study on Het Plein. In the beginning of 1687 he found a way of constructing tangents on rosettes. 'If I am to receive any fame for this discovery,' he wrote, 'then I wish to share it with him [Christiaan] or even give it to him entirely.'[87]

Christiaan's own notes show how justified this is:[88]

On 13 or 14 March Duillier told me about his way of finding tangents on the curves of Tschirnhaus. The following day, I showed my proof that this manner was exact and that one must

[84] Vollgraff 734   [85] OC 9, 109–111, 117–120   [86] OC 9, 108, 122
[87] OC 9, 175 (n)   [88] OC 9, 181

proceed step by step. On Sunday 16 I found that the perpendicular through ... On Monday 17 I said this to Duillier, who at first tried to deny it.

On these days he felt unwell, and Fatio, who wrote him a letter some weeks later when he was about to leave for England, blamed himself: 'I fear that my visit has contributed to preventing the full recovery of your health.'[89] The sensitive letter was a letter of condolence.

[89] OC 9, 134

14 Orphan

Father Constantijn died on 28 March 1687. It cannot have been unexpected – he was well over ninety – but the bereavement affected Christiaan deeply.

One month later, he wrote to La Hire in Paris on 'affairs that need to be settled upon such an occasion', mentioning somewhat coolly the funeral in the Jacobskerk, after a state ceremony by torchlight in the evening, with fifteen carriages. He also wrote about 'his poor state of health'.[1]

Behind the aloofness of these words lay concealed a genuine grief, for he dressed himself as an orphan, and had himself portrayed as such (Plate 7). We know of this 'orphan garb', and also the name of the painter: Pierre Bourguignon, from a piece of writing by Lodewijk's then twelve-year-old Constantijn.[2] With the same concern for convention as his father, Christiaan mourned the loss of both parents and desired to demonstrate it in some tangible form. But not only this.

The act of mourning spurred him on, whether or not under the influence of Fatio, to the strict discipline of a pure mathematical thought process. One year before, Leibniz had challenged the readers of *Acta Eruditorum* (Deliberations of Scholars) to prove his solution for the isochronous line. Christiaan proved this solution with all the virtuosity that a geometrical method demanded, sent this to Bayle, and saw it published six months later in his *Nouvelles*.[3] His proof was also published in the *Acta*. But this was not enough.

It was probably shortly after the funeral that Christiaan wrote: 'Heathens and barbarians attribute to God a human body, philosophers

---

[1] Strengholt 127; OC **9**, 130   [2] Vollgraff 754   [3] OC **10**, 224

a human soul and human feelings, only in greater perfection. They ascribe to him a mode of thinking, willing, understanding, loving. What else can they do?'⁴ This must surely be a reaction to words that were spoken in the Jacobskerk, namely that God is perfect love. But, he thinks, love is related to hate, just as good is to evil, and father to son. To love can be attributed no absolute meaning. This is why Christiaan answers those who ask what else they can do, 'Admit that it is beyond man to have an idea of God.'

He was inconsolable. The same intellect that rejected an anthropomorphic God because He, too, was unknowable, understood that something was needed to take His place. Christiaan set out to find it.

Apart from the letter we have mentioned to La Hire, concerning the delay in a despatch of documents (that was sent in June) and the proof of Leibniz's solution, nothing was heard from him until 11 July. He did nothing about the division of the estate, a task that fell to him as sole inhabitant of the ancestral home, as brother Constantijn was required by the Stadholder on the Veluwe. 'And how does it fare in the home of the deceased?' wrote Constantijn, at the end of the summer, 'Don't people talk about an inventory, plans for division, all that has to be done?'⁵ A short time later, on 4 September, he added the following lines:⁶

> Friends of ours have written to me that you have set your intentions upon B. in Rijswijk. Even matchmakers are said to concern themselves with the affair. Up until now I have not considered it seriously. As far as I am concerned, may you be happy with her. I wish at least that you will not regret it. But there is quite some fault to be found with the one to whom I allude, regarding background, money, and reputation of the person in question. If you wish to take a wife, you must, even at our age, consider carefully whether she, of all people, is to be the one.

⁴ OC 21, 341   ⁵ OC 9, 207   ⁶ OC 9, 210

Was he referring to Elizabeth Buat? Perhaps. Two years earlier she was still in Paris, where she was said to have converted to Roman Catholicism. Whether she belonged to the stream of refugees of 1686 is unknown, and also whether she was staying that summer with her family in The Hague. But we do know that, two years later, she left The Hague and went to Paris.[7] She is the subject of gossip in Constantijn's journal. She was said to be hot. Thus perhaps the discretion in his letter, and the spiteful tone. But there are other women by the name of B. The affair did not come to anything. And there was nothing in Christiaan's notes to set us upon the trail of any love affair.

What did take place was the division of the estate. As if he had been shaken awake from a dream, he set about making arrangements, consulted with Lodewijk and brother-in-law Philips Doublet, had catalogues made of the paternal library, and the extensive property valued. From the precious collection of books he chose those to his own taste, so that his own collection now included 3325 titles, of which 406 were to do with mathematics and physics, with, in addition, sets of illustrations and art books. He would also take charge of the hundreds of pieces of music either copied or composed by his father, a treasure that has disappeared with him.[8]

While taking stock of the past, Christiaan remembered the engraving that he had once commissioned in Paris, which was to depict him as the author of *Horologium oscillatorium*, together with an illustration of the clock. In October he wrote to the intermediary responsible: 'I find it tiresome to have to write so often concerning such a small matter. I request you, therefore, by whatever means, to have it settled. It is ridiculous to have to wait for two and a half years upon the whim of an engraver.'[9] This engraver was called Gerard Edelinck. He based his work on the drawing that, considering the striking resemblance, was based on Netscher's portrait from 1671. Christiaan would have to wait another six months before finally obtaining this engraving, which we know as the most frequently reproduced representation of him (Plate 8).

[7] OC 9, 335   [8] OC 9, 230–231, 241–246   [9] OC 9, 238–239

The division of the estate had far-reaching consequences for him, for he had to move out of the house on Het Plein. The father's will, made with characteristic care in 1682, specified that the house was to go to the oldest son. He was also to be given the manor Zuylichem ('our hapless property of Gelderland') and use of Hofwijck that was to remain the undivided property of the three sons.[10] Although Constantijn had a house in Korte Vijverberg that was to his liking, it never occurred to him to relinquish his right to the capital house. Around New Year, the Lord of Zuylichem moved in (again). By then the brothers had found a solution amongst themselves for Christiaan's accommodation.

'After our last discussion I was, as usual, unable to sleep,' he informed Lodewijk on 20 December, but, 'I have considered the proposal of Hofwijck from all sides and have decided to follow brother Zuylichem in his right to the enjoyment of the property. Then I shall not need to rent a house or rooms in The Hague. I plan to make some alterations in the building by enlarging the living quarters somewhat, to accommodate my library, and so that I may occupy them in comfort.'[11] He was allowed to stay on for a few more months, along with his servants, with Constantijn at Het Plein. And he regarded it as such a formal arrangement that he wanted to pay his sister-in-law Suzanna for the housekeeping, but she would not dream of accepting anything from him. In the spring he moved to Hofwijck.

He wrote to Constantijn on 4 May 1688:[12]

I have been in my new home now for five days, and am busy from morning till night putting matters in order, until the annex is ready that is to house my books and part of my furniture. During this time I have not yet been to The Hague, nor have I received any news from there. This provides me with some foretaste of the lonely life to which I must become accustomed. What grieves me is that I must take the midday and evening meals on my own, but as we know, this is the lot of crowned heads as well.

[10] Strengholt 128   [11] OC 9, 252   [12] OC 9, 295

In this loneliness he lost himself in a magnificent work, and for several months nothing was heard from him.

Isaac Newton's *Philosophiae naturalis principia mathematica* (Mathematical Principles of Natural Philosophy) had already come up for discussion on 11 July 1687, the only time in that eventful summer that he had actually got down to writing.[13] It had been a reply to Fatio, who had announced to him in June, shortly after his arrival in London, that a book was to be published by Newton that people said would radically change the science of physics. He had written something about the content, and also about the principle of attraction, which was not Cartesian. 'I do not care that he is no Cartesian,' Christiaan had replied, 'so long as he does not serve us up with such concoctions as attraction.'

Edmund Halley, Newton's keen patron, had taken care to present him a copy of the book as soon as it was printed.[14] It had been in his possession since August. First he seized upon a detail: the size of the flattening of the Earth's surface, for this greatly interested him. Already in November and December he calculated 1/578, where 1/229 was given in the *Principia mathematica* (and the actual figure is about 1/300).[15] He needed to know the figure to solve a practical problem for his marine timekeepers.

Also in that eventful summer of 1687, the *Alkmaar* returned to Texel from her voyage to the Cape. Thomas Helder, who had supervision of the clocks, was no longer on board, because he and fourteen others had died at sea. He had had problems with the clocks on the outward journey, because they did not stand up well to rough seas. His aid Johannes de Graaff, however, had continued to carry out measurements, and brought enough of these home for Christiaan to be able to reconstruct a course. Now, according to the clocks, the *Alkmaar* had sailed right through Ireland and Scotland.[16]

---

[13] OC 9, 190   [14] Westfall 472   [15] OC 19, 401
[16] OC 18, 636–642; OC 9, 273 (+map)

Although he wrote in his report to the East India Company that his work had given him pleasure, we can hardly imagine that he greatly contributed to his sister-in-law's enjoyment at table. As he was unable to discover any error in the logbooks, Christiaan was forced to admit that there was some kind of serious disturbance.

If the Earth was not completely round but flattened, then the second-pendulum must be shorter at the equator than at the pole. He had heard about this in 1685, but then he had thought that the effect was small, or even unproven. Apart from this, Jean Richer had already reported in 1679 to the Paris Academy that the second-pendulum was 5/4 line shorter in Cayenne than in Paris, and also that he considered this to be disputable.[17] Christiaan had to begin to take the flattening seriously.

His report to the East India Company was probably delayed, due to his continuous engrossment in the *Principia mathematica*. As we have mentioned, he wrote no letters for months (to be exact, between May and November 1688). He was deeply impressed, as were many of his contemporaries. The philosopher John Locke, political refugee from England,[18] visited him to ask whether the mathematics, which he was unable to follow, were correct. It was perfectly sound, Christiaan was said to have replied, and Newton, who heard this later from Locke, passed this on to various people.[19] Indeed, a long, admiring review of the book appeared in *Acta Eruditorum* in the summer of 1688.

In the autumn Christiaan examined Newton's calculation of the Moon's orbit around the Earth, with the supposition of an attracting force that is inversely proportional to the square of the distance. He did the same for the orbits of the planet around the Sun.[20] He found no flaw in Newton's proof that these orbits in which such a force prevailed must be ellipses, and that this force, moreover, explains the other laws of Kepler. And he found all this so astonishing that he placed

---

[17] Vollgraff 728; Mahoney 258   [18] Cranston 46–51
[19] Westfall 470   [20] OC 21, 408–412

his Paris notes (probably from 1680) about the planet orbits alongside, and added, in Latin: 'The famous Newton has dealt with all these difficulties, as he does with the Cartesian whirls. He has shown that the planets are maintained in their orbit by their gravitation towards the sun.'[21]

Let us, at this moment in Christiaan's life, add a reflection. When, in 1659, he had correctly expressed centrifugal force, speculation may have led him to the nature of a gravitational force. Expressed as a ratio, his discovery for this force f now comes to

$$f :: (ss : d)$$

in which s is the speed of the body in a circular path at a distance d around a middle point.

Kepler's third law of planetary motion, published in 1619, had stated that the square of the time taken to complete an orbit around the Sun is proportional to the cube of the distance of that planet to the Sun. Because this time t is found by dividing the circumference of the orbit by the orbital speed s, and because the circumference is proportional to the distance d to the middle point, we get

$$tt :: ddd$$
$$(dd : ss) :: ddd$$
$$ss :: (1 : d)$$

By filling in the last line of Christiaan's expression for force, we get

$$f :: (1 : dd)$$

therefore obtaining the result that the force is inversely proportional to the square of the distance.

We must take into consideration that this force is not centrifugal, but centripetal. This is the heart of such a speculation. The centrifugal force is neutralised by an equal, but counter-directional centripetal force, a gravitational force. Hooke saw something of this sort in 1673, shortly after Christiaan's results for centrifugal force, and he

[21] OC **21**, 143

never forgave Newton for carrying off the prize that, according to his own firm conviction, should have been his.[22]

Had Christiaan now, after so many years, become convinced of such a force? He could not believe it and kept fighting it off. 'Absurd,' he wrote two years later to Leibniz, the other great protester.[23] But he began to reflect upon an explanation for such a force of convex symmetrical whirls of particles, and regretted not yet having published his own theory of weight. On 8 November 1688 he noted down, in passing, in his calculations: 'Last night nearly choked on vomit that compressed the windpipe and hindered breathing. The evening before ate artichokes and grated apple with too much sugar, the beer was too immature and bland.'[24] The first sentence is half in French, half in Latin, the languages of his unpublished *Discours* and of Newton's *Principia mathematica*.

It was 12 November 1688 before he resumed his correspondence. Constantijn became involved in a military expedition, and Christiaan, filled with concern, wrote to him about it.[25] On 30 December he wrote again: 'How lamentable that we have been unable meanwhile to exchange any letters. Thank God that matters are now improving. You can imagine how happy we were with this great success, after all the worry and anxiety.'

Constantijn was now in London, in the retinue of the Stadholder, who had crossed over to England with an army and would be crowned King William there in February 1689. He continued:[26]

Many will be travelling to England this spring, including myself perhaps. I have now spent the entire winter at Hofwijck, with a few wretched evenings when the weather was bad. But one can become accustomed to anything . . . Have you met Boyle yet? I should like to be in Oxford [he must have meant Cambridge] just to meet Newton, whom I greatly admire for the discoveries in the work that he has sent to me.

[22] Westfall 451 (471) [23] OC 9, 538 [24] Vollgraff 764 (n114)
[25] Israel (1995) 849–852; OC 9, 303 (+n2) [26] OC 9, 304–305

He began to long for entertainment. Lodewijk's son Constantijn describes the behaviour of his learned uncle as a mixture of deep concentration on his work and light-hearted companionship with his family. 'He spoke with the same ease about matters of no great consequence as about matters on all kinds of art and science.'[27] Lodewijk himself, who would live in Hofwijck during his brother's absence, received a letter in April 1689 that was filled with doubts about the trip. On 25 May, however, Christiaan wrote to Hudde that he was about to leave for England.[28] He arrived in Harwich on 11 June, and in London on 16 June.

Before we set out this well-documented visit (both Christiaan and Constantijn kept a journal), we should mention Christiaan's activities to make amends for his absence. The despatch to La Hire of June 1687 contained a manuscript of his theory of weight, the *Discours de la cause de la pesanteur*. The aim was to have it published by the Academy, which is indeed what happened, six years later.[29]

Now he made a copy of the old text, with a slight alteration to the introduction, and brought it, in March, to Van der Aa, the printer in Leiden. He also brought him the ten-year-old text of his theory of light, *Traité de la lumière*. He wanted to have them both in one volume, with the *Traité* in front and the shorter *Discours* at the back, by way of an addition that merited little attention.[30] This was the material he wished to discuss at the Royal Society in London. Even before he had left for England, the setting and correcting had been largely completed.

Now for the visit, which we cannot represent better than by quoting from the diaries of Christiaan (Ch)[31] and Constantijn (Co):[32]

> 16 June (Co): While I was seated at table in Whitehall, my wife, son and brother arrived and, to my great joy, all in good health. In the afternoon we looked over one or two lodgings with them and

[27] Vollgraff 743 (n10)   [28] OC 9, 317, 319   [29] OC 9, 299 (+377); OC 19, 429
[30] OC 9, 276 (n8)   [31] OC 22, 743–749   [32] Huygens 23, 142–161

cousin Becker, took one with Mrs Row, widow of Sir Robert Row, and spoke with the daughter. Our rooms, together with those of brother Christiaan, cost 33 guilders per week. We moved in straightaway, after we had been out shopping.

20 June (Co): Brother visited Boyle in the morning, who informed him, amongst other things, that a man had been to see him who, with a bright red powder, had made an ounce of gold from lead, and that he had heard that this man was arrested in France.

21 June (Co): Rode with my wife and the others to Hampton Court, where we made our way as best we could. But Brother, who had reserved a room in Toy, could not get in, when he thought to be able to sleep there at half past midnight. He slept on chairs in my room.

22 June (Co): Brother went in a boat to London to be at the meeting that evening of the Royal Society. (Ch): Meeting in Gresham College in a small room, a collection of curiosities, very full but beautifully kept. Askin [Inchiquin] was chairman, Henshaw vice-chairman, Halley secretary. Handed over the letter from Leeuwenhoek. I was there together with Newton and Fatio.

30 June (Ch): At the concert we heard pieces sung from French operas, amongst others, by Bourguignon. Paisible played the flute excellently. It was a fine room in Italian style. (Co): My wife and I arrived in London at eight o'clock and waited until eleven o'clock to eat, because brother and Tien went to listen to music by Bourguignon. My wife informed me that the daughter of Mrs Row and another woman, who had little mouse pelts instead of eyebrows, had been up to our Tien's room in order to tickle him. When they got nowhere with him, they went to the servants' room, but they chased them away as well.

2 July (Co): My wife said to me that her maid had seen Miss Row on the staircase in front of our room with a man, she taller than he, and that he pulled up her skirts but she, caught in the act, let

them fall. The woman with the little mouse pelts was called Mrs Pamel.

3 July (Co): In the evening went for a stroll in the park and met, by chance, the ladies Row and Pamel. The latter had told brother Christiaan that that very morning, a woman, clothed all in white and with a veil over her head, had been sentenced in the church of St Margaret to atone in public for committing adultery.

8? July (Ch): Spoke with the King at Hampton Court, dined with Bentinck. Hamden proposed that I put in a good word for Newton, who submitted an application to His Majesty.

9 July (Co): The King went hunting, and I crossed over the water to London. My wife said that she had heard that the wife of the glazier, where 7 stays, had seen now and then, through a hole in the garret, that 8 swived her.

10 July (Co): Brother Christiaan went to London at seven o'clock in the morning with Mister Hamden, Fatio Duillier, and Newton to recommend the latter to the King for a vacant governor's post of a college at Cambridge.

20–22? July (Ch): Went to Epsom, where one may drink spring water, together with Cole and the ladies Pamel and Row. Lodged with Clinch, where we slept for three nights. Spent three pounds. We were close to the house of Lord Barclay with gardens and park, the house splendidly furnished. After that we visited the house of Lady Evelyn, which is flanked by a beautiful chapel containing fine woodcarving. The garden has three straight terraces, one above the other. The second day we visited Boxhill, a beech forest measuring eight or nine miles, with overgrown avenues and beautiful views towards the hills. We were drenched to the skin by the rain. In Epsom there are rooms where one can dance in the evening. In another room one may drink spring water, and by eleven o'clock in the morning one can see people taking a walk. There is an orchestra with little shops all around it for

sweets and other bits and bobs. In the evening, before our
departure, we dined in a tent on a raised platform in the garden
of Clinch.

23 July (Co): In the afternoon I visited 's Gravemoer with my wife
and brother, where we played for a silver jug, which brother
Christiaan won. (Ch): On our return from Epsom I accompanied
my brother and sister-in-law to Gravemoer, where I won a jug of
10 guineas. We were nine players together, amongst whom Solms,
Odijk, Ouwerkerk, 's Gravemoer, Wilderen and Marquet. Three or
four days earlier, we ate there.

27? July (Ch): Saw Boyle three times. The last time he carried out
an experiment for us with two cold liquids that, when mixed
together, caused a flame to appear. With one of them, which
smelled almost like aniseed, he had drenched some linen in a
silver spoon. The other, which he poured over it, was contained in
a small bottle that smoked when he opened it.

28 July (Ch): Met Hamden at the concert, who spoke with the
Duke of Somerset about the matter concerning Newton. Somerset
is chancellor of the college in Cambridge.

6 August (Co): Went with my wife, Tientje, brother, and
Oldersum, who joined us, in a carriage to Windsor and viewed the
castle. Tien rode Oldersum's horse, a grey. Ate at the White? (Ch):
Went from Hampton Court to Windsor with my brother, his wife
and Oldersum. There we saw seven or eight rooms with ceiling
paintings by Verrio. The painting in the chapel pleased me most.
We saw the bathtub of King Charles II, surrounded in white
marble, and a small room adjacent to it, his bed surrounded by
portraits of various women . . . We ate at the White Lion, bad
drink but for the rest quite pleasant.

10 August (Ch): With my sister-in-law, moved out of Row's
lodgings, and moved into another on the Pall Mall, my room
opposite hers.

13? August (Ch): With Pamel and Row, together with Cole, to the bleach-works at Monmouth, taking an indirect route along small roads. Festive meal on damask.

16 August (Co): Bade farewell to brother Christiaan, who thought that Captain Van der Holck would depart next Friday.

18 August (Ch): To the park with Pamel, to her mother. From there to the tailor, where the mistress of her husband lived. Spent the night, f.m.g.a.c.[33]

19 August (Co): Bade farewell to Boyle, Fatio and Locke. Boyle promised me the formula to make ice without ice or snow. To Pamel, f.c.f.

20 August (Ch): In the evening to Pamel, f.m.g.r.g. Then changed money.

21 August (Ch): Walk in the park with Fatio, in the evening to Pamel for the last time, b.d.p.a.l.

24 August (Ch): In the morning on board the *Brielle*, the same ship that transported the King of Holland to England, with Captain Van der Holck. As the wind was very favourable, we would be able to cross in twenty-four hours, but at the mouth of the Thames we were obliged to wait for an escort. Then the wind was against us. Early on the 27th three ships were seen that were thought to belong to the French, and the men prepared themselves for a fight. The women went into the cable locker, also Villers and Sablière. The rest armed themselves with rifle or gun, all except Geldermalsen and I. With my telescope, we gradually perceived that the three ships carried a Dutch flag.

30 August (Ch): We arrived at the roadstead of the Maas where, for 25 cents per person and 5 cents per case, we changed over to a pilot boat. Ate in *Brielle* with the King and Queen of England. From there we took two barques to Maassluis, where we left clothes and a few cases, and in three carriages, each for 7 pounds, we arrived that evening in The Hague.

[33] The meaning of this and other codes in the diary is unknown.

Added afterwards, in pencil:[34]

> Friendship is a real tresor.
> Ah houw vain is inclination.

On 20 November Constantijn wrote in his journal: this morning Mrs Pamel had a disreputable fellow inform me 'that she desired the honour of my company in her new lodgings near Saint James'.

The above account provides no grounds for the anecdote that Annie Romein tells in her biographical sketch of the meeting held on 22 June in Gresham College.[35] And neither do other, contemporary English sources. Christiaan would have spoken about gravity and Newton about the double refraction of light. They would have discussed just those matters about which, in hindsight, the other would prove to possess greater insight. But Newton, whose presence there was not accidental, listened to Christiaan's summary of his *Discours* and his *Traité*, which had now been printed as one volume. Perhaps he also took advantage of the opportunity to comment on it.

This account also provides no grounds for Bell's claim in his Huygens biography,[36] that the journey in the carriage on 10 July was from Cambridge to London, which would mean a distance of at least 100 kilometres and discussions between Christiaan and Newton that lasted a day. The unknown beginning of the journey was closer to London, because that is where Fatio was living, and Newton would have been living in the vicinity of the court for the purpose of his application, which, incidentally, would prove unsuccessful.

Indeed, it would be interesting to know what the men did talk about. They spoke at least once about light and colour, for Christiaan later informed Leibniz that Newton had told him about a couple of splendid experiments, probably on thin films.[37] They also spoke about motion: in August Christiaan received two articles from

---

[34] OC 22, 749   [35] Romein 420   [36] Bell 15   [37] OC 9, 471

Newton on motion in a medium with resistance.[38] They clearly thought differently about the relativity of motion, which Newton did not wish to give full significance and encumbered with an absolute space. We don't know exactly what passed between them during these discussions, but they were sufficiently important to move Christiaan to write a postscipt to his *Discours.*

Anyone who thinks that his first priority on arriving home would have been his book on light and weight will be quite wrong. He had barely got back when he wrote to Constantijn that he had heard that Simon van Petkum had died, the man who had succeeded his father as chairman of the council. Although he had previously refused the position, he now found it necessary to assure himself of extra income. The high salary of the chairman of the council moved him to make an ill-considered application: 'I would like to offer my services to His Majesty.' Within a week, when he had received no reply – which would not even have been possible – he wrote once more. And after ten days, yet again.[39]

Did the income that he received from the Stadholder, which amply provided for his needs, suddenly seem insufficient? In an earlier letter to Constantijn he had complained about 'the Gentleman States, with their continual demands for two percent', an extra levy that was necessary to pay for the English expedition.[40] Did he worry about Hofwijck, for which the cost of alterations had perhaps turned out to be higher than estimated? Was he filled with uncertainty about a life alone, without the caring father with his high salary? Regret at his earlier refusal of the post?

Constantijn had wished to discuss the application with him before he presented it to the king, but after the second letter he understood that Christiaan was in earnest. He noted in his journal that he appealed to the king 'about a second letter, with which brother Christiaan vexed me, with the request to be permitted to occupy the post of the deceased Van Petkum. He [the King] muttered something about

---

[38] OC 9, 321–327 & 328–330   [39] OC 9, 334, 336, 344   [40] OC 10, 3–5

not knowing whether he wished the post to be filled. When I pushed the matter a bit further, he said that he did not have any appropriate post for my brother, that he had a keen mind, which he put to good use, and that he had higher capacities than those suited to a steward. I left it at that.'[41] In similar words, but rather more diplomatically, he wrote to Christiaan who, deceived by the diplomatic tone, wrote back that he was happy that there remained still some prospect of gaining the post.[42]

In his second letter, of 9 September, Christiaan mentioned a meeting with Suzette Caron in The Hague. He found her greatly changed since their last meeting (when?). Did he compare his memory of her with that of Pamel when he added those lines in pencil to his English journal? Things were not going well for her. He also mentioned that her husband was not present, and 'perhaps he avoids me'. As we saw, the news of her arrival in The Hague four years earlier went accompanied by a gallant poem about Iris. This Iris was the Iris of another, and the other, her husband Chevalier de la Ferté, certainly had reason to avoid him.

Because Suzette would play a not unimportant part in his life, we shall recount some of what befell her.[43] She was imprisoned for her Protestant beliefs in the Paris Bastille and released after three months, on renouncing her religion. She immediately left, along with a group of refugees, for The Hague; her husband only followed later. He had lost his property. In 1688 Suzette resurfaced, according to Constantijn's journal, in a refugee milieu in London, where she became involved in an attempt to obtain an allowance from the king.

Her husband would die in or before 1698. In 1716 she would remarry, to a Bonaventure in Groningen, because 'an end had to be made to an irregular relationship'. She was loose, 'not unrelenting towards admirers', as Constantijn delicately puts it in his journal of 1694. Christiaan's remark about the evasive behaviour of her husband

---

[41] Huygens **23**, 180; OC **9**, 336 (n1)   [42] OC **9**, 348
[43] Vollgraff 386 (n10); Keesing (1983) 64

suggests that he, after the Paris period, too, between 1686 and 1688, received Suzette more than once. Iris.

There is a dark undercurrent in Christiaan's thoughts after the trip to England. Not only money and women preoccupied him. That autumn, he wrote curious Latin texts on fame, the unknowable, and death. In the first, he gives account of the sensual pleasure that fame can bring about; in the second, of the uncertain, angst-ridden paths that the mind is unable to follow. The third and briefest text we give below:[44]

On death

(1) Only if we join in our mind the past with the present do we remain ourselves. If every memory disappears irrevocably, we have become changed. We must not think that, to remain ourselves, the body is enough, for the mind is necessary in order to remember. Therefore, if we assume that in an afterlife we shall not remember what has happened in this life, or who we have been, then this second life will not concern us, though it last into eternity. Therefore it does not concern me either, however wondrous and joyous it may be, unless I possess consciousness of this, my present life.

(2) To think that our salvation depends upon the absence of memory! How many are the things that we would prefer to forget! But these matters are so bad that poets have thought up for us the Lethe, the river from which the spirits drink, so as to forget all that has passed in this life, in order to be blessed. Such happiness after death, if this be truly happiness, is what each of us may wish for.

(3) Suppose that all your wishes were fulfilled, that you understood everything, that you might converse with the greatest minds of the past, wander in the loveliest of landscapes, enjoy all that there is to enjoy! But if the memory of the life before has vanished, you will have no more happiness than all the others.

[44] OC 21, 522–523

(4) Nature allows us to truly experience all that happens in life, but cheats us in death. She appears as great evil, but ensures that man may survive. For sickness frequently goes hand in hand with great pain. What healthy person would deny that sickness is an evil? He who is privileged to die without pain should consider the misfortune of those who are denied euthanasia.

(5) He who is physically sick and withers away is also weak in mind. Because of this he must not rely upon his own judgement, and must regard that which is distressful as being only how it appears to his sick mind. Without doubt, the mind only functions well when the body is healthy.

(6) Would you wish to be immortal? Why not, if you are granted a strong, healthy body and a strong, healthy mind? But if old age comes upon you with physical decay and weakness of mind, would you not then rather die, or make your own exit [*vel eripi, vel excedere e vita*]?

In translating from the Latin we have tried to follow the train of thought as closely as possible, but we have shortened and smoothed out those lengthier sentences that are sometimes set out in a confusing and illogical manner. How harrowing this text is! Impersonal musings turn into personal questioning and end with thoughts of suicide: *vel eripi, vel excedere e vita.*

'It is not possible to spend the winter here [at Hofwijck], in such loneliness,' he wrote on 18 October to Constantijn in London.[45] Perhaps this brother did not require any explanation, or only this: 'I have no carriage. Somewhere to stay, a house in The Hague is almost indispensable, otherwise I must leave each time with the six-thirty boat.' We see that the proximity of family – his sister-in-law on Het Plein, where he still had his study, and his sister and her family on the Voorhout – had become a necessity of life for him. In November 1689

[45] OC 9, 349

he moved into rooms on the Noordeinde, 'next to De Crabbe' (that is, at number 6 or 10), close to them both.

Only then could he settle down to writing the postscript to his *Discours de la cause de la pesanteur*, to which the discussions with Newton, and his own intellectual integrity, committed him.[46] It would be longer than the work itself.[47] First he scrapped the sentence that concluded the *Discours* of 1669: 'As long as no phenomena are found that contest it, nothing stands in the way of the truth of our assumption.' Strictly speaking, Newton's explanation for the motion of the planets was not another natural phenomenon as such, but it appeared to demonstrate to him 'new characteristics of universal gravity'. 'I have nothing against centripetal force, as Newton calls it.'[48] The consequences of this force were clear to him, but not its cause, and thus, ultimately, explained nothing to him.

He tried to explain the motion of the planets as being related to whirls of fine particles, which would form almost unresisting ether. For him, the existence of such particles followed from 'the peculiar speed' of light, peculiar in the sense that it is not infinitely great (a central theme of his *Traité*). He went so far as to say that Newton's theory ought to be contested, unless 'it may be assumed that gravity be an inherent quality of natural bodies. But I do not believe that Newton is able to agree with this, because such an assumption would remove us far from mathematical or mechanical fundamental principles.'[49]

So he remained steadfast in his estimate of the speed of the terrestrial whirl (seventeen times the rotation speed of the Earth) and also his whirl model of weight that limits the falling speed (when a falling body reaches the whirl speed, its gravity must disappear). But he had to admit that the planet whirls must have an 'extreme agitation', and that seventeen times the rotation speed of the Earth for the Earth's whirl was rather low. Also, he had to settle for convex symmetrical whirls instead of cylindrical whirls, because the centripetal force, according to its name, works towards a point, not a line. But he

---

[46] Costabel (1982) 146–148   [47] OC 21, 466–488   [48] OC 21, 472   [49] OC 21, 474

could hardly have realised what the consequences of this geometry would mean for the behaviour of whirls.

This made the postscript almost a confession of faith, and it provided no more explanation of the force than the *Principia mathematica* had done. At any rate, he had defined his critical stance towards 'Newton's book, which is justifiably held in high esteem, because there exists nothing better or more astute upon these subjects'.[50]

He completed it shortly before Christmas, on 23 December 1689. Immediately after this, he continued working on the final edition of his foreword to the *Traité de la lumière*, which is dated 8 January 1690:[51]

> I wrote this treatise during my stay in France, twelve years ago.
> [As we have seen, he did not write it there, but in The Hague.] In
> the year 1678 I brought it to the notice of the learned gentlemen
> who, at the time, comprised the Royal Academy of Sciences. One
> may ask why I have waited so long with the publication.

He then mentions his plan to translate the work into Latin, and to add a description of his work on optics. 'But when pleasure in the novelty had worn off, I kept postponing the implementation of that plan. I do not know when I would ever have finished it.' This foreword concludes with an explanation of his view of the method of natural science. It is an epistemology that is so lucidly articulated, and so reminiscent of the ideas of Karl Popper, that we give the full text:[52]

> In this, a type of proof will be seen which is less certain than in
> mathematics, and which differs from it considerably.
> Mathematical proofs are based upon established, irrefutable
> principles, while here, because it cannot be otherwise, the
> principles are confirmed by the conclusions that are drawn from
> them. Even so, a degree of probability can often be achieved that is
> hardly surpassed by complete certainty. This is the case when that

[50] OC 21, 475–476   [51] OC 19, 453   [52] OC 19, 454–455

which is shown to be based upon assumed principles agrees with the result of experiments, especially when these are numerous. And this is certainly the case when people predict new phenomena, and then subsequently discover them as well. These must follow from the applied assumptions. In this manner we can establish that the result complies with our expectations. If all these proofs of probability point in the same direction, namely to that which I propose to be true, this must constitute an important confirmation of the result of my research. It is hardly to be imagined that the situation is not more or less as I propose it to be. I therefore readily believe that those of an inquiring mind, who hold in awe the miracle of light, will derive some satisfaction in my various speculations upon it, and especially in my new explanation of its distinctive property [of light propagation]. This forms the basis for the structure of our eyes, and for the inventions so conducive to the use of light. I hope, too, that new scientists, following this initiative, will penetrate deeper into this material than I have been able to do, for there still remains much to discover. That is proved by the matters that I have indicated, where I have left the problems as they are, without solving them. And this is even truer for those subjects that I have left wholly untouched, such as all kinds of luminous bodies, and all things to do with colour. Up until now there is no one who can pride himself on any result in these matters. Finally, there remains very much more to find out about the nature of light than I claim to have discovered, and I shall be greatly obliged to those who are able to fill the gaps in my knowledge.

15 Heaven

Consider that we are in the heavens, Christiaan noted down somewhere, '*penser que nous sommes dans le ciel,*' or elsewhere, '*in coelo sumus*'.[1] Consider that we recognise ourselves, 'that in these heavenly regions there are observers, not human beings like us, but all the same, reasonable creatures'.[2] It was a thought that had remained with him and that he would now pursue in a book on natural philosophy.

> All matter is a world in itself and at the same time a mirror of the cosmos, which manifests itself in its own fashion. In one way or another, the cosmos is multiplied as often as there is matter. Men may even say that every manifestation of matter, in one way or another, bears the infinite wisdom of the cosmos and mimics it in as much as it is susceptible to it. For it manifests, even in its confusion, all that occurs in the cosmos, past, present or future, and that resembles an infinite understanding or insight. And because all other matter expresses the same and conforms to it, one may say that it makes its impact upon all other matter . . . [3]

This is a kindred thought of Gottfried Leibniz, who immediately worked it out and wrote it down in ten days, in a metaphysics of thirty-seven articles in the winter of 1686, when frost and snow kept him from working on mine pumps in the Harz Mountains to earn his living.

In the brief ice age, then at its worst, the winters were severe, perhaps because the Sun was so inactive, almost without spots, without

---

[1] OC **21**, 351 (§11) & 362 (§37)   [2] OC **21**, 729 (lines 13–14)   [3] Leibniz 37–38

corona.[4] Let us imagine a blizzard blowing across the frozen Court Lake of The Hague and Christiaan striding along the edge. Beneath his tailored coat he is exquisitely dressed in a black suit of English cloth and a splendid silk waistcoat woven with gold, lined with East Indian crimson; the silk alone cost him 40 guilders and 95 cents.[5] 'In coelo sumus.' Cosmic whirls of ice-crystals around his head.

He has written to Fatio de Duillier that he has sent seven copies of his book to Constantijn in London, intended for him, Newton, Hamden, Halley, Locke and Flamsteed, although he has his doubts whether Newton will understand the French.[6] His servant, Hendrik, helps him out of his coat when he arrives home at Noordeinde. Now he will write to Leibniz. It is 8 February 1690:[7]

> It is rather late to say that I received your excellent letter of eight or ten months ago [in reality two years], on my solution to your problem [of the isochronous line]. I am no longer able to say why I have not replied sooner. As so often happens, I have postponed doing so. That is why I have waited until I could send you the book that I have just recently had printed. Laxity on the part of the printers, and my journey to England once publication had begun [sic] has delayed the answer until today. But see, the substantial volume is ready finally. An expert such as yourself can read it through easily in a few hours.

Leibniz, usually so prompt to reply, did not respond. Christiaan did receive an answer by return post from his other philosopher friend, John Locke, who sent him his Essay concerning Human Understanding.[8] 'It is astute and lucid in style, quite different from what one is accustomed to from all the others there,' he wrote to Fatio, 'and when I have read it further, I shall certainly write to him.'[9] Nothing came of it.

Newton did not respond either, at least not to him directly. His questions and criticism came concealed in the long commentary by

---

[4] Eddy 252–256   [5] OC 22, 371   [6] OC 9, 357   [7] OC 9, 366
[8] Broad 12–18   [9] OC 9, 392–393

Fatio on the *Discours* ('pity if it turns out to be untrue'), to which Christiaan replied with some irritation, 'Judge nothing before you have understood it'.[10] But where was the reaction from Leibniz, in whom he could expect an ally?

Spring came. He saw the full moon change colour from white to red as it calmly passed through the Earth's shadow: 'Why do whirls, which carry the moons, travel in the same direction as the great sun whirl? God may have ordered it so, but it is also an unchangeable law of nature.'[11]

He moved to Hofwijck, where Matthijs the gardener's apple orchard was pink with blossoms. 'The flow of water ensures the growth of trees, leaves and fruit. Also stones. It is ascertained that mountainous substances, crystals and rock grow out of sand with the flow of water, although people cannot see it because it happens very slowly.'[12]

He asked the East India Company to try out the sea clocks once more, 'this summer or in the autumn, upon the departure of the next fleet'.[13] He received letters from Guillaume de l'Hospital on the calculation of the centre of oscillation in *Horologium oscillatorium*, also from Papin with questions on the *Traité*, and from Fatio, who suddenly turned up in Utrecht and wished to visit.

Summer came, with 'work on the natural causes of thunder and the rainbow'.[14] Leibniz's letter was written on 15 and 25 July:[15]

> Only returned to Hanover five or six weeks ago after a journey of
> almost two years, in which I passed through large parts of
> Germany and Italy, in search of historical documents for the
> Duke.[16] I met few people with whom I could converse upon
> developments in science and in mathematics. I impatiently await
> your *Traité*, which I expect from Hamburg as soon as it shall have
> arrived there. It has been long-awaited. We need such books in

[10] OC 9, 381 & 392   [11] OC 21, 343 (lines 3–4, 6–7)
[12] OC 21, 705 (lines 8–11)   [13] OC 9, 418–419
[14] OC 21, 343 (lines 8–9)   [15] OC 9, 448–452   [16] Aiton 139–168

order to make real progress. I do not know if you have seen the mathematical method that I propose in *Acta Eruditorum*. Instead of, the calculation of quantities that, up until now, has been applied, such as roots and powers, I presently use sums $\int$ and differences dy, ddy, $dddy$, which is to say, differences of the quantity y, or differences of differences, or differences of differences of differences. And just as roots are the inverse of powers, sums are the inverse of differences. For example, just as we get $\sqrt{yy} = y$, we get also $\int dy = y$ . . . If the analysis of transcendental curves is implemented in this manner, one is able to find their properties, and I have many examples of this. I should like sometime to hear your opinion, the value of which I am familiar with.

This was how Christiaan replied on 24 August:[17]

I have received your kind letter of 15/25 July. I had already sent you a copy of my book, on 8 February. I had the parcel delivered by Van der Heck, agent to the Duke of Hannover, but as you had not yet returned from your Italian journey, it will have arrived at the wrong address. Make inquiries. From time to time I have seen your new arithmetic method in the *Acta*, but find it rather obscure, and am unable to understand it. I believe that I have something similar. If your method is as helpful as you say, then you may be assured of my favourable judgement. You would greatly oblige other mathematicians and myself by clearly explaining it in a separate tract.

Leibniz had begun with the publication of his differential calculus in 1684, without referring to related work by Newton. That would have had no point, for almost no one had ever heard of it. Newton had, in fact, found an arithmetic method in 1666 that he later called the fluxion method, but he had not published it. Ten years later Newton

[17] OC 9, 470–473

had let him guess at the rules for differentiation and integration, but Leibniz had already found them out himself, as a result of a mathematical education that had begun under Christiaan's guidance. He was fully entitled to publish his own findings, but would later regret not having made reference to Newton's letters of 1676. He was not allowed to forget it.[18]

The German and the Englishman maligned each other once more, thirty years later, to get sole honour, when old age had weakened their control over their emotions. The findings were great enough for the honour to be shared, but they could not bring themselves to do so. Fatio was the agitator. He had something against the abstruseness of Leibniz, and as Newton's confidant he became persuaded of his priority in the matter.

Fatio was back in the Netherlands and, during the latter part of his stay, from February until September 1691, in The Hague. There is nothing to indicate that he discussed Newton's rules with Christiaan, when the latter was occupied with the new arithmetic method on the instigation of Leibniz, while he, Fatio, was there and said he understood nothing![19] When he got back to England, his report must have been a blow to Newton, who had so far not expressed himself on the matter. Only then did he publish his letters of 1676, clarified them with arithmetic examples, and ignored the work of Leibniz. In 1699, the bombshell would be dropped.

Autumn 1690: sparks of ground amber against rainy windows, 'effects of whirling electric particles'.[20] Exploding grenades in the ammunition storehouse in The Hague. 'I did not know what to make of it at first, for the noise lasted a long time, as if a house in the neighbourhood were collapsing.'[21]

An intensive and rich exchange of letters started up between Christiaan and Gottfried Leibniz. On 9 October a long letter with

---

[18] Aiton 335–340; Westfall 698–780   [19] OC **10**, 74, 77, 145, 163; Westfall 516–518
[20] OC **19**, 615 (§17)   [21] OC **9**, 531

questions to Hannover, on 13 October a lengthy answer. Nineteen letters to and twenty-seven letters from Leibniz would follow, first about the differential calculus, in which the strength lay, amongst other things, in the solution to the problem of the catenary curve; followed by the theory of light, the structure of the cosmos, metaphysics and gnosiology. The letters were courteous, critical and, judging by the sheer amount, indispensable.[22] In short: 'I have met a person with whom I can converse upon developments in natural science and mathematics!'

The catenary curve was described through equations – in other words, inverses of logarithms. This follows out of Leibniz's arithmetic method, and Christiaan, who had proved in his youth that the curve could not be a parabola, was prepared to believe it. But did he not have his own method, 'something similar', to find this result? He grappled with it for a long time.

In 1693 he came up with his own solution, a paragon of the strength and style of his mathematics.[23] He conceived the chain as a set of equal weights, joined together by weightless cords of equal length. Balance demands that all four consecutive weights A, B, C and D (see Figure 6) hang in the chain in such a manner that, after they are extended, lines AB and CD intersect one another on the vertical line that divides BC through the middle. The angles of the cords to the horizontal line form an arithmetic set. This he had already discovered in his youth. Now he went further.

Christiaan invented a chain $C_1 C_2 C_3 C_4$ (see figure again), of which the lowest piece lies horizontally, which is then stretched horizontally along $C_1 D_2 D_3 D_4$. He chose a point P on the vertical line through C_1 for which the angle between $C_1 P$ and PD_2 is equal to the angle between $C_2 C_1$ and $C_1 B_1$. As he knew, it could be proved that $C_2 C_1$ approximates the radius of the chain at its lowest point. Because the tangents of the angles between $D_i P$ (where $i = 1, 2, 3 \ldots$) and $C_1 P$ form an arithmetical series, the angle between $D_i C_1$ and $C_1 P$ must be

[22] Heinekamp 100, 104–105   [23] OC **10**, 135–136, 414–416

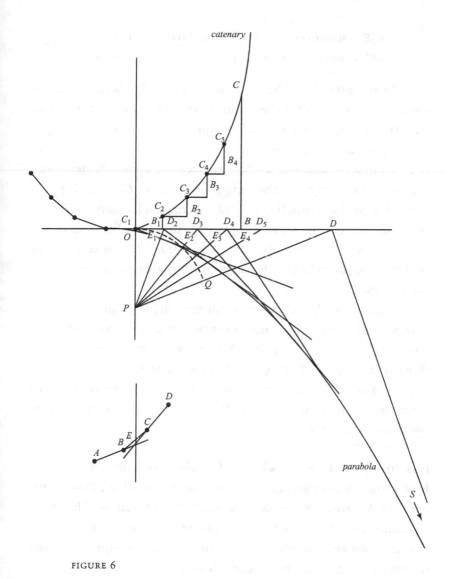

catenary

C

C_5
B_4
C_4
B_3
C_3
B_2
C_2
B_1 D_2
C_1 D_3 D_4 B D_5 D
O E_1 E_2 E_3 E_4

Q

P

D
C
E
B
A

parabola

S

FIGURE 6

equal to the angle between $C_{i+1}C_i$ and C_iB_1. When perpendicular lines D_iE_i on $D_{i+1}P$ are introduced, it follows that the triangles $D_iD_{i+1}E_i$ are identical to triangles $C_iC_{i+1}B_i$, so that the chain, as it were, is stretched out along the horizontal line in a set of distinctive triangles. When one considers further the abscissa C_1B and the ordinate BC from a point C on the chain, then

C_1B = sum over i of C_iB_i = sum over i of D_iE_i

BC = sum over i of B_iC_{i+1} = sum over i of E_iD_{i+1}

Next he imagined the cords to be infinitely short, so that C_1 coincides with point O of the chain, and took OD as equal to the arc OC. Then the sum over i of E_iD_{i+1} must equal QD, if $PQ = PO$, so that the ordinate BC equals QD.

To find the abscissa OB he extended the perpendiculars D_iE_i, and noticed that they are the tangents of a curve OS, which is distinguished by the perpendiculars PD_i meeting on its tangents D_iE_i at one single point. This makes the curve OS into a parabola. According to his theory of the unwinding, the sum over i of D_iE_i is equal to the arc OS of the parabola, less the tangent SD, so that the abscissa OB is equal to the arc OS minus SD.

Together with the earlier result that BC is equal to QD, this makes possible the geometrical construction of corresponding ordinates and abscissas of the chain. This construction assumes the rectification of the parabola that, as Christiaan knew, depends upon the logarithm or, as he called it, the quadrature of the hyperbola. This makes his geometrical solution to the problem of the catenary line of equal standing to the analytical solution.

It is difficult for us to say whether knowledge of Leibniz's result put him on the track of the parabola as 'provisional solution'. This knowledge did play a role, as he acknowledged in the publication,[24] but how? Certainly, though, this is the work of a virtuoso, who put into it all that he knew and saw. Perhaps that included the towline, on which he was working while finding the solution.

For in the autumn of 1692 he also happened to be occupied with devices by which the motion of towing, which could be seen daily in the barges along the Vliet, could be exactly mimicked. In the ideal

[24] OC **10**, 513 (lines 26–27)

sense, it is the motion of a heavy body towed across a horizontal surface with a cord by moving the loose end of this cord away along a straight line, which does not lie in the extension of the cord. Because the length of the cord does not change, and the body, by its weight and the resulting friction with the surface, follows the direction in which the cord moves, we get a curved path. This path, the towline, has a logarithmic form.[25]

Christiaan, who proposed the problem of the form and immediately solved it by means of geometry, recognised this. The quadrature of the hyperbola featured here as well. Leibniz got the same answer. The aim of the devices was to be able to draw logarithmic lines precisely. Christiaan envisaged a small, stiffly moving cart with a drawing pen, even a small boat in syrup.[26]

After the publication of the catenary, Johann Bernoulli came up with a related problem, which he managed to solve in 1694 with his sophisticated geometry.[27] But by that time he was almost the only mathematician who worked in this fashion.

In the differential calculus, all geometrical sophistication became reduced to a simple formula, and the brainwork of months became the manipulation of minutes. Christiaan had mastered the manipulations, but he did not care for them, because they robbed one of insight into the accuracy of the solution. This is what he wrote, on 24 December 1693, to De l'Hospital, who had changed to differential calculus, mainly on his instigation:[28]

> Leibniz has recently published an article in the *Acta* about curved lines, under a pompous title, as if a general method were being given, while it concerns merely tangents. I find it deficient and useless, considering the confusing and impracticable directions that would not enable one to draw with any certainty the simple line [the catenary line] that I have given.

[25] Bos (1988) 25–28 (OC **10**, 407–411)   [26] Bos (1988) 30–31
[27] Bos (1988) 32–43 (OC **10**, 512–515)   [28] OC **10**, 578–579

Leibniz was familiar with such criticism. Even so, he did not hold it against Christiaan. 'It has something wayward about it,' he wrote to Johann Bernoulli, 'not ill-natured.' And elsewhere he wrote, 'Such an exceptional man has almost the right to look down upon everything that he does not understand.'[29]

On 17 September 1693, Christiaan wrote to Leibniz that he admired the beauty 'of the progress to which you have so greatly contributed – that peculiar arithmetic method of yours'.[30] And in the conclusion of his publication on the catenary line, he acknowledged, 'In addition, it must be stated of this unusual problem that the geometrical finesses could hardly have been found without the directions of the famous inventor of the differential calculus.'[31] This made the wayward Christiaan a partisan after all, in the question of precedence with Newton, the question that would grow into a conflict.

Leibniz was fighting another battle. With good reason, he considered Christiaan to be an ally in his arguments against Cartesianism. In July 1692 he sent him his *Animadversiones*, a defence against the principles of Descartes, which had been raised almost to a dogma, and asked for his opinion.[32] But the man from The Hague did not draw any philosophical conclusions from the blunders made by the Frenchman in his physics, and instead, contented himself with finding the correct solutions. Leibniz found that the blunders eroded the whole Cartesian way of thinking and that it should therefore be rejected.[33]

He was unable to win Christiaan over in something so radical as this. (Later, in 1714, Leibniz would complain about this: 'He had no interest whatsoever in metaphysics.'[34]) In his answer, also in July 1692, Christiaan simply replied to him that he did not wish to stick to the dogma that the essence of a body lies concealed only in its extensiveness. This is why Leibniz thought, as he wrote to De l'Hospital, that 'Huygens, in my opinion, thinks just as I upon dynamics.'[35] But Christiaan was not thinking at all of a Leibnizian dynamics, in which

---

[29] Heinekamp 102–103   [30] OC **10**, 511   [31] OC **10**, 513   [32] OC **10**, 302 (n23)
[33] Heinekamp 105; Aiton 197   [34] Heinekamp 106 (lines 34–35)
[35] *Ibid.* 106 (lines 45–46)

bodies were not only extended, but also possessed a *potentia motrix*, a tendency to autonomous motion. This idea was totally alien to him. In fact, he thought it was absurd.

Was he, deep down, a Cartesian after all? The question has often been asked, and it is best to give his own answer, even though it is fairly long. Adrien Baillet had published a biography of Descartes, and Christiaan, who appeared in the book, found so many errors in it that he wrote a commentary. At the beginning of 1693 he sent this commentary to Pierre Bayle for publication. First the errors were put right, then came the following text:[36]

> Descartes wrote in such a way that people took his conjectures and fabrications to be the truth. And he achieved the same effect upon the readers of his *Philosophical principles* as upon the readers of enjoyable novels that are about seemingly true historical events. What is new, the representation of small particles in a whirl, is very attractive. When I first read this book about the principles, it seemed to me to be the best in the world and I thought, whenever I came across some difficulty, that it was my own incapacity that prevented me from understanding it. I was only fifteen or sixteen at the time. But later on, I found more and more things in it that are demonstrably wrong, or highly improbable, so that I have quite gone back upon my former preference. Now I find almost nothing in all his physics, his metaphysics, or his assertions about the weather that I hold to be true.
>
> What was initially pleasing about this philosophy, when it first appeared, was that people understood what Descartes was saying, contrary to what other philosophers said with their quite incomprehensible language on qualities, material forms, intentional sorts, and whatever else. More radically than anyone before him, he repudiated all this misplaced balderdash. But the greatest credit attributable to his philosophy is that he did not get

[36] OC 10, 403–406

bogged down in aversion for the old ways, but dared to replace this by understandable causes for all that existed in nature. For Democritus, Epicurus and various other earlier philosophers were unable to explain satisfactorily a single phenomenon, although they were convinced that everything must be explained by the form and the motion of particles in a void. This is apparent from their shadowy representation of vision, a constant process whereby minute skin particles of objects are detached and touch upon our eyes. They held that gravity was an inherent characteristic of all objects. They claimed that the sun was actually only two feet in diameter and that it renewed itself each night, to be reborn the next morning. In a nutshell, they understood nothing of what people wished to know.

Modern philosophers, such as Telesio, Campanella and Gilbert, hung onto various concealed qualities, as did the followers of Aristotle, and possessed too little ingenuity or mathematical knowledge to create a whole new system. Gassend, too, although he did unmask the Aristotelian humbug. Verulam [Francis Bacon] likewise recognised shortcomings of this peripatetic Aristotelian philosophy and, moreover, pointed to excellent methods towards building a better philosophy, based upon the good services of experiments. He has successfully given us an example of warmth in bodies, in which he concludes that warmth consists of a motion of particles that make up the bodies. But apart from this he understood nothing of mathematics, and his understanding of physics was inadequate. This was apparent from his incapacity to grasp the possibility of the movement of the Earth, a possibility he derided and called absurd. Galileo possessed all the acumen and mathematical knowledge necessary to make progress in the science of physics. One must concede that he made some fine discoveries on the nature of motion, although he overlooked a number of topics. He was not so overconfident or conceited as to wish to explain all natural causes, nor was he so vain as to wish to be head of a sect. He was modest and truth-loving. Though he did believe to have won eternal fame with his discoveries.

But Descartes, who in my opinion was very jealous of Galileo's fame, did have the strong desire to be regarded as the author of a new philosophy. This was apparent from his efforts to have it taught at academies instead of the Aristotelian, his desire that it be espoused by the Jesuits, and finally, from the obstinacy with which he clung, through thick and thin, to all that he had put forward, even if it was flawed. He replied to all objections – though in my opinion rarely satisfactorily, at variance with the usual practice in public debates at academies, where such objections had the last word. It would have been different if he could have explained clearly the truth of his dogmas. And he had that opportunity, if they indeed contained the truth. I have said that he held his conjectures to be true. This is apparent from his concoction of grooved particles that are meant to explain a magnet, from the ice arc in the sky that was meant to explain the Roman parhelia, and from hundreds of other things that he assumed to be true, without stopping to consider all the incongruities that came with them. He claimed things without providing proof, such as the laws of motion of colliding bodies. According to his belief, they had to be true, because if they were not, his whole physics would be untrue. It came down to his proving them by swearing on them. Nevertheless, only one of these laws is true, and I found it very simple to prove it so.

He presented his system to us as an attempt to say which is probable and which depends only on mechanical principles, and he invited the gifted to conduct their research in a similar manner. This is highly commendable. Yet by letting it be thought that he had discovered the truth, as he does each time, by relying upon what follows, which he even prides himself on, and by the beautiful coherence of his explanations, he has inflicted great damage on progress in philosophy. Those who adhere to it and have become members of his sect indeed imagine that they know the origin of all things, in so far as it is possible to know this. In this manner they lose time defending the teachings of their master, instead of probing into the real causes of the numerous natural phenomena upon which Descartes offered only statements.

His greatest finding in the field of physics, probably the only one that he fully understood, is the cause of double rainbows. That is to say, as far as the determination of their angles and apparent diameters is concerned, for regarding the origin of their colours, nothing is more improbable. What other writers before him have written on this subject is enough to move one to tears, for they knew too little geometry and nothing of the law of refraction, nor were they informed by experiments. Granted, in all probability, Descartes did not find the actual law of refraction himself. Certainly he saw [Willibrord] Snel's book while still in manuscript form. I have seen it too. The book was all about the nature of refraction, and it concluded with this law, for which he thanked God. Instead of considering the sine he took something that amounted to the same, the ratio of sides of a triangle. But he was mistaken in thinking that the light ray perpendicular to the water surface becomes shorter, so that the bottom of a ship appears higher than it actually is.

Despite the little that is true, in my opinion, in Descartes' *Philosophical principles*, I do not deny that he has shown great power of thought in actually setting up this new system, this monument of probability that satisfies and pleases masses of people. It could also be said that, by spreading these dogmas with great self-confidence and by becoming very famous, he has stimulated subsequent writers into examining the system anew, in an attempt to find something better. Has he gained all this respect, then, without actually deserving it? No. On the grounds of his geometry and algebra alone, he merits to be regarded as a great mind.

Let us return to the final month of 1690, for in our attempt to sketch the influence of Gottfried Leibniz, there is much that we have missed out on. In December we see Christiaan in Amsterdam, on board the *Brandenburg* to help Johannes de Graaff with the repair, regulation and placing of his clocks in a separate compartment in the latter's cabin.[37] Professor of mathematics Burchard de Volder from Leiden, whom the East India Company had asked for advice after the previous test, found a second test worthwhile.[38]

The ship was obliged to wait two weeks in the fleet at Texel, before a favourable wind allowed it sail out to the East, on New Year's Eve. De Graaff reported to him from the roadstead on various matters, including the impossibility of seeing the Sun rise.[39] Icy mist over the mud flats springs to mind.

On New Year's Day, Bayle, who was stranded in a theological groundswell and could have done with some assistance, asked for an explanation for measuring a star parallax. Christiaan's answer of 13 January 1691 was precise and extensive.[40] For us it seems odd that this triangulation of stars, which, in principle, enables the Earth to make its orbit around the Sun, should have had a theological significance. Robert Hooke claimed to have measured a star parallax, but Christiaan mistrusted him and maintained the parallax to be immeasurably small, because the stars stand at such a great distance.

Willem III returned to The Hague and crowds of people flocked to court, so many that one could not turn round.[41] On 3 February Christiaan welcomed Constantijn, who had been overseas for more than two years and, after the crossing, with floating ice at Hellevoetsluis, was glad to see his family once more. The welcome celebrations went on, and there was a Sunday coach trip through the snow. Is this why Christiaan fell ill?

He came down with flu, which kept him indoors for three weeks, and, as he wrote to Leibniz, for the first time in his life he suffered

[37] OC 9, 567   [38] OC 9, 343 (418)   [39] OC 9, 577–584
[40] OC 10, 3–5   [41] Keesing (1983) 136

from diarrhoea.[42] When he had recovered, he came to Constantijn's house regularly to dine and, we must conclude from the following, to fret about money.

Fatio de Duillier, who was accompanying a group of English students and was now in The Hague, found time now and again to speak with him. He also looked him up at Hofwijck. The few letters that document this bear witness to an agitated enthusiasm, 'In the confusion of yesterday evening, I forgot to ask . . . ,' Christiaan writes to him on 3 April. Promptly Fatio replied briefly, 'for I am too tired to write more'. On returning to England later that year, the native of Switzerland wrote: 'I departed from The Hague in such haste that I could not find a moment to acknowledge your orders.'[43]

Christiaan and Fatio flung themselves into Leibniz's new arithmetic method, which both excited and confused them: 'Do not imagine that Fatio and I understand your differential calculus.'[44] Christiaan must have grilled his young friend. And Fatio, who must have recognised something of Newton's method, was not able to say all.

What did the Swiss actually discover? Religious anxieties may have been part of his agitation, a turmoil that would practically destroy him two years later in his conflict with Newton.[45] Christiaan was susceptible to it, otherwise he would never have cursed the pastors who came to bring spiritual care.

During these months Bayle tried to involve him in his theological dispute with Pierre Jurieu, a Calvinistic quibbler, and sent a pamphlet. Christiaan felt obliged to bolster him up and wrote, on 6 June:[46]

> These unfortunate disputes [between Protestants] are spent effort and Catholic zealots just laugh at them. I am astonished that Jurieu does not seek in himself the error of attacking one's own supporters, for which he blames our religion. The administrators should call him to account.

---

[42] OC 10, 55   [43] OC 10, 74, 78, 145   [44] OC 10, 93
[45] Westfall 538   [46] OC 10, 103

And there was something else that played a part in his agitation at that time: his family, too, was putting him under pressure.

In Constantijn's journal we read that his brother-in-law visited Hofwijck, armed with a financial proposition: 'So as to be able to live more comfortably, it was suggested [to him] that he marry cousin Constantia de Wilhem. He was taken aback. Had she been twelve or fourteen years younger, he might have been able to consider it, but now he felt no inclination for it whatsoever.'[47]

Constantia was of the same age as Christiaan. She had money.[48] This bright idea had come from Constantijn's wife, Suzanna, who had not dared to come forward herself, but now that the matter had been raised, was not to be so easily put off. In the journal entry of 29 June we read: 'Had a letter from my wife. Wrote that she had dined with sister and brother and cousin Constantia, and that she thought that he paid more attention to Constantia than before. She thought that it would still go ahead.' And on 18 July, '[My wife] wrote that brother Christiaan still looked thin and feeble and that his love did not appear to be stirred.' Did he have flu again? Perhaps, but it had been hot and dry, with thundery showers in the evening.

Migraine? Could be, from thinking things over. That summer he found the provisional solution for the form of the catenary curve: his letters to Leibniz on the subject, on 1 and 4 September, are elated.[49] Melancholy? We may assume that he was secretly receiving the considerably younger Suzette de Caron; her new petition to Constantijn for a pension from the king shows her to be in the vicinity.

If there was any question of his being ill in July, then it cannot have been more severe than his collapse in September and October, which Christiaan himself mentions in his letters. On 2 November to Denis Papin: 'A long interruption in my studies for reasons of health...' On 16 November to Leibniz: 'I have not worked or studied these last two months, had difficulty in staying healthy at a time when almost everyone in this country has fallen ill.'[50]

[47] Huygens 23, 445–446   [48] Keesing (1983) 137   [49] OC 10, 129, 139
[50] OC 10, 175, 182

It could have been the epidemic, but it is remarkable that Constantijn, who noted down almost everything, never mentioned Christiaan's collapse in health. He dined with him at Hofwijck at the end of October; they went together to the fair at Voorburg, bought books, looked at Joliveau's maps at their brother-in-law's house, discussed grinding instruments, and visited (to discuss the matter once again?) Constantia de Wilhem. He withstood the pressure. On 27 October, the brother, head of family but also servant of the king, was obliged to leave once more for England.

On the same day De Graaff reported to him from Veere, where he had arrived on a ship of the return fleet. The clocks would be brought on a lighter to the Oostindisch Huis in Amsterdam. He would not impart anything to him about how they had worked. He had arrived at the Cape on 3 June, and had been obliged to wait there for a long time for a return fleet.[51]

Not until 16 November 1692 would De Blocquerie of the East India Company report that the clocks had not worked well.[52] Professor De Volder was sceptical, and sent him the journals of De Graaff for his comments. Christiaan leafed through them and discovered errors. On 10 February 1693 he wrote to De Graaff:[53]

> Having seen the Journal from São Tiago [on the Cape Verde Islands] to the Cape, I take it ill from you that the correction for the pendulum motion, which is proportional to the latitude, is everywhere added where it should have been taken away.

According to the inventor the clocks did, in fact, work quite well. He wished to know more. On 14 February he received the following reply from De Graaff:[54]

> With regard to their irregular running on the voyage to the Cape and on the return journey, you may find several proofs in the journal. If you wish to view these as belonging to the vicissitudes

[51] OC 10, 166   [52] OC 10, 340, 433–436   [53] OC 10, 389   [54] OC 10, 396

of clocks, then you will see just how often they have stopped, time and again, due to other causes. They stopped so often that I sometimes did not deem it necessary to make a note of it in the journal. Twice or thrice a piece of the spring of a clock broke off, and since the spring was then too short, it broke to pieces. To conclude, with regard to the 'adding up and taking away' that was incorrectly carried out by myself, I can offer no answer.

Christiaan turned to De Volder, who concluded on 6 April that the result of determining the longitude was untrustworthy by several degrees.[55] He had to resign himself to professional opinion, and wrote, 'have lain down flat upon my bed for several days in great pain, with a discharge on the hip'.[56]

So now he is chronically sick. 'I am not fully recovered from an illness that has plagued me these three weeks,' he reports to Guillaume de l'Hospital on 20 May 1693, and adds, 'with pain in the liver and in the spleen.'[57] Constantijn, who had visited him a week earlier, compared the illness with his complaints in Paris, although this time it was less severe.

In June Suzette was back at Hofwijck. While his sister-in-law made taunting remarks about how Christiaan was diverting himself with her,[58] we wonder if she wasn't nursing him as well. Their relationship was no longer a secret. Suzette might have had illusions. The year before she had stayed with him for two weeks,[59] together with her daughter, as if, for a short while, they were a little family there at Hofwijck. Iris –

L'on se trompe aisement lors que l'on aime bien
Je croiois vostre amour de la force du mien

Your love as brittle as my own,
You come, you kiss, you disappear once more

[55] OC 10, 435–436   [56] OC 10, 442   [57] OC 10, 451
[58] Keesing (1983) 143   [59] Huygens 25, 131

That year, 1692, also began with his being ill. In his letter of 4 February to Leibniz, Christiaan wrote of a prolonged cold and headache. And on 15 March he wrote, also to Leibniz: 'Thank you for your concern about my health, which since my last letter, has suffered greatly once again from migraine.'[60] The long frost was most unwelcome, for he wanted to get out of the house, had probably hired a carriage and was allowed to use his nephew Tien's horses, on condition that he did not leave them standing in the snow at the houses he was visiting, 'because there has been an offer for 850 pounds for the horses'.[61] On 4 April he rode to Delft for a visit to Leeuwenhoek. There he saw blood flow through the tail of a small eel,[62] as he looked through a microscope of which Leeuwenhoek knew the secret – the formula for dark-field illumination. But the reason for his visit to Delft was to look for lathes for Fatio, which were for sale there. He described them to him, but added 'I cannot imagine that you wish to waste your time behind a lathe.'[63]

That same year, 1692, was also the year in which he began his description of the heavens. Not earlier than in the summer, for on 11 July he wrote to Leibniz that he was still working on improvements to his dioptric writings, apparently because he wanted to publish them. We may regard his notes on colours in a soap bubble as one such improvement.[64]

Did he blow bubbles for Suzette's little daughter? True, Leibniz had written to him that he sorely missed a theory for colours in his *Traité de la lumière*, but such a simple explanation is perfectly conceivable. After all, his nephew tells how entertaining he could be with children. It is not unthinkable that Suzette's stay at Hofwijck partly contributed to the fact that he put aside his old dioptrics, and began on something new.

Then he must have ordered an enormous telescope, which certainly entailed almost a year's work. 'I have ordered a fine rectangular tube to be made for my lens [with a focal distance] of 45 feet, to

---

[60] OC **10**, 268   [61] OC **10**, 221   [62] OC **13**, 720 (523)
[63] OC **10**, 277   [64] OC **19**, 549–550

allow people of rank, who so desire, the pleasure of viewing the moon and the planets,' he wrote to Constantijn on 1 September 1693.[65] He continued: 'Cassini claims to view all five moons of Saturn with a shorter telescope. Why should I not be able to do the same? I regret not having used a tube these last six years, for this works better. But 45 feet is quite something. The tube weighs more than 200 pounds, and there is the same weight again on the other side to balance the frame.'

Surprisingly, he thought only of magnification and not of intensity of light as a condition for seeing the moons of Saturn. 'People of rank', unless it is intended as a joke, must be a code for Suzette, for nothing is known of other visitors. And after this order he began writing Cosmotheoros (The Celestial Worlds Discover'd).

This book kept him busy for one and a half years. On 19 March 1694 he wrote to Constantijn that it was finished, 'but half Latin, half French; there is still much to be translated and the drawings have still to be made, but these are not many'.[66]

We shall go into it in more detail, but first we need to say something about the drama that was played out during those eighteen months between Fatio and Newton. We begin at the end. 'I do not know whether you have heard that Newton has gone mad. It has gone on for eighteen months. People say that friends have treated him by force and locked him up. By now he is more or less recovered. This is the worst that can happen to a person.'[67] This is how Christiaan passed on the rumour to Leibniz on 8 June 1694, having had it from a certain Colm from Scotland.[68] But Leibniz did not respond.

Nor was it true. What is true is that in the summer of 1693 Newton went through a deep crisis.[69] Theology and alchemy, which possibly caused a fire amongst his papers, had disillusioned him, and everything that he had discovered appeared to be futile in his eyes. Nothing was heard from him for many months. Fatio, who also underwent a deep crisis, played an important role in this as well; he had

---

[65] OC 10, 488   [66] OC 10, 583   [67] OC 10, 618
[68] OC 10, 616 (n2)   [69] Westfall 531–540

urgently requested Newton's advice about an amalgamation of gold and mercury that grew, like something alive, to be a universal remedy against nearly every kind of complaint. What occurred between them in London is unknown, but ended in an irrevocable rift. Fatio more or less disappeared from the scene.

Christiaan wrote to the Swiss on the 30 November 1693: 'Sometimes I have news of you from De Quesne, but because he has not received any letters from you for a long time, I take it that you have fallen ill again because, as I imagine it, the air in London is less healthy than in The Hague.'[70] Constantijn did not manage to deliver this letter. On further inquiries he only managed to find out that Fatio, just as Constantijn had written to Christiaan on 5 March 1694, 'is now tutor to the children of a lord, whose name I have forgotten, for good fortune does not always come to those who deserve it'.[71] 'Fatio will be obliged to accompany the lord upon his travels,' Christiaan concluded dryly in the above-mentioned letter of 19 March.[72]

In *Cosmotheoros*, observations of celestial bodies and notes on their location and significance, which he had been collecting since 1686, were developed into a cosmology. In 1686 a similar book, Bernard de Fontenelle's *Entretiens sur la pluralité des mondes* (Talks on the Multitude of Worlds), was published and quickly became popular, and one may wonder who had the idea first.[73]

Ever since the Aristotelian cosmos had been thrown overboard, there were scores of writings on the subject. Apart from the books of Nicolaas van Cusa and Giordano Bruno, which Christiaan frequently cites, we also have Galileo's *Siderius nuncius* (Messenger from Heaven) from 1610, and Kepler's *Somnium* (Dream) from 1634, not to mention the *Principia philosophiae* by Descartes from 1644. All contain speculations as to man's place in the universe that could be observed through a telescope for the first time. Without being exhaustive, we recall the work of Pierre Gassend: *Sintne coelum et sidera*

---

[70] OC 10, 567   [71] OC 10, 581   [72] OC 10, 583   [73] OC 21, 655–656

habitabilia (Whether Heaven and the Stars Are Inhabitable, 1658),
*Discours nouveau prouvant la pluralité des mondes, que les astres
sont habités* (New Writing that Proves the Multiplicity of Worlds, that
the Stars Are Inhabited, 1658). Finally we have Athanasius Kircher's
Iter extaticum coeleste (Ecstasy of a Celestial Journey, 1656), which
Christiaan strongly criticised. We must also mention the men-on-the-
moon, which figured in science-fiction-like works by John Wilkins,
Francis Godwin and even a serious astronomer like Adrien Auzout,
which earned the latter a sharp rebuke from Hooke. So Fontenelle was
by no means the first.[74]

Christiaan had waited so long with his *Cosmotheoros* for sub-
jective – and objective – reasons. He considered a cosmological design,
which entailed so much guesswork, to be didactically unjustified if
no reasonable estimate were given of the distance to the stars. In 1686
he defined his didactic goal: 'This is an easily comprehensible summary of astronomy.'[75] And his objection to the guesswork of earlier
cosmological works, even then, was that they are sometimes unneces-
sarily or even completely wrong: 'Knowledge in these matters affords
us greater understanding of the greatness and majesty of the creator of
the world, and lack of knowledge must lead to a multitude of absurd
opinions.'[76]

Therefore, the order of magnitude of the universe, which people
then knew nothing about, was a central theme for him. The estimate
he is able to provide for the distance to the star Sirius, on the basis of
a photometric observation, makes *Cosmotheoros* revolutionary.

But first he had to carry out this observation, and in 1686 he
needed more time. Determination of the distance, which we shall
discuss later, is based upon a conjectured analogy between stars like
the Sun and Sirius. This sort of conjectured analogy between such
planets as Earth and Saturn led him to notions about life in the cosmos. We know now that in the case of the first, he was not far off,
but in the second, he was so far removed from the truth that he

[74] Seidengart 209–210   [75] OC **21**, 358 (§26)   [76] OC **21**, 358 (§25)

has been called naive, or worse. Annie Romein, for example, wonders here whether 'we are not up against a natural weakening of the faculties'.[77]

Christiaan presents his cosmology as a modern work in which estimations are made, not at random, but according to straight lines of probability. 'Cusa, Bruno and Kepler write that Tycho Brahe thought just so, having attributed inhabitants to the planets, but no one has investigated this seriously, not even the Frenchman [Fontenelle] who, in the meantime, has published an ingenious dialogue upon the abundance of worlds.'[78]

Since it was intended as an academic work, *Cosmotheoros* was written in Latin. He said straightaway: 'I would not want to tell this to everybody, but . . . I wish to choose readers of my own taste, to whom astronomy and philosophy are not yet unfamiliar.'[79] He wishes, therefore, to prevent it from becoming popular, like Fontenelle's *Entretiens*, and gives it a Greek title ('*theoros*' means 'one who observes', and sounds rather like theologian), and the subtitle *sive de terris coelestibus, earumque ornatu, conjecturae* (or conjectures about celestial worlds and their adornment).

But when it was published, it attracted wide interest and was quickly translated into one language after the other: English, Dutch, French, German and Russian.[80] Immanuel Kant would praise it.[81] How could it be otherwise, with a text in which he had taken such great pleasure? 'The writing of it entertained me, for what Archytas says is true: he who . . . is wise to the beauty of the stars, experiences no joy if there is no one to whom he can tell it.'[82] That someone, here, was brother Constantijn, to whom he dedicated the work. The book was actually written in letter form, to his 'dear brother'.

The first letter is about the Copernican world-view and inhabitants of the planet, the second about the view that these inhabitants must have of the heavens and of the scale of the cosmos. We note that

---

[77] Romein 422   [78] OC 21, 683 (lines 3–7)   [79] OC 21, 685 (lines 4–6)
[80] Rabus XII–XIV   [81] Seidengart 221   [82] OC 21, 685 (lines 2–4)

this view gives a physical interpretation of the 'insight' that, according to Leibniz's metaphysics from 1686, 'all matter, even if it is confused, is a manifestation of the cosmos'. We should mention here that the quotations at the beginning of this chapter come from the first part of *Cosmotheoros*.

The book begins, then, with an explanation of the world according to Copernicus, with the Sun in the centre and the planets revolving around it. This is no longer new, but even so, Christiaan manages to give a new twist to it, one he derived from the slow revolution of the equinoxes:[83]

> It is to explain why nowadays the star at the tip of the tail of the Little Bear orbits in a small circle, at $2\frac{1}{3}$ degrees from the pole, but in the time of Hipparchus, at 24 degrees from the same pole, for a few centuries at 45 degrees, and in twenty-five thousand years, in the same position as now. If people say that the whole firmament revolves [instead of the Earth] they must think that its axis is continually changing direction, which is disorderly and absurd. But with the hypothesis of Copernicus, it is easy to explain.

In this, he was also thinking of the immense swiftness that stars must have in this revolving firmament, so that the centrifugal force would cause them to fly outwards. After these Copernican didactics, he traces the magnitude of the planets in relation to the Sun.[84] Only Mercury is fairly accurate, which is logical because Mercury can be observed as a disc on the Sun. But he greatly overestimates the size of the other planets. We have explained why, in our discussion of his *Systema Saturnium*, which, in fact, he returns to.

Next he comes to the question of whether the planets are inhabited. Although it remains a matter of conjecture, he considers it highly probable. Does this put his reputation at stake?

[83] OC 21, 693 (lines 20–27)   [84] OC 21, 697 (& Fig.)

394 HUYGENS: THE MAN BEHIND THE PRINCIPLE

Not in the least. In all his work on physics, it is a matter of whether something is either more or less probable. To recall an earlier statement of his (to Perrault): 'There is nothing that we know with complete certainty, but everything with probability . . . and there are vast differences in measures of probability.'[85] This is why he writes here: 'We are able to rely upon our knowledge of one single planet, one which can be observed on the spot, and make excellent conjecture about other planets of the same family.'[86] But his decisive argument is not the comparability of their orbiting, clouds, mountains or oceans – it is a relativity principle.

'What truly causes me to believe in the existence of rational beings on the planets is this: If it is only the Earth that would have an animal that so far surpasses the other animals (and has something of the divine), it would have too great an advantage in comparison to the other planets, and be too noble.'[87]

What is clever about this argument is that it gives purpose to the multitude of worlds. In Christiaan's time, the multitude of worlds was mainly used to challenge the belief in a divine providence directed towards man on Earth. For that reason, Fontenelle's book ended up on the Roman Index. Christiaan turned it round: the multitude of worlds directs the divine providence upon the whole cosmos. He refers to Power, who writes in his publication on magnetic experiments, in 1664, 'The world is not first and foremost created for man, nor to serve Him.' In short, it is 'against reason' to deny inhabitants to other worlds.[88]

If it has been established that this adornment of the planets is intended as the manifestation of a cosmic plan, or a divine providence, then it only remains to define its nature with some measure of probability. Similarity and aim, analogy and teleology led Christiaan to state the following upon the inhabitants of the planets.

Their bodies are of another substance than our own, but have the same form. They have feet, hands, eyes, a face, and five senses,

---

85 OC 7, 300   86 OC 21, 699 (lines 19–21)
87 OC 21, 715 (lines 2–4)   88 Seidengart 215

for a sixth is superfluous. Some are gifted with the power of reason (of necessity the same as our own) and study the sciences in order to better admire the universe. They live in an organised society with rules and institutions. They protect themselves with houses, against bad weather. They sail the oceans with the aid of a compass to orient themselves, and calculate longitude by observing the moons. They entertain themselves with musical instruments, and know perhaps the (his!) solution to the problem of the circle of fifths. They make use of vegetable nourishment, metals, gunpowder, wind and water, books, and even complete pendulum clocks . . .[89]

The second part is more interesting, although Christiaan first goes into Kircher's *Iter extaticum coeleste,* which adds nothing new. Perhaps just for those who enjoy caustic criticism, 'And then he lapses into even greater drivel . . . long outdated poppycock of astrologers . . . of sweet-seeming light and fresh running water . . . that a Jew or a heathen may easily become baptised.'[90] Only after a couple of pages does he come to the conclusion: 'Let us leave this celebrity for what he is.' In the meantime he has ensured that the reader will not become too attached to earthly examples.

There is now a strong resemblance to Fontenelle's *Entretiens,* but the tone in *Cosmotheoros* is different, and also its precision. He discusses how inhabitants of the planets view the skies.[91] On Mercury the Sun is three times closer by, and this is why the inhabitants get nine times as much light and warmth. Water must have other properties from our own, so that the inhabitants were of another substance. For them, Venus is a kind of moon that, on approach, gives off six times as much light as our Moon. On Venus the Sun is one and a half times closer, and the inhabitants get twice as much light and warmth. From its smooth appearance it follows that the atmosphere is thick and covers the mountains and oceans. Christiaan has no knowledge of the two (small) moons of Mars, and concludes that of the four

[89] Rabus 85–122   [90] OC 21, 767 (lines 15–18 & 22); 769 (lines 4–5)
[91] OC 21, 771–775

innermost planets the Earth is the only one with a moon, which is significant.

Then it becomes truly fascinating.[92] Jupiter and Saturn are gigantic planets with multiple moons. On Jupiter, which he estimates to be twenty times as large as Earth, the innermost planets must be invisible. The inhabitants can only see Saturn, and four moons that are subject to numerous eclipses and conjunctions. Days and nights are very short, but the equinox is everlasting. When he arrives at Saturn, Christiaan recalls his discovery of Titan:[93] 'In 1655 I sighted one of the moons, the clearest and second outermost, and I was the first to do so, with my telescope of under twelve feet.' He backs the view that Saturn is the outermost planet, but has doubted whether this is so. In his notes from 1686 we find 'Possibly planets beyond Saturn. Observe stars with a telescope on the ecliptic.'[94] The inhabitants of the planets see the stars just as we do, for although the distance to the Sun is great, the distance to the stars is far greater. They see Jupiter move no further than 37 degrees from the Sun.

And finally the ring:[95]

> In the middle of the night they see a part of it as a luminous arc from the one horizon to the other, interrupted in the middle by Saturn's shadow. But after midnight, for an inhabitant of the Northern Hemisphere, this shadow moves slightly to the right; for an inhabitant of the Southern Hemisphere, slightly to the left. And this shadow disappears in the morning, while the arc can be seen for the entire day, but shining less brightly than the moon appears to us at daytime. . . . Moreover, because they see it turn upon its surface, the spectacle of the ring is adorned by the movement of spots and unevenly illuminated patches.

Next Christiaan considers the second sort of celestial body, the moons of Saturn, Jupiter and Earth. Because our Moon has no

---

[92] OC **21**, 775–791   [93] Rabus 140; OC **21**, 777 (lines 12–13)
[94] OC **21**, 362 (§39)   [95] OC **21**, 789

atmosphere, he believes that the other moons also have no atmosphere, and that, therefore, they can have no inhabitants. But this does not stop him from saying that the inhabitants can see stars at full daylight, and that they can even observe these more easily than we are able to at night.

There remains only the third sort of celestial body, the Sun and the stars. This closing section is by far the most interesting and, as we have said, revolutionary. Having ruled out life on the Sun, Christiaan places the Sun, without any hesitation, as a star amongst the other stars, revealing a whole new multitude of worlds.

He blames Johannes Kepler for having given the Sun too important a place among the stars, stars that, according to Kepler, are fixed to the inner surface of a celestial globe. The distance from the Sun to this celestial globe would be 600 000 diameters of the Earth (nearly 8 billion kilometres), because this is 2000 times the distance from Saturn to the Sun, while Saturn resides at 2000 times as far from the Sun as the Sun is large. Christiaan is thoroughly indignant at the inappropriate lack of seriousness of this rating: 'Astonishing that such ideas should come from such a great man, the great innovator in astronomy.'[96] He finds it necessary to give a sound estimate of the distance from the Sun to the closest star, so it can be proved that the stars are illuminated not by the Sun, but that, just as the Sun, they too radiate light. This he did as follows:[97]

> But the stars shine so minutely, even those greatest in size, that even in a telescope they flicker like luminous specks, without perceptible width. From such observations one is unable to deduct any measurements. So when this proved impossible, I conducted an experiment to reduce the dimension of the Sun so that it would bring to the eye the same amount of light as Sirius, or some of the other, clearer stars. I covered the aperture of a telescope with a lens of twelve feet with a small thin plate, in the middle of which

---

[96] OC 21, 813 (lines 3–4)   [97] OC 21, 815 (line 14) – 817 (line 1)

I had made a minute hole, not above a twelfth part of a line or the 144th part of an inch. I directed the telescope towards the Sun and held the other end before my eye. And then in it I saw a minute part of the Sun that gave 182 times as little light. But I discovered that this part shone much brighter than Sirius at night. Therefore, when I saw that this minute hole must be made much narrower, I remedied this by placing in the hole of this plate a tiny glass ball that I had first used as a magnifying glass. When I then looked through the telescope at the Sun, with my head wrapped on all sides so that I would not be hindered by the daylight, the Sun shone no less brightly than Sirius. According to the rules of dioptrics, the size of the Sun was now a 152th part of that 182th part that I had seen first. Now, we know that by multiplying 152 by 182, we get 27 664. Therefore, if the Sun is reduced, or brought to a great distance (which amounts to the same), so that its diameter is the 27 664th part of the Sun that we see in the sky, then it gives light that does not deviate from Sirius. The distance from the Sun, so far away, must necessarily be 27 664 times the distance that it now has.

It only remains for us to add that Christiaan's distance is 538 times as great as that 'calculated' by Kepler and that this distance amounts to 0.46 light years, while the real distance of Sirius is now put at 8.5 light years. He had brought the heavens closer, to subsequently to place them at an unprecedented distance.

He never saw the publication. *Cosmotheoros* appeared post-humously.[78] In 1694 his working capacity clearly deteriorated. He lost weight. Did he have cancer? The question arises as we describe the end of his life, but remains unanswerable.

When he was once again at Hofwijck after 22 May, his pulse was weak and irregular.[98] Without checking first, he spread the improbable rumour of Newton's mental illness, 'the worst that can happen to a person'. In July, studying became too much for him. He fretted.

We do not know whether he saw his niece Suzette again. 'Without satisfactory occupation, the mind surrenders to transient desire that often harms only another,' he noted in a margin.[99] Did he blame himself for something? Did he have feelings of guilt?

We can only answer that he went regularly to church, and to communion, and that he was sincere when he wrote to Bayle about 'our religion'.[100] But the doctrine of reconciliation can hardly have accorded with all that he knew, and with his doubts. He often lay awake at night, brooding, watching the shadows sharpen above the village of Nootdorp and the sky take on its morning hue. After one such night he wrote:[101]

> *Ut valeas sit cura, minantemque effuge morbum*
> *Nam ratio atque animi languent cum corpore vires*
> *Tristitia quodcumque agitat mens inficit aegri,*
> *Nec tibi iudiciis proprius tunc fidere fas est.*

> Be strong, surrender not to grief.
> A soul that suffers even as the flesh,
> With bitter gloom deceives
> The mind that on the soul relies.

'I shall console myself with all reasons for your silence,' Leibniz wrote to him on 26 April, 'as long as they are not sickness or anger . . .'[102] He replied that it was neither the one nor the other, but that he was waiting for information.

After the letter of 8 June, containing the rumour about Newton, he answered his avid correspondent, who continued to pose mathematical problems, only twice, and he ended by sending his good wishes for the New Year.[103] He must have still been working on the

---

[99] OC **22**, 493   [100] OC **10**, 103   [101] OC **10**, 719 (n1)   [102] OC **10**, 600
[103] OC **10**, 664 & 696

translation of his book until well into the autumn of 1694. Was he distracted by money worries?

Rather precipitously, and despite objections from his family, he decided to move to Utrecht, where taxes were supposed to be 800 guilders lower. That was in September. He terminated his lease of Noordeinde, but discovered in October that taxes differed far less than he had thought, and reactivated the lease.[104] In November, when he moved back into his residence in The Hague, it would be for good.

Printing of *Cosmotheoros* began in 1695, according to his letter to Constantijn on 4 March. The letter ended as follows:[105]

> I have read an English poem in praise of the deceased Queen, which apparently earned the writer 100 guineas. I do not know whether you consider it to be a fine one. I do not. What are such names as Pastora, Astrofel and others doing in it? Why the inappropriate exaggeration? But people there will lap it up, for they adore all that is English. Did you, by any chance, write something for the occasion?

These are the last lines from his hand. Meanwhile he was already dangerously ill and had his will drawn up:[106]

> I, the undersigned, do reflect upon the certainty of death and the uncertainty of the moment in time and the manner in which this shall occur. Bearing this in mind, I should not wish to depart from this world without having made my will regarding those temporal goods that have been bestowed to me by God Almighty, and do declare without influence or persuasion from any person:
>
> Firstly: to bequeath a preference legacy to Christiaan Huygens, the son of my brother, the Delegate Council in the

---

[104] Huygens 25, 422–423   [105] OC 10, 710   [106] OC 22, 775–778

Admiralty, above his inheritance, one of my silver jugs, because he did not receive any christening gift from me;

To my cousin, Madame de la Ferté, I bequeath the sum of two thousand guilders, and another five hundred guilders to her oldest daughter, whom I held up for holy baptism;

To my cousin, Miss Ida van Dorp, I bequeath five hundred guilders;

To Monsieur Johan Wiljeth, two hundred guilders;

To Hendrick, my servant, because he has served me well, I bequeath one hundred two-and-a-half guilder coins;

To Anna, my serving maid, I bequeath one hundred and fifty guilders;

To Matthijs, my gardener, one hundred guilders, the same sum to Greetje, his sister;

My manuscripts of mathematics, most of which lie in the lowest drawer of my largest study at Hofwijck, consisting of nine bound volumes, marked with the letters A until and including I, and in addition, many dissertations, in which I was engaged, I bequeath to the Academy or Library of Leiden, and I request the professors De Volder from Leiden and Fullenius from Franeker to look these through, and what may be appropriate for publication, to carry this out to the best of their ability.

The completed treatise *Cosmotheoros*, which lies in my study in The Hague, except for three or four pages that I had sent to the printer Rammazeijn, I entrust to my brother, the Lord of Zuijlichem, to whom it has been explained, and who should arrange that it be further printed, as it was begun by Mr. Moetjes; the same I entrust to the executors, hereafter to be named, my last will and testament.

. . . And I have, to ratify this document, signed it with my customary signature in 's Gravenhage, the 23rd of March 1695, signed

Chr. Huygens

402 HUYGENS: THE MAN BEHIND THE PRINCIPLE

'He who is privileged to die without pain, should consider the misfortune of those who are denied euthanasia.' We end with Constantijn's sober journal:[107]

> 16 April: Christiaan is badly indisposed. He cannot sleep and fears to lose his mind. His room is darkened. He allows no one in, for if he speaks, he immediately suffers greater pain.
>
> 26 April: His health continues to fail. He suffers harrowing thoughts, whatever our sister and my wife say to divert him. By great good fortune he does not have a fever, otherwise he would not last long, for he is very weak.
>
> 25 May: I arrive to find brother Christiaan sorely ill. He complains of pain and of bedsores. Anything with which he may harm himself has been removed. He began to cut himself with broken glass and prick himself with pins. Also he stuck a marble down his throat. His servant heard it rattle and managed to extricate it by slapping upon his back. He cried out then, 'Slap hard'. Sometimes dreams and hears people speak who are not there. Says that people would tear him apart if they heard his view on religion. Hopes that he will not be held to blame for this view, for he is out of his senses. Sometimes screams loudly and curses.
>
> 3 June: Slept reasonably well this past night and did not rave. The doctors Van Lieberghen and Van Wouw find him weak, say that he cannot endure any heavy food. In the evening I came to his door [to bid farewell] and his servant said that I could come in if I wished to see him one more time in his last suffering. If not, he wishes me happiness upon my journey.
>
> 13 June (from this point on, based on news from Constantijn's wife): Since he began drinking goat's milk, there is some improvement.

[107] Huygens 25, 472–504

17 June: Worse again than he was before. He now imagines that his food is being poisoned, and therefore will not eat; is so thin that it is incomprehensible that he is able to stay alive.

20 June: Became angry when she [Constantijn's wife] asked permission to call a clergyman, cursed and raged.

24 June: No sign of improvement.

27 June: Has not tossed and turned thus since three or four days.

5 July: The doctors say that he can live like this for yet another year.

11 July: Last Thursday [that is, 7 July] a sudden change. Christiaan agreed that the clergyman Olivier, an acquaintance, be called. Olivier spoke long with him and prayed for him. But his mind is not to be changed, whatever one may say. Great sorrow for all.

During the night he lost consciousness. At half past three in the morning the family was notified.

On Friday 8 July 1695, he fell asleep forever.

Bibliography

OC = *Oeuvres complètes de Christiaan Huygens*, D. Bierens de Haan, J. Bosscha, D. J. Korteweg & J. A. Vollgraff (eds.) on behalf of the Hollandsche Maatschappij van Wetenschappen (Martinus Nijhoff, La Haye/The Hague, 1888–1950); by far the most important source for this book, with 22 volumes to which will be referred by a bold number (for instance: OC **16**, 259 refers to page 259 in volume 16 of these *Oeuvres*)

Acloque, Paul, 'L'oculair de Huygens, son invention et sa place dans l'instrumentation', pp. 177–185 in: R. Taton (ed.), *Huygens et la France* (Paris, Vrin, 1982)

Aiton, Eric J., *Leibniz: A Biography* (Bristol, Adam Hilger, 1985)

Andriesse, C. D., 'The melancholic genius', in: L. Palm (ed.), *De zeventiende eeuw*, vol. 12 (Hilversum, Verloren, 1996)

Arago, François, 'Huygens', pp. 319–322 in: J.-A. Barral (ed.), *Biographies des Principaux Astronomes* (Paris, Gide & Baudry, 1855)

Bachrach, A. G. H., 'The role of the Huygens family in seventeenth-century Dutch culture', pp. 27–52 in: H. J. M. Bos, M. J. S. Rudwick, H. A. M. Snelders & R. P. W. Visser (eds.), *Studies on Christiaan Huygens* (Lisse, Swets & Zeitlinger, 1980)

Barbour, Julian B., *Absolute or Relative Motion? – A Study from a Machian Point of View of the Discovery and the Structure of Dynamical Theories*, vol. 1, *The Discovery of Dynamics* (Cambridge, Cambridge University Press, 1989)

Beaulieu, Armand, 'Christiaan Huygens et Mersenne l'inspirateur', pp. 25–32 in: R. Taton (ed.), *Huygens et la France* (Paris, Vrin, 1982)

Bell, Arthur E., *Christian Huygens and the Development of Science in the Seventeenth Century* (London, Edward Arnold, 1947)

Blom, F. R. E., Bruin, H. G. & Ottenheym, K. A., *Domus – Het huis van Constantijn Huygens in Den Haag* (Zutphen, Walburg, 1999)

Boissonnade, Prosper, *Colbert et la dictature du travail 1661–1683* (Paris, Marcel Rivière, 1932)

Bos, H. J. M., 'Christiaan Huygens – a biographical sketch', pp. 7–16; & 'Huygens and mathematics', pp. 126–146 in: H. J. M. Bos, M. J. S. Rudwick, H. A. M. Snelders & R. P. W. Visser (eds.), *Studies on Christiaan Huygens* (Lisse, Swets & Zeitlinger, 1980)

Bos, H. J. M., 'Tractional motion and the legitimation of transcendental curves', *Centaurus* **31** (1988) 9–62

Bosscha, Johannes, *Christiaan Huygens – Rede op den 200sten gedenkdag van zijn levenseinde* (Haarlem, J. Enschedé & Zonen, 1895)

Bost, Hubert, *Un 'intellectuel' avant la lettre: Le journalist Pierre Bayle (1647–1706)* (Amsterdam-Maarssen, Academic Publishers Associated, 1994)

Bots, Hans, 'Constantijn Huygens een wetenschapsbeoefenaar?', pp. 149–160 in: *De zeventiende eeuw*, vol. 3 (Hilversum, Verloren, 1987)

Broad, Charlie Dunbar, 'John Locke', pp. 1–24 in: J. S. Yolton (ed.), *A Locke Miscellany: Locke Biography and Criticism for All* (Bristol, Thoemmes, 1990)

Brugmans, Henri L., *Le séjour de Christian Huygens à Paris et ses relations avec les milieux scientifiques français* (Paris, E. Droz, 1935)

Cohen, H. F., 'Christiaan Huygens on consonance and the division of the octave', pp. 271–301 in: H. J. M. Bos, M. J. S. Rudwick, H. A. M. Snelders & R. P. W. Visser (eds.), *Studies on Christiaan Huygens* (Lisse, Swets & Zeitlinger, 1980)

Cohen, H. F., *Over aard en oorzaken van de 17de eeuwse wetenschapsrevolutie* (Amsterdam, Van Oorschot, 1983)

Cohen, H. Floris, *The Scientific Revolution: A Historiographical Inquiry* (Chicago, University of Chicago Press, 1994)

Costabel, Pierre, 'Huygens et la mécanique de la chute des corps à la cause de la pesanteur', pp. 139–152 in: R. Taton (ed.), *Huygens et la France* (Paris, Vrin, 1982)

Costabel, Pierre, 'Le père Marin Mersenne 1588–1648', pp. 3–19; & 'Gilles Personne de Roberval 1602–1675', pp. 20–31 in: P. Costabel & M. Martinet (eds.), *Quelques savants et amateurs de science au XVIIe siècle* (Paris, Société française d'histoire des sciences et des techniques, 1986)

Coumet, Ernest, 'Sur "Le calcul ès jeux de hazard" de Huygens – Dialogues avec les mathématiciens français', pp. 123–137 in: R. Taton (ed.), *Huygens et la France* (Paris, Vrin, 1982)

Cranston, Maurice, 'John Locke's exile', pp. 46–54 in: J. S. Yolton (ed.), *A Locke Miscellany: Locke Biography and Criticism for All* (Bristol, Thoemmes, 1990)

Crommelin, Claude A., *Descriptive Catalogue of the Huygens Collection* (Leiden, Boerhaave Museum, 1949)

De Lang, Hendrik, 'Christiaan Huygens, originator of wave optics', pp. 19–30 in: H. Blok, H. A. Ferwerda & H. K. Kuiken (eds.), *Huygens' Principle 1690–1990: Theory and Applications* (Amsterdam, North-Holland, 1992)

Dijksterhuis, Eduard Jan, *De mechanisering van het wereldbeeld* (Amsterdam, Meulenhof, 1950)

Dijksterhuis, Eduard Jan, 'Christiaan Huygens – Address to the Holland Society of Sciences in Haarlem on the occasion of the completion of Huygens's collected works', *Centaurus* 2 (1951/53) 265–282

Dijksterhuis, Fokko Jan, *Lenses and Waves: Christiaan Huygens and the Mathematical Science of Optics in the Seventeenth Century* (Enschede, Ipskamp, 1999)

Duistermaat, J. J., 'Huygens' principle for linear partial differential equations', pp. 273–297 in: H. Blok, H. A. Ferwerda & H. K. Kuiken (eds.), *Huygens' Principle 1690–1990: Theory and Applications* (Amsterdam, North-Holland, 1992)

Eco, Umberto, *Foucault's Pendulum* (New York, Harcourt, 1995)

Eddy, John A., 'The "Maunder Minimum"': Sunspots and climate change in the reign of Louis XIV', pp. 226–268 in: G. Parker & L. M. Smith (eds.), *The General Crisis of the Seventeenth Century* (London, Routledge & Kegan Paul, 1978)

Eringa, Dieuwke, *Christiaan Huygens: Verhandeling over het licht* (Utrecht, Epsilon Uitgaven, 1990)

Eyffinger, Arthur, 'Constantijn Huygens beschrijft de jeugd van zijn kinderen', pp. 89–154 in: A. Eyffinger (ed.), *Huygens herdacht – Catalogus herdenkingstentoonstelling* ('s Gravenhage, Koninklijke Bibliotheek, 1987)

Flaubert, Gustave, *L'Education sentimentale*, R. Dumesnil (ed.) (Paris, Société les belles lettres, 1942)

Frankfourt, Usher I. & Frenk, Aleksandr M., *Christiaan Huygens* [French translation of *Khristian Gyuigens* by I. Sokolov] (Moscou, Editions Mir, 1976)

French, A. P., *Einstein: A Centenary Volume* (London, Heinemann, 1979)

Freud, Sigmund, *Rouw en melancholie* [Dutch translation by Th. Graftdijk & W. Oranje of *Trauer und Melancholie* in: *Psychoanalytische theorie*, vol. 1] (Boom, Meppel, 1985)

Gabbey, Alan, 'Huygens and mechanics', pp. 166–199 in: H. J. M. Bos, M. J. S. Rudwick, H. A. M. Snelders & R. P. W. Visser (eds.), *Studies on Christiaan Huygens* (Lisse, Swets & Zeitlinger, 1980)

Gabbey, Alan, 'Huygens et Roberval', pp. 69–83 in: R. Taton (ed.), *Huygens et la France* (Paris, Vrin, 1982)

Hahn, Roger, 'Huygens and France', pp. 53–65 in: H. J. M. Bos, M. J. S. Rudwick, H. A. M. Snelders & R. P. W. Visser (eds.), *Studies on Christiaan Huygens* (Lisse, Swets & Zeitlinger, 1980)

Hatch, Robert A., 'Between friends: Huygens and Boulliau', pp. 106–116 in: L. Palm (ed.), *De zeventiende eeuw*, vol. 12 (Hilversum, Verloren, 1996)

Heijbroek, J. Frederik, 'Het geheimschrift van Huygens ontcijferd', pp. 167–172 in: A. Eyffinger (ed.), *Huygens herdacht – Catalogus herdenkingstentoonstelling* ('s Gravenhage, Koninklijke Bibliotheek, 1987)

Heinekamp, Albert, 'Christiaan Huygens vu par Leibniz', pp. 99–114 in: R. Taton (ed.), *Huygens et la France* (Paris, Vrin, 1982)

Hunten, D. M., Tomasko, M. G., Flasar, F. M., Samuelson, R. E., Strobel, D. F. & Stevenson, D. J., 'Titan', in: T. Gehrels & M. S. Matthews (eds.), *Saturn* (Tucson, University of Arizona Press, 1984)

Huygens, Constantijn Jr., *Journaal van Constantijn Huygens, den zoon* (Utrecht, Historisch Genootschap, 1876–1888) [early journals are edited in **46** (1888); those of the 1670s in **32** (1881); those of 1688–1691 in **23** (1876); his last journals of 1692–1696 in **25** (1877)]

Israel, Jonathan I., *The Dutch Republic: Its Rise, Greatness and Fall, 1477–1806* (Oxford, Clarendon Press, 1995)

Israel, Jonathan I., *Radical Enlightenment: Philosophy and the Making of Modernity, 1650–1750* (Oxford, Oxford University Press, 2001)

Keesing, Elisabeth, *Constantijn en Christiaan – Verhaal van een vriendschap* (Amsterdam, Querido, 1983)

Keesing, Elisabeth, 'Les frères Huygens et Spinoza', *Cahiers Spinoza* **5** (1984/85) 109–128

Keesing, Elisabeth, *Het volk met lange rokken – Vrouwen rondom Constantijn Huygens* (Amsterdam, Querido, 1987)

Kirk, G. S., *The Nature of Greek Myths* (Harmondsworth, Penguin Books, 1974)

Kossmann, Ernst H., 'Rowen's De Witt', pp. 190–192 in: *Politieke theorie en geschiedenis* (Amsterdam, Bert Bakker, 1987)

Kossmann, Ernst H., 'Koning-stadhouder Willem III', pp. 87–101 in: *Vergankelijkheid en continuiteit – Opstellen over geschiedenis* (Amsterdam, Bert Bakker, 1995)

Koyré, Alexandre, *Newtonian Studies* (London, Chapman & Hall, 1965)

Kubbinga, H. H., 'Christiaan Huygens' wetenschappelijke opleiding', pp. 161–170 in: *De zeventiende eeuw*, vol. 3 (Hilversum, Verloren, 1987)

Leeuwenhoek, Antoni, *Alle de brieven – Collected Letters*, vol. 2, G. Van Rijnberk (ed.) (Amsterdam, Swets & Zeitlinger, 1941)

Leibniz, Gottfried Wilhelm, *Discours de métaphysique*, H. Lestienne (ed.) (Paris, Vrin, 1990)

Leopold, Jan Hendrik, 'Christiaan Huygens and his instrument makers', pp. 221–233 in: H. J. M. Bos, M. J. S. Rudwick, H. A. M. Snelders & R. P. W. Visser (eds.), *Studies on Christiaan Huygens* (Lisse, Swets & Zeitlinger, 1980)

Mahoney, Michael S., 'Christiaan Huygens – the measurement of time and of longitude at sea', pp. 234–270 in: H. J. M. Bos, M. J. S. Rudwick, H. A. M. Snelders & R. P. W. Visser (eds.), *Studies on Christiaan Huygens* (Lisse, Swets & Zeitlinger, 1980)

Manuel, Frank E., *A Portrait of Isaac Newton* (Cambridge, Belknap Press of Harvard University Press, 1968)

Mesnard, Jean, 'Les premières relations Parisiennes de Christiaan Huygens', pp. 33–40 in: R. Taton (ed.), *Huygens et la France* (Paris, Vrin, 1982)

Parker, Geoffrey & Smith, Lesley M., *The General Crisis of the Seventeenth Century* (London, Routledge & Kegan Paul, 1978)

Pascal, Blaise, *Pensées et opuscules*, M. L. Brunschvicq (ed.) (Paris, Hachette, 1968)

Payen, Jacques, 'Huygens et Papin – Moteur thermique et machine à vapeur au XVIIᵉ siècle', pp. 197–208 in: R. Taton (ed.), *Huygens et la France* (Paris, Vrin, 1982)

Ploeg, Willem, *Constantijn Huygens en de natuurwetenschappen* (Rotterdam, Nijgh & Van Ditmar, 1934)

Rabus, Pieter, *Christiaan Huigens – Wereldbeschouwer, of onderzoek over de hemelsche aard-klooten en derzelver cieraad* [Dutch translation of Κοσμωθεωρος 's Gravenhage, Adriaan Moetjens, 1698] (Utrecht, Epsilon Uitgaven, 1989)

Rasch, Rudolf, *Christiaan Huygens' cycle harmonique & novus cyclus harmonicus* (Utrecht, Diapason Press, 1986)

Rasch, R., 'Constantijn Huygens: Een muzikale heer van stand', pp. 99–114 in: *De zeventiende eeuw*, vol. 3 (Hilversum, Verloren, 1987)

Rasch, R., 'Constantijn en Christiaan Huygens' relatie tot de muziek', in: L. Palm (ed.), *De zeventiende eeuw*, vol. 12 (Hilversum, Verloren, 1996)

Rietbergen, P. J. A. N., 'Den Haag, 20 april 1660: de bruiloft van Susanna Huygens', pp. 181–189 in: *De zeventiende eeuw*, vol. 3 (Hilversum, Verloren, 1987)

Roger, Jacques, 'La politique intellectuelle de Colbert et l'installation de Huygens à Paris', pp. 41–47 in: R. Taton (ed.), *Huygens et la France* (Paris, Vrin, 1982)

Romein, Annie, 'Christiaen Huygens – Ontdekker der waarschijnlijkheid', in: J. & A. Romein, *Erflaters van onze beschaving* (Amsterdam, Querido, 1938)

Rowen, Herbert A., *John de Witt: Statesman of the 'True Freedom'* (Cambridge, Cambridge University Press, 1986)

Schama, Simon, *The Embarrassment of Riches: An Interpretation of Dutch Culture in the Golden Age* (London, Collins, 1987)

Schopenhauer, Arthur, 'Parerga und Paralipomena' in: W. Brede (ed.), *Selection of Schopenhauer's Works*, vol. 2 (München, Carl Hanser, 1977)

Seidengart, Jean, 'Les théories cosmologiques de Christiaan Huygens', pp. 209–222 in: R. Taton (ed.), *Huygens et la France* (Paris, Vrin, 1982)

Shapiro, Alan E., 'Huygens' kinematic theory of light', pp. 200–220 in: H. J. M. Bos, M. J. S. Rudwick, H. A. M. Snelders & R. P. W. Visser (eds.), *Studies on Christiaan Huygens* (Lisse, Swets & Zeitlinger, 1980)

Smit, Jacob, *De grootmeester van woord- en snarenspel – Het leven van Constantijn Huygens* ('s Gravenhage, Martinus Nijhoof, 1980)

Snelders, H. A. M., 'Christiaan Huygens and the concept of matter', pp. 104–125 in: H. J. M. Bos, M. J. S. Rudwick, H. A. M. Snelders & R. P. W. Visser (eds.), *Studies on Christiaan Huygens* (Lisse, Swets & Zeitlinger, 1980)

Sobel, Dava, *Longitude: The True Story of a Lone Genius who Solved the Greatest Scientific Problem of his Time* (London, Fourth Estate, 1996)

Sonnino, Paul, *Louis XIV and the Origins of the Dutch War* (Cambridge, Cambridge University Press, 1988)

Sparnaay, Marcus Johannes, *Adventures in Vacuums* (Amsterdam, North-Holland, 1992)

Strengholt, Leendert, *Constanter – Het leven van Constantijn Huygens* (Amsterdam, Querido, 1987)

Taton, René, 'Huygens et l'académie royale des sciences', pp. 57–68 in: R. Taton (ed.), *Huygens et la France* (Paris, Vrin, 1982)

Van Berkel, Klaas, *Citaten uit het boek der natuur* (Amsterdam, Bert Bakker, 1998)

Van der Stighelen, Katlijne, 'Constantijn Huygens en Anna Maria van Schurman: veel werk, weinig weerwerk', pp. 138–148 in: *De zeventiende eeuw*, vol. 3 (Hilversum, Verloren, 1987)

Van Helden, Albert, 'Huygens and the astronomers', pp. 147–165 in: H. J. M. Bos, M. J. S. Rudwick, H. A. M. Snelders & R. P. W. Visser (eds.), *Studies on Christiaan Huygens* (Lisse, Swets & Zeitlinger, 1980)

Van Helden, Albert, 'Contrasting careers in astronomy: Huygens and Cassini', pp. 96–105 in: *De zeventiende eeuw*, vol. 12 (Hilversum, Verloren, 1996)

Van Lieburg, Mart J., 'Constantijn Huygens en Suzanna van Baerle – Een pathobiografische bijdrage', pp. 171–180 in: *De zeventiende eeuw*, vol. 3 (Hilversum, Verloren, 1987)

Van Strien, Ton & Van der Leer, Kees, *Hofwijck – Het gedicht en de buitenplaats van Constantijn Huygens* (Zutphen, Walburg Pers, 2002)

Verbeek, Theo, *Descartes and the Dutch: Early Reactions to Cartesian Philosophy 1637–1650* (Carbondale, Southern Illinois University Press, 1992)

Vollgraff, Johan A., [*Fragments d'une*] *Biographie de Christiaan Huygens*, in: OC **22**, 383–778 (La Haye, Martinus Nijhoff, 1950)

Westfall, Richard S., *Never at Rest: A Biography of Isaac Newton* (Cambridge, Cambridge University Press, 1980)

Worp, Jacob Adolf, *De gedichten van Constantijn Huygens* [the father; edited in nine volumes to which will be referred by a bold number] (Groningen, Wolters, 1892–1899)

Yoder, Joella G., *Unrolling Time: Christiaan Huygens and the Mathematization of Nature* (Cambridge, Cambridge University Press, 1988)

Further reading

More literature exists on Huygens than has been used for this biography. We list some 400 further titles, without claiming that this list is exhaustive. The titles are ordered in sections on the man and on his works, the latter again ordered by subject (Astronomy, Horology, Instruments, Mathematics, Optics, Physics and Varia).

ON THE MAN

Aris, Daniel, 'La découverte de la France par Christiaan Huygens', pp. 58–72 in: J. M. Pastré (ed.), *Les Récits de voyage* (Paris, Nizet, 1986)

Bachelard, Suzanne, 'L'influence de Huygens au XVIIIe et au XIXe siècle', pp. 241–257 in: R. Taton (ed.), *Huygens et la France* (Paris, Vrin, 1982)

Bachrach, Alfred Gustave Herbert, 'Christiaan Huygens en de invloed van het ouderlijk huis', *Spiegel Historiael* **14** (1979) 201–209

Bachrach, A. G. H., 'Les Huygens entre la France et l'Angleterre', pp. 17–24 in: R. Taton (ed.), *Huygens et la France* (Paris, Vrin, 1982)

Bachrach, A. G. H. & Collmer, R. G. (eds.), *Lodewijck Huygens: The English Journal (1651–1652)* (New York, E. J. Brill, 1997)

Barchilon, Jacques, 'Les frères Perrault à travers la correspondence et les oeuvres de Christian Huygens', *Dix-septième siècle* **56** (1962) 19–36

Bell, Arthur E., 'Christian Huygens', *Nature* **146** (1940) 511–514

Bernés, Anne-Catherine, 'Christiaan Huygens, les Pays-Bas et la principauté de Liège', *De zeventiende eeuw* **12** (1996) 37–51

Biermann, Kurt-Reinhard, 'Ch. Huygens im Spiegel von Alexander von Humboldts Kosmos', *Janus* **66** (1979) 241–247

Bogazzi, Riccardo, 'Aspetti dell'opera filosofica e scientifica di Christiaan Huygens', *Bolletino della Società filosofica italiana* **146/147** (1992) 17–24

Bos, Hendrik J. M., 'Huygens, Christiaan', pp. 597–613 in: C. C. Gillespie (ed.), *Dictionary of Scientific Biography* vol. 6 (1972)

Bos, H. J. M., 'The influence of Huygens on the formation of Leibniz' ideas', in: A. Heinekamp & D. Mettler (eds.), *Leibniz à Paris (1672–1676), Studia Leibnitiana* (Supplementa 17) **1** (1978) 59–68

Bos, H. J. M., 'Christiaan Huygens', in: E. Pauly (ed.), *Célébrations nationales* (Paris, Direction des archives de France, 1995)

Bosscha, Johannes, 'Les œuvres complètes de Christiaan Huygens', *Bibliotheca Mathematica: Zeitschrift für Geschichte der mathematischen Wissenschaften*, serie 3, **1** (1900) 93–96

Bosscha, J., 'Huygens', pp. 1180–1186 in: *Nieuw Nederlandsch Biographisch Woordenboek*, vol. 1 (Leiden, Sijthoff, 1911)

Cantor, Max, 'Huygens', *Allgemeine Deutsche Biographie* **13** (1881) 480–486

Chareix, Fabien, 'Experimenta ac ratio – L'oeuvre de Christiaan Huygens', *Revue d'histoire des sciences* **56** (2003) 5–13

Cohen, H. F., 'Wie was Christiaan Huygens?', *Spiegel Historiael* **14** (1979) 210–214

Cohen, H. F., *Christiaan Huygens en de wetenschapsrevolutie van de 17de eeuw*, Mededeling van het Museum Boerhaave no. 267 (Leiden, Museum Boerhaave, 1996)

Crommelin, Claude A., 'Christiaan Huygens', *Christiaan Huygens: Internationaal Mathematisch Tijdschrift* **17** (1938/39) 247–270

Dekker, Elly, 'Christiaan Huygens 1629–1695', *Zenit* **6** (1979) 234–238

D'Elia, Alfonsina, *Christiaan Huygens – Una biografia intellettuale* (Milano, Franco Angeli, 1985)

De Pater, Cornelis, 'In de schaduw van Newton: Het Huygensbeeld bij enkele Nederlandse newtonianen in de achttiende eeuw', pp. 64–73 in: L. Palm (ed.), *De zeventiende eeuw*, vol. 12 (Hilversum, Verloren, 1996)

De Vleeschauwer, Herman J., *De briefwisseling van Ehrenfried Walther von Tschirnhaus met Christiaan Huygens*, Mededelingen van de Koninklijke Vlaamsche Academie voor Wetenschappen, Letteren en Schoone Kunsten van België, Klasse der Letteren, 3 (Brussels, 1941) no. 6

Dijksterhuis, Eduard Jan, 'Christiaan Huygens en Frankrijk', *De Gids* **99/4** (1935) 240–250

Dijksterhuis, Fokko Jan, 'Titan en Christiaan: Huygens in werk en leven', *Tijdschrift voor de geschiedenis der geneeskunde, natuurwetenschappen, wiskunde en techniek* **23** (2000) 56–68

Dijksterhuis, F. J., 'Huygens', pp. 364–365 in: A. Hessenbruch (ed.), *Reader's Guide to the History of Science* (London, Fitzroy Dearborn, 2001)

Elzinga, Aant, *Notes on the Life and Works of Christiaan Huygens*, Göteborg University Reports no. 88 (Göteborg, Department of the Theory of Science, 1976)

Favaro, Antonio, 'Intorno alle opera complete di Christiano Huygens pubblicate dalla Società olandese delle scienze', *Atti del Reale istituto Veneto di scienze, lettere ed arti*, serie 6, **7** (1888/89) 403–421

Feingold, Mordechai, 'Huygens and the Royal Society', pp. 22–36 in: L. Palm (ed.), *De zeventiende eeuw*, vol. 12 (Hilversum, Verloren, 1996)

Forti, Umberto, 'Nel terzo centenario di Christiano Huygens', *Nuova antologia* **1386** (1929) 500–512

Frege, Gottlob, *Über den Briefwechsel Leibnizens und Huygens mit Papin*, (Jena, 1881; reprinted on pp. 93–96 in: I. Angelelli (ed.), *Gottlob Frege – Begriffsschrift und andere Aufsätze* (Hildesheim, Georg Olms, 1964)

Gerland, Ernst, *Leibnizens und Huygens' Briefwechsel mit Papin, nebst der Biographie Papins und einigen zugehörigen Briefen und Actenstücken* (Berlin, 1881; reprinted Wiesbaden, Sändig, 1966)

Gerland, E., 'Nachtrag zu Leibnizens und Huygens' Briefwechsel met Papin', *Sitzungsberichte der Königlich Preußischen Akademie der Wissenschaften zu Berlin, Philosophisch-historischen Classe* (1882) 979–984

Harting, Pieter, *Christiaan Huygens in zijn leven en werken geschetst* (Groningen, Hoitsema, 1868)

Harting, P., 'Christiaan Huygens in de Parijsche Akademie van Wetenschappen', pp. 16–20 in: *Album der Natuur: Een werk ter verspreiding van natuurkennis onder beschaafde lezers van allerlei stand* (Haarlem, Kruseman, 1869)

Henry, C., *Huygens et Roberval: Documents nouveaux* (Leiden, E. J. Brill, 1880)

Holleman, A. F., De Sitter, Willem & Zeeman, Pieter (eds.), *Christiaan Huygens 1629–14 april-1929: Zijn geboortedag, 300 jaar geleden, herdacht* (Amsterdam, H. J. Paris, 1929)

Jöcher, Christian Gottlieb (ed.), 'Huygens', pp. 1792–1794 in: *Allgemeines Gelehrten-Lexicon*, vol. 2 (Leipzig, 1750)

Keesing, Elisabeth, 'Wanneer was wie de heer van Zeelhem?', *De zeventiende eeuw* 9 (1993) 63–65

Keesing, E., 'De samenwerking van de broers Huygens', pp. 14–21 in: L. Palm (ed.), *De zeventiende eeuw*, vol. 12 (Hilversum, Verloren, 1996)

Klever, Wim, 'Spinoza en Huygens: Een geschakeerde relatie tussen twee fysici', *Tijdschrift voor de geschiedenis der geneeskunde, natuurwetenschappen, wiskunde en techniek* 20 (1997) 14–31

Korteweg, Diederik Johannes, *Een en ander over de Huygens-uitgave en over den invloed van Descartes op Christiaan Huygens*, Jaarboek der Koninklijke Akademie van Wetenschappen (Amsterdam, Johannes Müller, 1909)

Korteweg, D. J., 'Christian Huygens' wissenschaftliche Lehrjahre', *Internationale Wochenschrift für Wissenschaft, Kunst und Technik* 3 (1909) 1391–1396 & 1411–1426

Kox, Anne J. & Polak, P. H., 'Christiaan Huygens 1629–1695', *De Gids* 142 (1979) 279–291

Kox, A. J. & Polak, P. H., 'Christiaan Huygens – Door ervaring en rede', in: Anne Kox & Margot Chamalaun (eds.), *Van Stevin tot Lorentz – Portretten van Nederlandse natuurwetenschappers* (Amsterdam, Intermediair Bibliotheek, 1980)

Land, J. P. N., *Over papieren van Constantijn Huygens en zijne zonen, Verslagen en Mededeelingen der Koninklijke Akademie van Wetenschappen, Afdeeling Letterkunde, derde reeks*, vol. 2 (Amsterdam, Akademie van Wetenschappen, 1884)

Lemans, Moses, *Levensbeschrijving van Christiaan Huygens* (redevoering in het Letteroefenend Genootschap tot Nut en Beschaving, 1820)

Loria, Gino, 'La vita scientifica di Christiano Huygens quale si desume dal suo carteggio', *Commentationes pontificiæ academiæ scientiarum* **6** (1942) 1079–1138

Moll, Gerrit, 'Over de uitgave van de handschriften van Ch. Huygens', *Algemene Konst- en Letterbode* **22** (1825)

Moll, Konrad, 'Von Erhard Weigel zu Christiaan Huygens: Feststellungen zum Leibnizens Bildungsweg zwischen Nürnberg, Mainz und Paris', *Studia Leibnitiana* **14** (1982) 56–72

Monchamp, Georges, 'Les correspondants belges du grand Huygens', *Bulletin de l'Académie royal des sciences, des lettres et des beaux-arts de Belgique*, serie 3, **27** (1894) 255–308

Mormino, Gianfranco, 'Christiaan Huygens e il problema della comunicazione scientifica', pp. 167–189 in: M. Galuzzi, G. Micheli & M. T. Monti (eds.), *Le forme della comunicazione scientifica* (Milano, Franco Angeli, 1998)

Mormino, G., 'Sur quelques problèmes éditoriaux concernant l'oeuvre de Christiaan Huygens', *Revue d'histoire des sciences* **56** (2003) 145–151

Mormino, G. & Chareix, Fabien, 'Bibliographie huguenienne', *Revue d'histoire des sciences* **56** (2003) 153–190

Mundt, C. S., 'Notable astronomers of past ages: VI – Huyghens', *Publications of the Astronomical Society of the Pacific* **39** (1927) 354–356

Nellen, Henk J. M., 'Editing 17th-century scholarly correspondence: Grotius, Huygens, and Mersenne', *Lias* **17** (1990) 9–20

Picard, Charles-Émile, 'Le troisième centenaire de Huygens', *Revue générales des sciences pures et appliqués* **40** (1929) 321–323

Picard, C.-É., 'Christiaan Huygens', *Notices et discourses de l'Académie des sciences* **1** (1935) 236–241

Picolet, Guy, 'Un génie mal connu: Christiaan Huygens', *La Recherche* **103** (1979) 906–908

Picolet, G., 'Huygens et Picard', pp. 85–97 in: R. Taton (ed.), *Huygens et la France* (Paris, Vrin, 1982)

Poggendorff, J. C. (ed.), 'Huygens', pp. 1164–1165 in: *Biographisch-Literarisches Handwörterbuch zur Geschichte der exacten Wissenschaften*, vol. 1 (Leipzig, J. A. Barth, 1863)

Radelet-de Grave, Patricia, 'Huygens et les Bernoulli', *De zeventiende eeuw*, **12**, (1996) 253–273

Robinet, André, 'Huygens et Malebranche', pp. 223–239 in: R. Taton (ed.), *Huygens et la France* (Paris, Vrin, 1982)

Schinkel, A. D., *Opgave der handschriften van Constantijn en Christiaan Huygens, benevens die handschriften welke tot hen betrekkelijk zijn en voor weinig jaren onder derzelver nakomelingen nog berustende waren* ('s Gravenhage, 1840)

Schinkel, A. D., *Nadere bijzonderheden betrekkelijk Constantijn Huygens en zijne familie* [. . .] in 2 deelen ('s Gravenhage, 1851–1856)

Schuh, Frederick, 'Christiaan Huygens (14 april 1629–9 juli 1695)', *Christiaan Huygens: Internationaal Mathematisch Tijdschrift* **1** (1921/22), 1–28

Stein, J., 'Christiaan Huygens en de Jezuïeten', *Bijdragen van de Philosophische en Theologische Faculteiten der Nederlandsche Jezuïeten* **4** (1941) 166–191

Stroup, Alice, 'Christian Huygens et l'Académie royale des sciences', *La vie des sciences* **13** (1996) 333–341

Struik, Dirk J., *The land of Stevin and Huygens – A sketch of science and technology in the Dutch Republic during the Golden Century* (Dordrecht, Reidel, 1981)

Tannery, Paul, 'À propos de la correspondence de Huygens', pp. 83–90 in: *La correspondance de Descartes dans les inédits du fonds libre, étudiée pour l'histoire des mathématiques* (Paris, Gautier-Villard, 1893)

Uylenbroek, Pieter J., *Oratio de fratribus Christiano atque Constantino Hugenio, artis dioptricoe cultoribus* [. . .] ('s Gravenhage, Staatsdrukkerij, 1840)

Van Geer, Pieter, 'De briefwisseling van Christiaan Huygens', *De Nederlandsche Spectator* **23** (1905) 1–8

Van Geer, P., 'Christiaan Huygens' leerjaren', *Tijdspiegel* **3** (1906) 1–22; – 'reis- en studiejaren', *Tijdspiegel* **1** (1907) 37–64; – 'verblijf te Parijs', *Tijdspiegel* **3** (1907) 24–46; – 'en Isaac Newton', *Tijdspiegel* **3** (1907) 24–46; '– en Gottfried Wilhelm Leibniz', *Tijdspiegel* **1** (1908) 17–42; – 'laatste levensjaren', *Tijdspiegel* **1** (1908) 367–395

Van Maanen, Johannes Arnoldus, 'Hendrick van Heuraet (1634–1660?): His life and mathematical work', *Centaurus* **27** (1984) 218–279 [contains an unpublished letter of Christiaan Huygens]

Van Maanen, J. A., 'Unknown manuscript material of Christiaan Huygens', *Historia Mathematicae* **12** (1985) 60–65

Van Vloten, Johannes, 'Kristiaan Huygens en Spinoza', *De Levensbode* **3** (1869) 252–254

Vermij, Rienk, H., *Huygens – De mathematisering van de werkelijkheid: Wetenschappelijke biografie* (Diemen, Veen, 2004)

Vermij, Rienk H. & Van Maanen, Johannes Arnoldus, 'An unpublished autograph by Christiaan Huygens: His letter to David Gregory of 19 January 1694', *Annals of Science* **49** (1992) 507–523

Vollgraff, Johan A., 'Christiaan Huygens (1629–1695) et Jean le Rond d'Alembert (1715–1783)', *Janus* **20** (1915) 269–313

Vollgraff, J. A., 'Christiaan Huygens: eenige citaten en beschouwingen naar aanleiding van den driehonderdsten geboortedag', *Christiaan Huygens: Internationaal Mathematisch Tijdschrift* **7** (1928/29) 181–191

Vollgraff, J. A., 'Deux pages consecutives du manuscrit G de Chr. Huygens', *Janus* **44** (1940) 10–23

Vollgraff, J. A., 'Christiaan (ou Chistiaen) Huygens 1629–1695', *Archives internationales d'histoire des sciences* **28** (1948) 165–179

Vollgraff, J. A., 'Deux lettres de Christiaan Huygens', *Archives internationales d'histoire des sciences* **16** (1951) 634–637

Vollgraff, J. A., 'Een en ander over Christiaan Huygens', pp. 42–52 in: *Verslag over 1956 van het Provinciaal Utrechts Genootschap van Kunsten en Wetenschappen* (Utrecht, Kemink, 1956)

Vollgraff, J. A., 'Waar overleed Christiaan Huygens?', pp. 22–26 in: *Jaarboek van 'Die Haghe' 1956* ('s Gravenhage, Trio, 1956)

Wieleitner, Heinrich, 'Forschung und Schule: Christian Huygens zu seinem 300. Geburtstag', *Unterrichtsblätter für Mathematik und Naturwissenschaft* **35** (1929) 107–117

Wolf, Emil, 'The life and works of Christiaan Huygens', pp. 3–17 in: H. Blok, H. A. Ferwerda & H. H. Kuiken (eds.), *Huygens' Principle 1690–1990 – Theory and Applications* (Amsterdam, North-Holland, 1992)

Yoder, Joella G., 'Christiaan Huygens' Great Treasure', *Tractrix* **3** (1991) 1–13

Yoder, J. G., 'The archives of Christiaan Huygens and his editors', pp 91–107 in: M. Hunter (ed.), *Archives of the Scientific Revolution and Exchange of Ideas in Seventeenth-Century Europe* (Woodbridge, Boydell Press, 1998)

Yoder, J. G., 'The letters of Christiaan Huygens', *Revue d'histoire des sciences* **56** (2003) 135–143

Zedler, Johann Heinrich (ed.), 'Huygens', p. 1109 in: *Grosses volständiges Universal-Lexicon aller Wissenschaften und Künste*, vol. 13 (Leipzig, 1739)

ON HIS WORKS

Astronomy

Ashbrook, Joseph, 'Astronomical scrapbook: The long night of selenography', *Sky and Telescope* **29** (1965) 92–94

Ashbrook, J., 'Astronomical scrapbook: the visual Orion nebula', *Sky and Telescope* **50** (1975) 299–301

Bednarczyk, Andrzej, 'Z dziejów idei zycia we wszechswiecie: Epoka Oswiecenia (Fontenelle, Huygens, Kant)', *Kwartalnik historii nauki i techniki* **40/3** (1995) 7–48

Bogazzi, Riccardo, 'Il *Kosmotheoros* di Christiaan Huygens', *Physis* **19** (1977) 87–109

Chapman, Allan, 'Christiaan Huygens (1629–1695): astronomer and mathematician–mechanician', *Endeavour* **19** (1995) 140–145

Cohen, I. Bernard, 'Perfect numbers in the Copernican system: Rheticus and Huygens', pp. 419–425 in: E. Hilfstein (ed.), *Science and History: Studies in Honor of Edward Rosen* (Wrocław, Ossolineum, 1978)

Débarbat, Suzanne, 'A la rencontre de Christiaan Huygens', *Bulletin de la Société astronomique de France* **112** (1998) 148–151

Defossez, Leopold, 'Le planétaire de Christiaan Huygens', *Journal suisse de l'horlogerie* **72** (1947) 404–416

Dekker, Elly, 'Sterrenkunde in de zeventiende eeuw', pp. 84–100 in: *De zeventiende eeuw*, vol. 2 (Hilversum, Verloren, 1986)

De Sitter, Willem & Stokley, James, 'The date of Huygens' announcement of the discovery of the rings of Saturn', *Journal of the British Astronomical Association* **42** (1932/33) 223–224

Dubois, Pierre, *Histoire et traité de l'horlogerie* [avec une traduction du] *Descriptio automatii planetarii* (Paris, 1849)

Dugas, René, 'Huygens devant le système du monde, entre Descartes et Newton', *Comptes rendus hebdomadaires des scéances de l'Académie des sciences* **237** (1953) 1477–1478

Goldstein, S. J., Jr, 'Christiaan Huygens' measurement of the distance to the sun', *The Observatory* **105** (1985) 32–33

Harting, Pieter, Kaiser, Friedrich & Bosscha, Johannes, *On the discovery of the satellite of Saturn by Christiaan Huygens*, Rapport fait à l'Académie Royale des Sciences des Pays-Bas, Section Physique, au 25 janvier 1868

Havinga, Egbertus, Van Wijk, Walter Emile & D'Aumerie, J. F. M. G., *Planetarium-boek Eise Eisinga* [met een vertaling van] *Christiaan Huygens' Automaton planetarii* (Arnhem, Van Loghum Slaterus, 1928)

Herczeg, Tibor, 'The Orion nebula: A chapter of early nebular studies', *Acta Historica Astronomiae* **3** (1998) 246–258

Jozeau, M. F. & Hallez, Maryvonne, 'Planétarium et fractions continues: Lecture d'une texte de Huygens', pp. 164–181 in: *Histoire et epistémologie des mathématiques* (Strasbourg, Université Louis Pasteur, 1988)

Kaiser, Frederik, 'Iets over de sterrekundige waarnemingen van Christiaan Huygens, naar aanleiding van zijne onuitgegeven handschriften', *Tijdschrift voor de wis- en natuurkundige wetenschappen, uitgegeven door de eerste klasse van het Koninklijk-Nederlandsch Instituut van Wetenschappen, Letterkunde en Schoone kunsten* **1** (1848) 7–24

Kaiser, F., 'De stelling van Otto Struve, omtrent het breeder worden van den ring van Saturnus, getoetst aan de handschriften van Huygens en de naauwkeurigheid der latere waarnemingen', *Verslagen en Mededeelingen der Koninklijke Akademie van Wetenschappen, Afd. Natuurkunde* **3** (1855) 186–232

Kaiser, F., 'Briefwechsel zwischen Herrn Staatsrath Otto Struve, Astronom an der Sternwarte zu Pulkowa, und F. Kaiser, Direktor der Sternwarte in Leiden, über die Änderungen in den Dimensionen des Saturnringes', *Verslagen en Mededeelingen der Koninklijke Akademie van Wetenschappen, Afd. Natuurkunde* **5** (1857) 150–172

King, Henry C. & Milburn, John R., *Geared to the Stars: The Evolution of Planetariums, Orreries, and Astronomical Clocks* (Bristol, Adam Hilger, 1978)

Knight, David M., 'Celestial worlds discover'd', *Durham University Journal* **58** (1965) 23–39

Mascart, Jean, *La découverte de l'anneau de Saturne par Huygens* (Paris, Gautiers-Villar, 1907)

Minnaert, Marcel Gilles Jozef, 'De briefwisseling tussen Rømer en Huygens betreffende de snelheid van het licht', *Hemel en Dampkring* **44** (1946) 51–57

Minnaert, M. G. J., 'Galilei and Huygens', pp. 362–376 in: C. Maccagni (ed.), *Saggi su Galileo Galilei*, vol. 1 (Firenze, Barbèra, 1972)

Moll, Gerrit, 'Schreiben des Herrn Professors und Ritters Moll an den Herausgeber', *Astronomische Nachrichten* **10** (1832) 201–208 [on the transit of Mercury across the Sun on 3 May 1661, observed by Christiaan Huygens in London]

Nijland, Albert A., 'De drievoudige conjunctie van Jupiter en Saturnus in 1683', *Hemel en Dampkring* **18** (1921) 142

Nijland, A. A., 'Christiaan Huygens, in het bijzonder als astronoom', *Christiaan Huygens: Internationaal mathematisch tijdschrift* **7** (1929) 192–208

Olmsted, J. W., 'The "application" of telescopes: 1667 or 1668?', *Sky and Telescope* **8** (1948) 7 & 19

Pedersen, Kurt Møller, 'Jacques Cassini, James Gregory, Christiaan Huygens et leur déterminations de la distance entre Sirius et la Terre', pp. 137–154 in: P. Brouzeng & S. Débarbat (eds.), *Sur les traces des Cassini* (Paris, Éditions du CTHS, 2001)

Radelet-de Grave, Patricia, 'L'univers selon Huygens, le connu et l'imaginé', *Revue d'histoire des sciences* **56** (2003) 79–112

Reverchon, Léopold & Ditisheim, Paul, 'La machine planétaire et l'oeuvre astronomique de Huygens', *Bulletin de la Société astronomique de France* (1950) 57–76

Shapley, D., 'Pre-Huygenian observations of Saturn's ring', *Isis* **40** (1949) 12–17

Van de Sande Bakhuyzen, Hendrik Gerard, 'Christiaan Huygens als sterrekundige', *Hemel en Dampkring* **27** (1929) 113–121 & 162–170

Van Helden, Albert, 'The Accademia del Cimento and Saturn's ring', *Physis* **15** (1973) 237–259

Van Helden, A., 'Saturn and his anses', *Journal for the History of Astronomy* **5** (1974) 105–121

Van Helden, A., '*Annulo cingitur*: The solution of the problem of Saturn', *Journal for the History of Astronomy* **5** (1974) 155–174

Van Helden, A., 'A note about Christiaan Huygens's De Saturni luna observatio nova', *Janus* **62** (1975) 13–15

Van Helden, A., 'Saturn through the telescope: a brief historical survey', pp. 23–43 in: T. Gehrels & M. S. Matthews (eds.), *Saturn* (Tucson, University of Arizona Press, 1984)

Van Helden, A., 'Rings in astronomy and cosmology, 1600–1900', pp. 12–22 in: R. Greenberg & A. Brahic (eds.), *Planetary Rings* (Tucson, University of Arizona Press, 1984)

Van Helden, Albert & Righini Bonelli, L., 'Divini and Campani: a forgotten chapter in the history of the Accademia del Cimento', *Annali dell'Instituto e Museo di Storia della scienza di Firenze* **6** (1981) 3–176

Wolf, Rudolf, 'Über die Nebelfleck im Orion: Huygens ist nicht der erste Entdecker des grossen Nebels im Orion, sondern der Schweizer Johann Baptist Cysat', *Astronomische Nachrichten* **38** (1854) 109–110

Horology

Ariotti, Piero E., 'Aspects of the conception and development of the pendulum in the 17th century', *Archive for History of Exact Sciences* **8** (1971/72) 329–410

Baillie, Granville Hugh, 'Huygens' pendulum clock', *Nature* **148** (1941) 412–415

Ball, M. F., 'Galileo Galilei and Christiaan Huygens: Addendum', *Antiquarian Horology and the Proceedings of the Antiquarian Horological Society* **15** (1984/85) 373–374 [cf. Dobson (1984/85)]

Barbin, E., 'Le secret des longitudes et le pendule cycloidal de Huygens', pp. 143–163 in: *Histoire et epistémologie des mathématiques: Les mathématiques dans la culture d'une époque* (Strasbourg, Université Louis Pasteur, 1988)

Bell, Arthur E., 'The *Horologium oscillatorium* of Christian Huygens', *Nature* **148** (1941) 245–248

Bennett, Matthew, Schatz, Michael F., Rockwood, Heidi & Wiesenfeld, Kurt, 'Huygens' clocks', *Proceedings of the Royal Society of London*, series A, **458** (2002) 563–579

Blackwell, Richard J., *Christiaan Huygens'* [treatise about] *the Pendulum Clock, or Geometrical Demonstration concerning the Motion of Pendula as Applied to Clocks* (Ames, Iowa State University Press, 1986)

Bosmans, Henri, 'Galilée ou Huygens? A propos d'un épisode de la première application du pendule aux horloges', *Revue des questions scientifiques*, série 3, **22** (1912) 573–586

Bruna, P. P., 'Christiaen Huygens' tabel voor de tijdsvereffening', *Hemel en Dampkring* **33** (1935) 56–59

Chareix, Fabien, 'Vaincre la houle: les horloges maritimes de Christiaan Huygens', pp. 169–202 in: V. Julien (ed.), *Le calcul des longitudes: Enjeu pour les mathématiques, l'astronomie, la mesure du temps et la navigation* (Rennes, Presses Universitaires de Rennes, 2002)

Crommelin, Claude A., 'De isochrone conische slinger van Christiaan Huygens', *Physica* **11** (1931) 359–364

Crommelin, C. A., 'De klok van Huygens met den isochronen conischen slinger', *Nederlandsch Tijdschrift voor Natuurkunde* **3** (1936) 273–280

Crommelin, C. A., 'Het uurwerk met den balansslinger van Christiaan Huygens', *Nederlandsch Tijdschrift voor Natuurkunde* **4** (1937) 172–180

Crommelin, C. A., 'Pendulum cylindricum trichordon van Christiaan Huygens', *Nederlandsch Tijdschrift voor Natuurkunde* **5** (1938) 314–318

Crommelin, C. A., 'Uurwerken van Christiaan Huygens uit zijn laatste levensjaren', *Nederlandsch Tijdschrift voor Natuurkunde* **7** (1940) 321–328

Crommelin, C. A., 'Les horloges de Christiaan Huygens', *Journal suisse d'horlogerie* **72** (1947) 189–204

Crommelin, C. A., 'The clocks of Christiaan Huygens', *Endeavour* **9** (1950) 64–69

Crommelin, C. A., 'Huygens' pendulum experiments, successful and unsuccessful', pp. 23–27 in: *Huygens' Tercentenary Exhibition at the Science Museum* (London, Antiquarian Horological Society, 1957)

Crommelin, C. A., 'Christiaan Huygens' invention of the pendulum clock, three hundred years ago', *Janus* **46** (1957) 79–80

De Fleury, P., 'Note sur une horloge à pendule régulateur construite à Angoulême quatorze ans avant la naissance d'Huygens', *Bulletin et mémoires de la Société archéologique et historique de la Charente*, série 6, **2** (1892)

Defossez, Léopold, *Les Savants du XVIIe siècle et la mesure du temps* (Lausanne, Edition du Journal suisse d'horlogerie et de bijouterie, 1946)

De Kock, Adrianus C., 'De uitvinding van het slingeruurwerk', *Hemel en Dampkring* **31** (1933) 393–404 & 417–429

Ditisheim, Paul, *Le Spiral réglant et le balancier depuis Huygens jusqu'à nos jours* (Lausanne, Edition du Journal suisse d'horlogerie et de bijouterie, 1945)

Dobson, Richard D., 'The development of the pendulum clock 1656–1659', *Antiquarian Horology and the Proceedings of the Antiquarian Horological Society* **13** (1982) 270–281

Dobson, R. D., 'Galileo Galiei and Christiaan Huygens', *Antiquarian Horology and the Proceedings of the Antiquarian Horological Society* **15** (1984/85) 261–270

Dobson, R. D., *De slinger als tijdmeter: Een nieuwe visie op de ontwikkeling van de slinger als tijdmeter in de periode 1602–1660: Galileo Galilei, Ahasuerus, Fromanteel, Christiaan Huygens* (Aalten, Achterland Verlagscompagnie, 1999)

Edwardes, Ernest L., *The Story of the Pendulum Clock* (Altrincham, John Sherrat, 1977)

Edwardes, E. L., 'The suspended foliot and new light on early pendulum clocks', *Antiquarian Horology and the Proceedings of the Antiquarian Horological Society* **12** (1981) 614–634

Ellery, R. L. J., 'The results of some experiments with Huygens' parabolic pendulum for obtaining uniform rotation', *Monthly Notices of the Royal Astronomical Society* **36** (1875) 72–76

Favaro, Antonio, 'Galileo Galilei e Christiano Huygens: Nuovi documenti sull'applicazione del pendolo all'orologio', *Rivista di fisica, matematica e scienze naturali* **13** (1912) 3–20

Favaro, A., 'Scampoli Galileiani 23–147: Galilei oppure Huygens?', *Atti e memorie della Accademia di Scienze, Lettere ed Arti in Padova* **30** (1913/14) 61–66

Feldhaus, Gilbert W., 'Christian Huygens', *Deutsche Uhrmacher Zeitung* **53** (1929) 273–275

Gerland, Ernst, 'Zur Geschichte der Erfindung der Penduluhr', *Annalen der Physik*, Folge 3, **4** (1878) 585–613

Gerland, E., 'Die Erfindung der Penduluhr', *Zeitschrift für Instrumentenkunde* **8** (1988) 77–83

Gould, Rupert T., *The Marine Chronometer: Its History and Development* (London, Holland Press, 1923)

Günther, Siegmund, 'Geschichte der Penduluhr vor Huygens', *Sitzungsberichte der Physikalisch-medizinischen Societät zu Erlangen* **6** (1873) 12–27

Günther, S. 'Quellenmässige Darstellung der Erfindungsgeschichte der Penduluhr bis auf Huygens', pp. 308–344 in: S. Günther, *Vermischte Untersuchungen zur Geschichte der mathematischen Wissenschaften* (Leipzig, Teubner Verlag, 1876)

Heckscher, A. & Von Öttingen, Arthur (eds.), *Christiaan Huygens: Die Pendeluhr* [Horologium oscillatorium] (Leipzig, Wilhelm Engelmann Verlag, 1913)

Jonkers, Art Roeland Theo, 'Finding longitude at sea: early attempts in Dutch navigation', pp. 186–197 in: L. Palm (ed.), *De zeventiende eeuw*, vol. 12 (Hilversum, Verloren, 1996)

Kluiver, J. H., 'De ontwikkeling van de vormgeving van het Nederlands uurwerk als gevolg van Huygens' uitvinding van het slingeruurwerk in 1657', pp. 141–150 in: L. Palm (ed.), *De zeventiende eeuw*, vol. 12 (Hilversum, Verloren, 1996)

Korteweg, Diederik Johannes, 'Huygens' sympatische uurwerken en verwante verschijnselen, in verband met de principale en de samengestelde slingeringen die zich voordoen wanneer aan een mechanisme met een enkele vrijheidsgraad twee slingers bevestigd worden', *Verslagen van de Gewone Vergaderingen der Wis- en Natuurkundige Afdeeling van de Koninklijke Akademie van Wetenschappen te Amsterdam* **13/14** (1905) 413–432

Leopold, Jan Hendrik, 'L'invention par Christiaan Huygens du ressort spiral réglant pour les montres', pp. 153–157 in: R. Taton (ed.), *Huygens et la France* (Paris, Vrin, 1982)

Leopold, J. H., 'Christiaan Huygens, the Royal Society and Horology', *Antiquarian Horology* **21** (1993) 37–42

Mahoney, Michael S., 'Huygens and the pendulum: from device to mathematical relation', pp. 17–39 in: E. R. Grosholz & H. Breger (eds.), *The Growth of Mathematical Knowledge* (Dordrecht, Kluwer, 2000)

Mesnage, Pierre, 'Les inventions d'Huygens', pp. 161–175 in: *Les chefs-d'oeuvre de l'horlogerie* (Paris, Edition de la revue française des bijoutiers–horlogiers, 1949)

Mongruel, A., 'Christiaan Huygens en enige Franse klokken uit zijn tijd', *Vakblad voor Uurwerkmakers* **60** (1949) 27–29

Pighetti, Clelia (ed.), *Christiano Huygens: Horologium oscillatorium & Traité de la lumière* (Firenze, G. Barbèra, 1963)

Plomp, Reinier, 'The Dutch origin of the French pendulum clock: what we learn from Christiaan Huygens' correspondence', *Antiquarian Horology and the Proceedings of the Antiquarian Horological Society* **8** (1971/72) 24–41

Plomp, R., 'Christiaan Huygens: Man van de klok', *Spiegel Historiael* **14** (1979) 234–241

Plomp, R., 'A longitude timekeeper by Isaac Thuret with the balance spring invented by Christiaan Huygens', *Annals of Science* **56** (1999) 379–394

Reverchon, Léopold, 'Huygens, horloger', *Revue générale des sciences pures et appliquées* **27** (1916) 105–112

Robertson, John Drummond, *The Evolution of Clockwork, with a Special Section on the Clocks of Japan, Together with a Comprehensive Bibliography of Horology* (London, Cassell, 1931)

Schliesser, Eric, 'Van Nierops paskaart van Europa in een rapport van Christiaan Huygens', *Caert-Tresoor* **16** (1997) 93–96

Schliesser, E., 'Een kaart van Visscher en de globe van Blaeu in de bewijsvoering van een proef met slingeruurwerken van Huygens', *Caert-Tresoor* **19** (2000) 51–55

Schliesser, Eric & Smith, George E., 'Huygens's 1688 report to the directors of the Dutch East India Company on the measurement of longitude at sea and its implications for the non-uniformity of gravity', pp. 198–214 in: L. Palm (ed.), *De zeventiende eeuw*, vol. 12 (Hilversum, Verloren, 1996)

Servaas van Rooyen, A. J., 'Een mededinger van Christiaan Huygens', *Album der Natuur: Een werk ter verspreiding van natuurkennis* (1884) 25–31

Speiser, David, 'Le "Horologium oscillatorium" de Huygens et les "Principia"', *Revue philosophique de Louvain* **86** (1988) 485–504

Tyler, E. J., *Three Hundred Years of Pendulum Clocks: The Huygens Exhibition at the Science Museum* (London, Horological Journal, 1957)

Van Gent, Robert Harry & Leopold, Jan Hendrik, *De tijdmeters van de Leidse Sterrewacht* (Leiden, Museum Boerhaave, 1992) [pp. 12–14 on a timekeeper by Isaac Thuret, once owned by Christiaan Huygens]

Van Swinden, Jan Hendrik, 'Verhandeling over Huijgens, als uitvinder der slingeruurwerken', *Verhandelingen der eerste klasse van het Koninklijk-Nederlandsche Instituut van Wetenschappen, Letterkunde en Schoone Kunsten te Amsterdam* **3** (1817) 27–168

Vollgraff, Johan A., 'Heeft Vincenzio Galilei op zijn sterfdag zijne uurwerken vernield?', *Hemel en Dampkring* **32** (1934) 5–8

Vollgraff, J. A., 'Heeft Prins Leopold gezegd dat in 1656 te Florence een slingeruurwerk is geconstrueerd?', *Hemel en Dampkring* **32** (1934) 56–63

Vollgraff, J. A., 'Chr. Huygens en het ankeréchappement', *Hemel en Dampkring* **32** (1934) 90–91

Vollgraff, J. A., 'Over het slingeruurwerk van Huygens', pp. 107–110 in: *Handelingen van het 25ᵉ Nederlandsch Natuur- en Geneeskundig Congres* (Haarlem, Ruijgrok, 1935)

Vollgraff, J. A., 'Het zee-horologie van Christiaan Huygens', pp. 59–64 in: *Jaarverslag van de Vereeniging Nederlandsch Historisch Scheepvaartmuseum te Amsterdam* (1935/36)

Voorbeijtel Cannenburg, Willem, 'Het zee-horologie van Christiaan Huygens', *De Zee* **5** (1936) 238–245

Yoder, Joella G., 'Christiaan Huygens' theory of evolutes: the background to the *Horologium oscillatorium*', Doctoral thesis, University of Wisconsin, Madison (1985)

Instruments

Andriesse, Cornelis Dirk, 'Das Unterste zuoberst: Vakuum, Luftdruck und Luftpumpe bei Christiaan Huygens', *Monumenta Guerickiana [51]* **6** (1999) 24–29

Ariotto, Piero E., 'Christiaan Huygens: aviation pioneer extraordinary', *Annals of Science* **36** (1979) 611–624

Ashbrook, Joseph, 'Astronomical scrapbook: some Huygens' telescopes', *Sky and Telescope* **18** (1959) 559–560

Bacchus, P., 'L'oculair de Huygens', *L'Astronomie: Bulletin de la Société astronomique de France* **112** (1998) 130–133

Bacchus, P., Biraud, F., Bottard, R., Daversin, B., Dollfus, A., Farroni, G., Fort, J., Lecleire, J.-M., Moathy, P., Philippon, G. & Thiot, A., 'Reconstution de la lunette "Astroscope" de Huygens: Une convergence des bénévoles', *L'Astronomie: Bulletin de la Société astronomique de France* **112** (1998) 134–138

Bedini, Silvio A., 'Lens making for scientific instrumentation in the seventeenth century', *Applied Optics* **5** (1966) 687–694

Bedini, Silvio A., 'The aerial telescope', *Technology and Culture* **8** (1967) 395–401

Christiani Hugenii, '*Astroscopia compendaria, Tubi optici molimine liberate*, Or the description of an aerial telescope', *Philosophical Transactions* **14** (1684) 668–670

Crommelin, Claude A., *Het lenzen slijpen in de 17de eeuw* (Paris, Amsterdam, 1929)

Dollfus, Audouin, 'Christiaan Huygens et la lunette sans tuyau "Astroscope" ', pp. 127–140 in: L. Palm (ed.), *De zeventiende eeuw*, vol. 12 (Hilversum, Verloren, 1996)

Dollfus, A., 'Les frères Huygens et les grandes lunettes sans tuyau', *L'Astronomie: Bulletin de la Société astronomique de France* **112** (1998) 114–129

Dollfus, A., 'La lunette de Huygens reconstitute, premières observations', *L'Astronomie: Bulletin de la Société astronomique de France* **112** (1998) 139–146

Dobson, Richard D., 'Huygens, the secret in the Coster–Fromanteel contract, the thirty-hour clock', *Antiquarian Horology and the Proceedings of the Antiquarian Horological Society* **12** (1980/81) 192–196

Edwardes, Ernest L. & Dobson, Richard D., 'The Fromanteels and the pendulum clock', *Antiquarian Horology and the Proceedings of the Antiquarian Horological Society* **14** (1983) 250–265

Fournier, Marian, 'Huygens' design for a simple microscope', *Annals of Science* **46** (1989) 575–596

Fournier, M., *Early Microscopes: A Descriptive Catalogue* (Leiden, Museum Boerhaave, 2003)

Gregory, David, *Elements of Catoptrics and Dioptrics* [with] *A Particular Account of Microscopes and Telescopes from Mr. Huygens* (London, Curll & Pemberton, 1715)

Harting, Pieter, 'Oude optische werktuigen, toegeschreven aan Zacharias Janssen, en een beroemde lens van Christiaan Huygens teruggevonden', *Album der Natuur: Een werk ter verspreiding van natuurkennis* (1867) 313

Hasimoto, Takehiko, 'Huygens, dioptrics, and the improvement of the telescope', *Historia Scientiarum: International Journal of the History of Science Society of Japan* **37** (1989) 51–90

[H.], 'Huyghenian eyepiece diaphragms', *English Mechanic and World of Science* **70** (1900) 536

Jansen, P., 'Les carrosses à cinq sols de Christian Huygens', *Revue d'histoire des sciences* **4** (1951) 171–172

Kaiser, Frederik, 'Iets over de kijkers van de gebroeders Christiaan en Constantijn Huygens', *Verslagen en Mededeelingen [van het] Koninklijk Nederlandsche Instituut van Wetenschappen, Letterkunde en Schoone Kunsten* 6 (1846) 396–429

Kaiser, F., 'On Huygens's telescope lenses in relation to the Mars-rotation', *Astronomische Nachrichten* 25 (1847) 245–250

Michel, Henri, 'Sur trois objectifs de Huygens', *Ciel et Terre* 67 (1951) 128–130

Mills, Allan A. & Jones, M. L., 'Three lenses by Constantine Huygens in the possession of the Royal Society of London', *Annals of Science* 46 (1989) 173–182

Monaco, Giuseppe, 'Sulle prime livelle a cannocchiale', *Rivista di storia della scienza* 3 (1986) 385–407

Nijland, Albert A., 'Eine Prüfung alter Fernröhrobjektive van Huygens und Campani', *Sirius* 32 (1899) 277–280

Nijland, A. A., 'Over eenige lenzen van Christiaan Huygens', *Hemel en Dampkring* 20 (1922) 241–246

Oudemans, Jean Abraham Chrétien, 'Über die Erfinder des negativen Oculars', *Astronomische Nachrichten* 95 (1879) 323–330

Oudemans, J. A. C., 'Over het vermogen van den 10-voets kijker van Huygens', *Verslagen en Mededeelingen der Koninklijke Akademie van Wetenschappen, Afd. Natuurkunde*, tweede serie, 20 (1884) 290–296

Raveau, C., 'Sur l'histoire des procédés mis en oeuvre par Foucault pour l'étude des miroirs et des objectifs', *Journal de Physique*, série 4, 1 (1902) 115

Sampson, Ralph Allen & Conrady, Alexander Eugen, 'On three Huygens lenses in the possession of the Royal Society of London', *Proceedings of the Royal Society of Edinburgh* 49 (1928/29) 289–299

Schneider, Ivo, 'Technik in der Sicht der exakten Naturwissenschaften am Beispiel von Archimedes, Christiaan Huygens und Carl Friedrich Gauss', *Kultur und Technik* 6 (1982) 21–24

Sparnaay, Marcus J., 'De betekenis van Christiaan Huygens voor de ontwikkeling van de vacuümtechniek', *Histechnicon* 16/2 (1990) 16–29

Stroup, Alice, 'Christiaan Huygens and the development of the air pump', *Janus* 68 (1981) 129–158

Terrier, Max, 'L'invention des ressorts de voiture', *Revue d'histoire des sciences* 39 (1986) 17–30

Van Cittert, P. H., 'Een historische lens', *De Natuur* 49 (1929) 76–78

Van Gent, Robert H. & Van Helden, Anne C., *Een vernuftig geleerde: De technische vondsten van Christiaan Huygens* (Leiden, Museum Boerhaave, 1995)

Van Heel, A. C. S., 'Over eenige metingen gedaan aan een objectief van Huygens', *Hemel en Dampkring* 27 (1929) 324–326

Van Heel, A. C. S., 'De "Memorien aengaende het slijpen van glasen to verrekykers" van Christiaan Huygens', *Hemel en Dampkring* **61** (1963) 189–193

Van Helden, Albert, 'Eustachio Divini versus Christiaan Huygens: a reappraisal', *Physis* **12** (1970) 36–50

Van Helden, A., 'The development of compound eyepieces, 1640–1670', *Journal for the History of Astronomy* **8** (1977) 26–37

Van Helden, Anne C. & Van Gent, Robert H., 'The lens production by Christiaan and Constantijn Huygens', *Annals of Science* **56** (1999) 69–79

Van Nooten, Sebastiaan I., 'Contributions of Dutchmen to the early history of film technology', *Janus* **58** (1971) 81–100

Van't Veer, Frans, 'Christiaan Huygens: des lunettes de plus en plus longues', *L'Astronomie: Bulletin de la Société astronomique de France* **112** (1998) 110–113

Vollgraff, Johan A., 'De rol van den Nederlander Caspar Calthoff bij de uitvinding van het moderne stoomwerktuig', *Physica* **12** (1932) 257–268

Wagenaar, Willem A., 'The true inventor of the magic lantern: Kircher, Walgenstein, or Huygens?', *Janus* **66** (1979) 193–207

Wolf, Rudolf, 'Die Verbesserung der Instrumente durch Tycho, Bürgi, Morin, Gascoigne, Picard, Vernier, Thévenot und Huygens', *Vierteljahrsschrift der Naturforschende Gesellschaft in Zürich* (1873) 108–126

Mathematics

Arnol'd, Vladimir I., *Huygens and Barrow, Newton and Hooke: Pioneers in Mathematical Analysis and Catastrophe Theory, from Evolvents to Quasi-Crystals* (Basel, Birkhäuser Verlag, 1990)

Bos, Hendrik J. M., 'L'élaboration du calcul infinitesimal: Huygens entre Pascal et Leibniz', pp. 115–121 in: R. Taton (ed.), *Huygens et la France* (Paris, Vrin, 1982)

Bos, H. J. M., 'Recognition and wonder: Huygens, tractional motion and some thoughts on the history of mathematics', *Euclides* **63** (1987) 65–76 & *Tractrix* **1** (1989) 3–20

Bosmans, Henri, 'Sur un point de l'histoire du calcul des probabilités (Pascal et Huygens)', *Annales de la Société scientifique de Bruxelles* **43** (1924) 318–326

Bottema, O., 'Verscheidenheden LXVII: Frans van Schooten aan Christiaan Huygens', *Euclides* **42** (1966/67) 204–208 & **43** (1967/68) 164–166

Boyer, Carl B., 'Note on an early graph of statistical data (Huygens 1669)', *Isis* **37** (1947) 148–149

Bruins, Evert M., 'On curves and surfaces in the 17th–19th century', *Physis* **12** (1970) 221–236

Bruins, E. M., 'Computation of logarithms by Huygens', *Janus* **65** (1978) 97–104

Bruins, E. M., 'On the history of logarithms: Bürgi, Napier, Briggs, De Decker, Vlacq, Huygens', *Janus* **67** (1980) 241–260

Conte, Luigi, 'Il "De circuli magnitudine inventa" di Christiano Huygens', *Archimède* 9 (1957) 140–142, 224–227, 267–270 & 10 (1958) 44–47

Crommelin, Claude A., 'Sur l'attitude de Huygens envers le calcul infinitesimal et sur deux courbes intéressantes du même savant', *Simon Stevin* 31 (1956) 5–18

Dijksterhuis, Eduard Jan, 'De ontdekking van het tautochronisme der cycloïdale valbeweging', *Euclides* 5 (1928/29) 193–227

Dupont, Pascal & Roero, Carla Silvia (eds.), *Il trattato 'De ratiociniis in ludo aleae' di Christiaan Huygens con le 'Annotationes' di Jacob Bernoulli 'Ars conjectandi'* (Torino, Academia delle Scienze, 1984)

Fenaroli, Giuseppina & Penco, Maria A., 'Su alcune lettere di Christiaan Huygens in relazione al suo "De ratiociniis in ludo aleae"', *Physis* 21 (1979) 351–356

Fenaroli, G. & Penco, M. A., 'Le prime analisi di problema di mortalità in terme probabilistici', *Physis* 23 (1981) 115–134

Fenaroli, G. & Garibaldi, Ubaldo & Penco, M. A., 'Foundations of Huygens' theory of probability', *Epistemologia* 6 (1983) 293–322

Freudenthal, Hans, 'Huygens' foundations of probability', *Historia Mathematica* 7 (1980) 113–117

Grimm, G., 'Zur Kreisberechnung von Huygens', *Elemente der Mathematiken* 4 (1949) 78–85

Hadamard, J., 'Le principe de Huyghens pour les équations à trois variables indépendantes', *Journal de mathématiques pures et appliquées*, 9ᵉ série, 8 (1929) 197–228

Hallez, Maryvonne, 'Teaching Huygens in the rue Huygens: introducing the history of 17th-century mathematics in a junior secondary school', *Science and Education* 1 (1992) 313–328

Hofmann, Joseph E., 'Über die ersten logarithmischen Rektifikationen: Eine historisch-kritische Studie in vergleichender Darstellung', *Deutsche Mathematik* 6 (1941) 283–304

Hofmann, J. E., 'Neues über die näherungsweise Kreisquadratur bei Huygens (1654)', *Der mathematisch-naturwissenschaftliche Unterricht* 4 (1952) 321–323

Hofmann, J. E., 'Über die Quadrisectio trianguli', *Mathematische Zeitschrift* 74 (1960) 105–118

Hofmann, J. E., 'Über die Kreismessung von Chr. Huygens, ihre Vorgeschichte, ihre Inhalt, ihre Bedeutung und ihr Nachwirken', *Archive for the History of Exact Sciences* 3 (1966) 102–136

Hofmann, J. E. & Hofmann, Josepha, 'Erste Quadratur der Kissoide', *Deutsche Mathematik* 5 (1940) 571–584

Kleijne, Wim, *Christiaan Huygens: Van rekeningh in spelen van geluck* (Utrecht, Epsilon Uitgaven, 1998)

Kiessling, H., *Chr. Huygens De circuli magnitudine inventa: Als ein Beitrag zur Lehre vom Kreise für die Lehrbücher elementar entwickelt* (Flensburg, 1868)

Korteweg, Diederik Johannes, 'La solution de Christiaan Huygens du problème de la chaînette', *Bibliotheca Mathematica: Zeitschrift für Geschichte der mathematischen Wissenschaften*, serie 3, 1 (1900) 97–108

Loria, Gino, 'Curve piane speciali nel carteggio di Huygens', *Bibliotheca Mathematica: Zeitschrift für Geschichte der mathematischen Wissenschaften*, serie 3, **7** (1906) 270–281

Meusnier, N., 'Huygens – De Witt: Un modèle mathématique de calcul de la valeur des événéments uncertains', pp. 192–205 in: *Les mathématiques dans la culture d'une époque* (Strasbourg, Université Louis Pasteur, 1988)

O'Hara, James G., 'Huygens, Leibniz and the "petit demon": agreement and dissension in their mathematical correspondence', pp. 151–160 in: L. Palm (ed.), *De zeventiende eeuw*, vol. 12 (Hilversum, Verloren, 1996)

Parrochia, Daniel, 'Optique, mécanique et calcul des chances chez Huygens et Spinoza (sur quelques paradigms possible du discourse philosophique)', *Dialectica* **38** (1984) 319–345

Riersøl, Olav, 'Notes on some propositions of Huygens in the calculus of probability', *Nordisk Matematisk Tidskrift* **16** (1968) 88–91

Rohrbasser, Jean-Marc & Véron, J., 'Les frères Huygens et "le calcul des ages": l'argument du pari équitable', *Population* **54** (1999) 903–1011

Rudio, F., *Archimedes, Huygens, Lambert, Legendre: 4 Abhandlungen über die Geschichte des Problemes von der Quadratur des Zirkels von den ältesten Zeiten bis auf unsere Tage* (Niederwalluf bei Wiesbaden, Sändig, 1971)

Schneider, Ivo, 'Christiaan Huygens's contribution to the development of a calculus of probabilities', *Janus* **67** (1980) 269–279

Schneider, I., 'Christiaan Huygens' non-probabilistic approach to a calculus of games and chance', pp. 171–185 in: L. Palm (ed.), *De zeventiende eeuw*, vol. 12 (Hilversum, Verloren, 1996)

Schuh, Frederick, 'Deux demonstrations dues à Huygens de son théorème concernant des quatre points d'intersection de deux coniques à axes parallèles', *Christiaan Huygens: Internationaal Mathematisch Tijdschrift* **1** (1921/22) 96–101

Schuh, F., 'De eerste uitingen van het genie van Christiaan Huygens', *Christiaan Huygens: Internationaal Mathematisch Tijdschrift* **7** (1928/29) 214–217

Scriba, Christoph J., 'Gregory's converging double sequence: a new look at the controversy between Huygens and Gregory over the "analytical" quadrature of the circle', *Historia Mathematica* **10** (1983) 274–285

Sheynin, O. B., 'Early history of the theory of probability', *Archive for the History of Exact Sciences* **17** (1977) 201–259

Shoesmith, Eddie, 'Expectation and the early probabilists', *Historia Mathematica* **10** (1983) 78–80

Shoesmith, E., 'Huygens' solution to the "gambler's ruin" problem', *Historia Mathematica* **13** (1986) 157–164

Sloth, Flemming, 'Chr. Huygens' rectification of the cycloid', *Centaurus* **12/13** (1968/69) 278–284

Stamhuis, Ida H., 'Christiaan Huygens correspondeert met zijn broer over levensduur: Hoe wetenschappelijke bronnen kunnen ontstaan', pp. 161–170 in: L. Palm (ed.), *De zeventiende eeuw*, vol. 12 (Hilversum, Verloren, 1996)

Van Geer, Pieter, 'Hugeniana Geometrica', *Nieuw Archief voor Wiskunde*, tweede reeks, **7** (1907) 215–226 & 438–454; **8** (1909) 34–63, 145–168, 289–314 & 444–464; **9** (1911) 6–38, 202–230 & 338–358; **10** (1913) 39–60, 178–198 & 370–395

Van Maanen, Johannes Arnoldus, 'Facets of seventeenth century mathematics in the Netherlands', Doctoral thesis, Mathematical Institute, Utrecht (1987)

Vermij, Rienk, Van Dijk, Hanne & Reus, Carolien, *'Wiskunde beter bekeken: Christiaan Huygens* (Utrecht, Epsilon Uitgaven, 2004)

Vollgraff, Johan A., *De kromme van Johann Bernoulli volgens Christiaan Huygens, of zijn en worden in de wiskunde en in het leven* (Lochem, De Tijdstroom, 1945)

White, Colin & Hardy, Robert J., 'Huygens' graph of Graunt's data', *Isis* **61** (1970) 107–108

Yoder, Joella G., 'Following in the footsteps of geometry: the mathematical world of Christiaan Huygens', pp. 83–95 in: L. Palm (ed.), *De zeventiende eeuw*, vol. 12 (Hilversum, Verloren, 1996)

Ziggelaar, August, 'Les premières démonstrations du tautochronisme de la cycloide', *Centaurus* **12** (1967) 21–37

Optics

Albury, William R., 'Halley and the *Traité de la lumière* of Huygens: New light on Halley's relationship with Newton', *Isis* **62** (1971) 445–468

Baker, B. B. & Copson, E. T., *The Mathematical Theory of Huygens' Principle* (Oxford, Clarendon Press, 1939)

Blay, Michel, 'Christiaan Huygens et les phénomènes de la couleur', *Revue d'histoire des sciences* **37** (1984) 127–150

Blay, M. (ed.), *Christiaan Huygens: Traité de la lumière* (Paris, Dunod, 1992)

Bruins, Evert M., 'Problema Alhaseni: at the tercentenary of Huygens' solution', *Centaurus* **13** (1969) 269–277

Buchwald, Jed Z., 'Experimental investigations of double refraction from Huygens to Malus', *Archive for History of Exact Sciences* **21** (1979/80) 311–373

Costabel, Pierre, 'Matière et lumière au XVIIe siècle', *Acta historiae rerum naturalium necnon technicarum* **3** suppl. (1967) 115–130

Costabel, P., 'La propagation de la lumière sans transport de matière de Descartes à Huygens', pp. 83–91 in: R. Taton (ed.), *Rømer et la vitesse de la lumière* (Paris, Vrin, 1976)

Crew, Henry (ed.), *The Wave Theory of Light: Memoirs by Huygens, Young and Fresnel* (New York, American Book Co., 1900)

Crommelin, Claude A., 'Het optische werk van Christiaan Huygens', *Nederlandsch Tijdschrift voor Natuurkunde* **9** (1942) 298–310

Dijksterhuis, Fokko Jan, 'Huygens' dioptrica', pp. 117–126 in: L. Palm (ed.), *De zeventiende eeuw*, vol. 12 (Hilversum, Verloren, 1996)

Dijksterhuis, F. J., 'Christiaan Huygens en de mechanica van het licht', pp. 55–80 in: M. Keestra (ed.), *Doorbraken in de natuurkunde* (Amsterdam, Nieuwezijds, 2001)

Ferraz, Antonio, 'Le *Traité de la lumière* de Huygens comme synthèse historique', pp. 159–164 in: R. Taton (ed.), *Huygens et la France* (Paris, Vrin, 1982)

Klever, Wim, 'Insignis opticus', *De zeventiende eeuw* **6** (1990) 47–63

Lochak, Georges, 'Huygens, Newton et la lumière: La naissance de l'optique au XVIIe siècle', *Revue de Palais de la découverte* **24** no. 40 (1996) 41–51

Marek, Jiří, 'Les notions de la théorie ondulatoire de la lumière chez Grimaldi et Huygens', *Acta historiae rerum naturalium necnon technicarum* **1** (1965) 131–147

Pighetti, Clelia, 'C. Huygens di fronte al problema del colore', pp. 3–8 in: *Actes des VIIe journées internationales de la couleur* (Florence, 1964)

Rosenfeld, Léon, 'Le premier conflit entre la théorie ondulatoire et la théorie corpusculaire de la lumière', *Isis* **11** (1928) 111–122

Rosmorduc, Jean, 'Le modèle de l'éther lumineux dans le *Traité de la lumière* de Huygens', pp. 165–176 in: R. Taton (ed.), *Huygens et la France* (Paris, Vrin, 1982)

Sabra, Abdelhamid L., *Theories of Light from Descartes to Newton* (London, Oldbourne, 1967)

Sakellariadis, Spiros, 'Descartes' experimental proof of the infinite velocity of light and Huygens' rejoinder', *Archive for the History of Exact Sciences* **26** (1982) 1–12

Shapiro, Alan E., 'Kinematic optics: a study of the wave theory of light in the seventeenth century', *Archive for the History of Exact Sciences* **11** (1973) 134–266

Shapiro, A. E., 'Newton and Huygens' explanation of the 22° halo', *Centaurus* **24** (1980) 273–287

Shapiro, A. E., 'Huygens's *Traité de la lumière* and Newton's *Opticks*: Pursuing and eschewing hypotheses', *Notes and Records of the Royal Society of London* **43** (1989) 223

Southall, James, 'Some of Huygens' contributions to dioptrics, with notes', *Journal of the Optical Society of America* **6** (1922) 461–475

Sparberg, Esther B., 'Misinterpretations of theories of light', *American Journal of Physics* **34** (1966) 377–389

Van Broekhoven, R., 'De kleurentheorie van Christiaan Huygens', *Scientiarum Historia* **12** (1970) 143–158

Verdet, Jean-Pierre, 'La théorie de la lumière et la découverte de Rømer', pp. 169–178 in: R. Taton (ed.), *Rømer et la vitesse de la lumière* (Paris, Vrin, 1976)

Visser, Simon Willem, 'Christiaan Huygens en de Stockholmse halo van 20 april 1535', *Hemel en Dampkring* **58** (1960) 89–94

Westfall, Richard S., 'Huygens's rings and Newton's rings: periodicity and seventeenth-century optics', *Ratio* 10 (1968) 64–77

Wiener, Otto H., 'Der Wettstreit der Newtonschen und Huygensschen Gedanken in der Optik', *Berichte über die Verhandlungen der Sächsischen Akademie der Wissenschaften zu Leipzig, Mathematisch-physische Klasse* 71/72 (1919) 240–254

Ziggelaar, August, 'How did the wave theory of light take shape in the mind of Christiaan Huygens?', *Annals of Science* 37 (1980) 179–187

Physics

Aisa-Moreu, D., 'La filosofia meccanica de Descartes, Boyle y Huygens', pp. 83–131 in: J. Arana (ed.), *La filosofia de los cientificos* (Sevilla, Themata, 1996)

Bernstein, Howard R., 'Leibniz and Huygens on the "relativity" of motion', pp. 85–102 in: A. Heinekamp (ed.), *Leibniz' dynamica – Symposium* (Stuttgart, Steiner, 1984)

Blackwell, Richard J., 'Christiaan Huygens: the motion of colliding bodies', *Isis* 68 (1977) 574–597

Blay, Michel, *Les raisons de l'infini: Du monde clos à l'univers mathématique* (Paris, Gallimard, 1993) [translated into English by M. B. DeBoise as *Reasoning with the Infinite: From the Closed World to the Mathematical Universe* (Chicago, University of Chicago Press, 1998)]

Blay, M., 'L'organisation déductive de la science du movement: Descartes, Galilée, Huygens', pp. 325–336 in: J. Montesinos & C. Solis (eds.), *Largo campo di filosofare: Eurosymposium Galileo* (Orotave, Fundación Canaria de historia de la cienza, 2001)

Burch, Christopher B., 'Huygens' pulse models as a bridge between phenomena and Huygens' mechanical foundations', *Janus* 68 (1981) 53–64

Burch, C. B., 'Christiaan Huygens: the development of a scientific research program in the foundation of mechanics', Doctoral thesis, University of Pittsburgh, Pittsburgh (1981)

Burke, Vincent I., 'The writings of Christiaan Huygens on the problem of colliding bodies', Doctoral thesis, Queen's University, Belfast (1964)

Chareix, Fabien, 'La pesanteur dans l'univers méchanique de Christiaan Huygens', pp. 244–252 in: L. Palm (ed.), *De zeventiende eeuw*, vol. 12 (Hilversum, Verloren, 1996)

Chareix, F., 'La découverte des lois des choc par Christiaan Huygens', *Revue d'histoire des sciences* 56 (2003) 15–58

Costabel, Pierre, 'La "loi admirable" de Christiaan Huygens', *Revue d'histoire de sciences* 9 (1956) 208–220

Costabel, P., 'La septième règle de choc élastique de Christian Huygens', *Revue d'histoire des sciences* 10 (1957) 120–131

Costabel, P., 'Isochronisme et accélération 1638–1687', *Archives internationales d'histoire des sciences* 28 (1978) 3–20

Dijksterhuis, Eduard Jan, 'Over de ontwikkeling der valwetten', *Christiaan Huygens: Internationaal Mathematisch Tijdschrift* 1 (1922) 238–262, 296–318, 355–379 & 2 (1923) 88–123

Dijksterhuis, E. J., *Val en worp: Een bijdrage tot de geschiedenis der mechanica van Aristoteles tot Newton* (Groningen, P. Noordhoff, 1924)

Dijksterhuis, E. J., 'Over een mechanisch axioma in het werk van Christiaan Huygens', *Christiaan Huygens: Internationaal Mathematisch Tijdschrift* **7** (1929) 161–180

Dugas, René, *Histoire de la mécanique* (Neuchâtel, Editions du Griffon, 1950)

Dugas, R., 'Sur le cartésianisme de Huygens', *Revue d'histoire des sciences* **7** (1954) 22–23

Dugas, R., *La mécanique au XVIIe siècle – Des antécédents scolastiques à la pensée classique* (Neuchâtel, Editions du Griffon, 1954)

Elena, Alberto, 'Huygens y el cartesianismo: A propósito de la noción de gravedad', *Llull* **5** (1982) 5–16

Elzinga, Aant, 'Huygens' theory of research and Descartes' theory of knowledge', *Zeitschrift für allgemeine Wissenschaftstheorie* **2** (1971) 174–194; **3** (1972) 9–27

Elzinga, A., *On a Research Program in Early Modern Physics: Studies in the Theory of Science* (New York, Humanities Press, 1972)

Elzinga, A., 'Christiaan Huygens' theory of research', *Janus* **67** (1980) 281–300

Erlichson, Herman, 'Huygens and Newton on the problem of circular motion', *Centaurus* **37** (1994) 210–229

Erlichson, Herman, 'Christiaan Huygens' discovery of the center of oscillation formula', *American Journal of Physics* **64** (1996) 571–574

Erlichson, H., 'The young Huygens solves the problem of elastic collisions', *American Journal of Physics* **65** (1997) 149–154

Finocchiaro, Maurice A., 'The concept of judgement and Huygens' theory of gravity', *Epistemologia* **3** (1980) 185–216

Forti, Umberto, 'La teoria della gravitazione di Huyghens', *Periodico di matematiche*, serie 4, **6** (1926) 305–313

Gabbey, Alan, 'Force and inertia in seventeenth-century dynamics', *Studies in History and Philosophy of Science* **2** (1971) 1–68

Gagnebin, Bernard, 'De la cause de la pesanteur: Mémoire de Nicolas Fatio de Duillier présenté à la Royal Society le 26 février 1690', *Notes and Records of the Royal Society of London* **6** (1949) 105–160

Hall, A. Rupert, 'Huygens and Newton', pp. 45–59 in: Ch. Wilson *et al.* (eds.), *The Anglo-Dutch Contribution to the Civilization of Early Modern Society: A Symposium* (London, Oxford University Press, 1976)

Halleux, Robert, 'Huygens et les théories de la matière', pp. 187–195 in: R. Taton (ed.), *Huygens et la France* (Paris, Vrin, 1982)

Hausdorff, Felix (ed.), *Christiaan Huygens' nachgelassene Abhandlungen: Über die Bewegung der Körper durch den Stoss & Über die Centrifugalkraft* (Leipzig, Engelmann Verlag, 1903)

Henrici, Julius, *Die Erforschung der Schwere durch Galilei, Huygens, Newton als Grundlage der rationellen Kinematik und Dynamik historisch-didaktisch dargestellt* (Leipzig, Teubner Verlag, 1885)

Klever, Wim, 'Zwaarte: Een polemiek in de zeventiende eeuw', *Tijdschrift voor filosofie* 52 (1990) 280–314

Kuhn, Wilfried, 'Das Problem der Relativbewegung bei Huygens', Doctoral thesis, Goethe Universität, Frankfurt am Main (1970)

MacLean, Johannes, 'De historische ontwikkeling der stootwetten van Aristoteles tot Huygens', Doctoral thesis, Vrije Universiteit, Amsterdam (1959)

Maltese, G., *Introduzione alla storia della dinamica nei secoli XVII e XVIII* (Genova, Brigati, 1996)

Martins, Roberto de A., 'Huygens' reaction to Newton's gravitational theory', pp. 203–213 in: J. V. Field & F. A. J. L. James (eds.), *Renaissance and Revolution: Humanists, Scholars, Craftsmen and Natural Philosophers in Early Modern Europe* (Cambridge, Cambridge University Press, 1993)

Mignard, F., 'The theory of the figure of the Earth according to Newton and Huygens', *Vistas in Astronomy* 30 (1987) 291–311

Mormino, Gianfranco, 'La relatività del movimento negli scritti sull'urto di Christiaan Huygens', pp. 107–138 in: E. I. Rambaldi (ed.), *De motu: Studi di storia del pensiero su Galileo, Hegel, Huygens e Gilbert* (Milano, Cisalpino-Golia, 1990)

Mormino, G., *Penetralia motus: La fondazione relativistica della meccanica in Christiaan Huygens, con l'edizione del Codex Hugeniorum 7A* (Milano, Facoltà di lettere e filosofia, Università di Milano, 1993)

Mormino, G., 'The philosophical foundations of Huygens' atomism', pp. 74–82 in: L. Palm (ed.), *De zeventiende eeuw*, vol. 12 (Hilversum, Verloren, 1996)

Mormino, G., 'Atomismo e meccanicismo nel pensiero di Christiaan Huygens', *Rivista di storia della filosofia* 51 (1996) 829–863

Nauenberg, Michael, 'Huygens and Newton on curvature and its application to dynamics', pp. 215–234 in: L. Palm (ed.), *De zeventiende eeuw*, vol. 12 (Hilversum, Verloren, 1996)

Plana, J., 'Note sur la figure de la terre et la loi de la pesanteur à sa surface, d'après l'hypothèse d'Huygens, publiée en 1690', *Astronomische Nachrichten* 35 (1853) 371–378

Reichenbach, Hans, 'Die Bewegungslehre bei Newton, Leibniz und Huygens', *Kantstudien* 29 (1929) 416–438

Schouten, Jan A., 'Die relative und absolute Bewegung bei Huygens', *Jahresbericht der Deutschen Mathematiker Vereinigung* 29 (1920) 136–144

Slowik, Edward, 'Huygens' center-of-mass space-time reference frame: constructing a cartesian dynamics in the wake of Newton's *De gravitatione* argument', *Synthese: International Journal for Epistemology, Methodology and Philosophy of Science* 112 (1997) 247–269

Snelders, Harry A. M., 'Christiaan Huygens and Newton's theory of gravitation', *Notes and Records of the Royal Society of London* 43 (1989) 209–222

Spranzi Zuber, Marta, 'Huygens' rejection of absolute motion: some aspects of the dilemma', *Physis* 36 (1999) 55–71

Strauss, Martin, 'Die Huygens-Leibniz-Machse Kritik im Lichte heutiger Erkenntniss', *Deutsche Zeitung für Philosophie* **16** (1968) 117–120

Vilain, Christiane, *La mécanique de Christian Huygens: La relativité du movement au XVIIe siècle* (Paris, Blanchard, 1996)

Vilain, C., 'Espace et dynamique chez Christiaan Huygens', pp. 235–243 in: L. Palm (ed.), *De zeventiende eeuw*, vol. 12 (Hilversum, Verloren, 1996)

Vilain, C., 'La loi galiléenne et la dynamique de Huygens', *Revue d'histoire des mathématiques* **2** (1996) 95–117

Vollgraff, Johan A., 'De relativiteit der beweging volgens Chr. Huygens', *Hemel en Dampkring* **32** (1934) 195–200

Westfall, Richard S., *Force in Newton's Physics: The Science of Dynamics in the Seventeenth Century* (New York, MacDonald, 1971)

Westman, Robert S., 'Huygens and the problem of Cartesianism', pp. 83–103 in: H. J. M. Bos, M. J. S. Rudwick, H. A. M. Snelders & R. P. W. Visser (eds.), *Studies on Christiaan Huygens* (Lisse, Swets & Zeitlinger, 1980)

Yokoyama, Masahiko, 'Origin of the experiment of impact with pendulums', *Japanese Studies in the History of Science* **11** (1972) 67–72

Yokoyama, M., 'Huygens and the times-squared law of free fall', *Proceedings of the 14th International Congress of the History of Science* **2** (1974) 349–352

Ziggelaar, August, 'Christiaan Huygens' dynamiske grundprincipper og de anvendelser han gjorde deraf', *Fysisk Tidsskrift* (1958) 66–76, 102–125, 156–182

Varia

Barth, Michael, 'Huygens at work: annotations in his rediscovered personal copy of Hooke's *Micrographia*', *Annals of Science* **52** (1995) 601–613

Bosscha, Johannes, *Memorie betreffende het ontwerp van dr P. J. H. Cuypers voor een standbeeld van Christiaan Huygens* (Haarlem, Hollandsche Maatschappij der Wetenschappen, 1905)

Charrak, André, 'Huygens et la théorie musicale', *Revue d'histoire des sciences* **56** (2003) 59–78

Cohen, Hendrik Floris, *Quantifying Music: The Science of Music at the First Stage of the Scientific Revolution, 1580–1650* (Dordrecht, Reidel, 1984)

Crommelin, Claude A., 'Hugeniana', *Oud Holland: Tweemaandelijks Tijdschrift voor de Nederlandsche Kunstgeschiedenis* **6** (1943) 1–6

De Vries, Philip, 'Christiaan Huygens tussen Descartes en de Verlichting', *De Gids* **142/5** (1979) 293–305

De Vries, P., 'Christiaan Huygens entre Descartes et le siècle des Lumières', *Theoretische Geschiedenis* **6** (1979) 7–19

Farrar, L. P., 'Christiaan Huygens: his musical contributions to seventeenth century science' Doctoral thesis, University of Texas, Austin (1962)

Fokker, Adriaan D., 'Christiaan Huygens' oktaafverdeling in 31 gelijke diëzen', *Caecilia en de Muziek* **98** (1941) 149–152

Hess, Heinz-Jürgen, 'Bücher aus dem Besitz von Christiaan Huygens (1629–1695) in der Niedersächsischen Landesbibliothek Hannover', *Studia Leibnitiana* **12** (1980) 1–51

Jansen, P., 'Une tractation commerciale au XVIIᵉ siècle', *Revue d'histoire des sciences* **4** (1951) 173–176

Land, Jan P. N., 'Het toonstelsel van Christiaan Huygens', *Tijdschrift voor de Noord-Nederlandsche Muziekgeschiedenis* **3** (1891) 197–203

Mormino, Gianfranco, 'Ammirare e comprendere: la concezione del sapere di Christiaan Huygens', pp. 495–511 in: G. Canzani, M. A. Granada & Y. Ch. Zarka (eds.), *Potentia Dei* (Milano, Franco Angeli, 2000)

Mormino, G., 'Le role de Dieu dans l'oeuvre scientifique et philosophique de Christiaan Huygens', *Revue d'histoire des sciences* **56** (2003) 113–133

Nuchelmans, Gabriël, 'A 17th-century debate on the consequentia mirabilis', *History and Philosophy of Logic* **13** (1992) 43–58

Rasch, Rudolf A., 'Six seventeenth-century Dutch scientists and their knowledge of music', pp. 185–210 in: V. Coelho (ed.), *Music and Science in the age of Galileo* (Dordrecht, Reidel, 1992)

Van de Craats, J., 'Christiaan Huygens en de muziek', *De zeventiende eeuw* **7** (1991) 7–16

Van Gelder, Hendrik E., *Ikonografie van Constantijn Huygens en de zijnen* ('s Gravenhage, Martinus Nijhoff, 1957)

Van Helden, Anne & Van Gent, Robert, *De Huygenscollectie & The Huygens Collection* (Leiden, Museum Boerhaave, 1995)

Vuillemin, Jules, 'Physique panthéiste et détermisme: Spinoza et Huygens', *Studia Spinozana* **6** (1990) 231–250

Walker, D. P., '17th -century scientists' view on intonation and the nature of consonance', *Archives internationales d'histoire des sciences* **27** (1977) 263–273

Wolloch, Nathaniel, 'Christiaan Huygens' attitude toward animals', *Journal of the History of Ideas* **61** (2000) 415–432

Index